FOUNDATION
SCIENCE
FOR ENGINEERS

Other titles of interest to engineers

Dynamics
  G.E. Drabble

Electric Circuits
  P. Silvester

Electromagnetism
  R.G. Powell

Engineering Mathematics, fourth edition
  K.A. Stroud

Fluid Mechanics
  M.B. Widden

Foundation Mathematics for Engineers
  John Berry and Patrick Wainwright

Structural Mechanics
  J.A. Cain and R. Hulse

Thermodynamics
  J.R. Simonson

Understanding Hydraulics
  Les Hamill

Understanding Structures
  Derek Seward

FOUNDATION SCIENCE FOR ENGINEERS

Keith L. Watson

Second Edition

First edition 1993
Reprinted 1994
Second edition 1998

Published by
PALGRAVE
Houndmills, Basingstoke, Hampshire RG21 6XS and
175 Fifth Avenue, New York, N. Y. 10010
Companies and representatives throughout the world

PALGRAVE is the new global academic imprint of
St. Martin's Press LLC Scholarly and Reference Division and
Palgrave Publishers Ltd (formerly Macmillan Press Ltd).

ISBN 0–333–72545–X

This book is printed on paper suitable for recycling and
made from fully managed and sustained forest sources.

A catalogue record for this book is available
from the British Library.

10   9   8   7   6   5   4   3
08   07   06   05   04   03   02

Printed and bound in Great Britain by
Antony Rowe Ltd, Chippenham and Eastbourne

CONTENTS

Preface   vii

PART 1: FORCE, MATTER AND MOTION

Topic   1   Quantities   1
Topic   2   Forces and Matter   9
Topic   3   Equilibrium   19
Topic   4   Pressure and Upthrust   28
Topic   5   Displacement, Velocity and Acceleration   37
Topic   6   Force and Motion   47
Topic   7   Momentum and Impulse   54
Topic   8   Work, Energy and Power   60
Topic   9   Motion in a Circle   68
Topic 10   Rotation of Solids   78
Topic 11   Simple Harmonic Motion   86
Topic 12   Mechanical Waves   96
Topic 13   Electromagnetic Waves   108

PART 2: STRUCTURE AND PROPERTIES OF MATTER

Topic 14   Atomic Structure and the Elements   118
Topic 15   The Nucleus   129
Topic 16   Chemical Bonding   141
Topic 17   Heat and Temperature   151
Topic 18   Heat Transfer   163
Topic 19   Gases   174
Topic 20   Liquids   184
Topic 21   Solids   197
Topic 22   Structure of Solids   206
Topic 23   The Nature of Ceramics   221
Topic 24   The Nature of Metals   230
Topic 25   The Nature of Polymers   241
Further Questions   252

PART 3: ELECTRICITY AND MAGNETISM

Topic 26   Electric Charge   256
Topic 27   Electric Field   261
Topic 28   Capacitance   270
Topic 29   Electric Current   279

Topic 30   Resistance   287
Topic 31   Some Simple Circuits   302
Topic 32   Magnetic Fields   316
Topic 33   Electromagnetic Induction   324
Topic 34   Magnetic Behaviour of Materials   333
Topic 35   Alternating Current   338

Appendix: Calculation Technique   347
Answers to Questions   365
Index   373

PREFACE

This second edition includes five new topics: Topic 15 deals with the nucleus, and Topics 22–25, together with a selection of descriptive questions, extend the materials content into the first year degree and diploma context. An appendix on calculation technique has been included for the benefit of readers who wish to improve their confidence and fluency in solving numerical problems.

The same general approach has been adopted with the new material as with the first edition, with emphasis on clarity and crispness of presentation. Again, the data supplied are approximate and for illustrative purposes only, and practical work has not been included. The inclusion of safe and effective laboratory exercises over the whole range of topics covered by the book would have made it too long – furthermore, the needs of individual courses and the resources of individual institutions tend to determine their particular selection of practical work.

I am indebted to colleagues here at Portsmouth for their advice and comments on the new material, particularly Michael Devane, Dr. Simonne Mason and Dr John Tsibouklis; also to Elizabeth Brookfield of Coventry University who reviewed the material and made many useful suggestions. I would like to thank Malcolm Stewart and his colleagues at the publishers for their valuable support.

Portsmouth, 1998   KLW

Part 1: Force, Matter and Motion

TOPIC 1 QUANTITIES

COVERING:

- SI units;
- base and derived units;
- scalar and vector quantities;
- vector addition.

1.1 SI UNITS

Engineering quantities (pressure, temperature, power, and so on) need to be expressed in terms of an agreed system of units. SI units (Système International d'Unités) have been adopted in the UK and in many other countries, so we shall use them in this book. The system is founded on seven base units and two supplementary units from which all the others are derived.

The base units which are going to be of most interest to us are shown in Table 1.1 together with some of the derived units. We shall add to the list as we go along.

Derived units can be construed in terms of independent dimensions (such as length, mass and time) that are provided by the base units. Let us consider the unit of power as an illustration. Don't worry if you are unable to follow the scientific arguments too well at this stage. We shall go over them much more thoroughly later. The important thing to appreciate is that we can analyse the relationship between quantities in terms of their constituent base units.

First, velocity is a measure of change of position in unit time and its magnitude is given in metres per second ($m\ s^{-1}$). Acceleration is a measure of the rate at which velocity changes and its magnitude is given in metres per second per second ($m\ s^{-2}$).

The unit of force is called the *newton* (N). As we shall see later, it is defined as the force needed to give a mass of 1 kg an acceleration of $1\ m\ s^{-2}$. The relationship between these quantities is given by

Table 1.1

	Name	Symbol
Some base units:		
Length	metre	m
Mass	kilogram	kg
Time	second	s
Electric current	ampere	A
Temperature	kelvin	K

(For our purposes °C = K − 273: see Topic 17)

	Name	Symbol
Some derived units:		
Force	newton	N
Energy	joule	J
Power	watt	W
Pressure	pascal	Pa
Electric charge	coulomb	C

$F = ma$, where F represents force, m represents mass and a represents acceleration. The constituent base units of force are therefore those of mass times acceleration and

$$1 \text{ N} = 1 \text{ kg m s}^{-2}$$

Taking this a step further, the *joule* (J) is the SI unit of energy. This is equal to the amount of work done when the point of application of a force of 1 newton moves through a distance of 1 metre in the direction of the force. If W is the work done and s is the distance moved in the direction of the force F, then $W = Fs$. The constituent base units of energy are therefore those of force times distance and

$$1 \text{ J} = 1 \text{ kg m}^2 \text{ s}^{-2}$$

Finally, power is the rate at which work is done or energy expended. The SI unit of power is called the *watt* (W), which corresponds to a rate of 1 joule per second. The base units of power are therefore those of energy divided by time and

$$1 \text{ W} = 1 \text{ kg m}^2 \text{ s}^{-3}$$

So, starting with the base units of mass, length and time, we have derived the unit of power.

Units often require prefixes to adjust them to an appropriate scale of magnitude for a particular measurement. For instance, we use *kilo*metres (km, a thousand metres) for large distances and *milli*metres (mm, a thousandth of a metre) for small ones. Table 1.2 shows the prefixes that we shall be using in this book.

Table 1.2

Prefix	Symbol	Factor	
giga-	G	$\times 10^9$	(a thousand million)
mega-	M	$\times 10^6$	(a million)
kilo-	k	$\times 10^3$	(a thousand)
deci-	d	$\times 10^{-1}$	(a tenth)
centi-	c	$\times 10^{-2}$	(a hundredth)
milli-	m	$\times 10^{-3}$	(a thousandth)
micro-	μ	$\times 10^{-6}$	(a millionth)
nano-	n	$\times 10^{-9}$	(a thousand millionth)
pico-	p	$\times 10^{-12}$	(a million millionth)

1.2 SCALAR AND VECTOR QUANTITIES

Scalar quantities are those that are completely specified by their magnitude (e.g. length, speed and mass) and can be manipulated by simple arithmetic operations such as addition and multiplication. *Vector* quantities have both magnitude and direction (e.g. displacement, velocity and force). This makes them more complicated to handle, because angles are involved.

Let us start with displacement, i.e. change of position. This can be defined in terms of distance in a particular direction. But, as we can see from the example in Figure 1.1, there is another approach. Figure 1.1(a) shows a displacement of 5 units at an angle of 37° measured anticlockwise from the positive *x*-axis (i.e. from the 3 o'clock direction). Figure 1.1(b) shows the same displacement as the *resultant* of moving a distance of 4 units in the 0° direction, then 3 units in the 90° direction. Since tan 37° = 3/4 and, from Pythagoras' theorem, $5 = \sqrt{4^2 + 3^2}$, we can see that both methods give the same result.

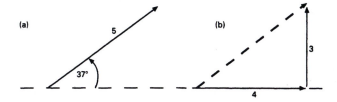

Figure 1.1

Vector quantities are often expressed in terms of distances and angles as in Figure 1.1(a), but sometimes as we need to *resolve* them into perpendicular components as in Figure 1.1(b). From the figure we can see that the vertical component, equal to 3, is given by 5 sin 37° and the horizontal component, equal to 4, is given by 5 cos 37°. This is useful when we need to find the resultant of two vector quantities.

For example, Figure 1.2(a) shows a displacement OA at an angle α

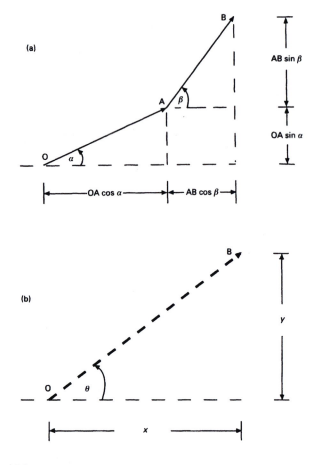

Figure 1.2

followed by a second displacement AB at an angle β. Figure 1.2(b) shows the resultant OB in the direction θ. The total vertical displacement y is equal to (OA sin α + AB sin β) and the total horizontal displacement x is equal to (OA cos α + AB cos β). So, knowing the magnitude and direction of the two original displacements, we can calculate y and x. We can then calculate the magnitude of OB from Pythagoras' theorem and its direction from tan $\theta = y/x$.

Clearly we can use this method to obtain the resultant of as many displacements as we wish by plotting vectors head to tail. If the vectors form a closed loop, then the resultant (hence, the net displacement) is zero. Taking Figure 1.2 as an illustration, if we travelled from O to A, then from A to B, and finally from B back to O, the three displacement vectors would form a closed triangle and we would end up where we started. This is an obvious but important idea that we shall need to use later.

Note that velocity is a measure of displacement in unit time and that it is a vector quantity that can be manipulated in the same way as displacement. Later we shall use similar methods to handle forces.

Before tackling the questions at the end of the topic, make sure that you understand the following worked examples. And remember that, unless stated otherwise, angles are measured anticlockwise from the positive x-axis.

Worked Example 1.1

Find the magnitude and direction of the resultant of a displacement of 103 m at 62° followed by another of 59 m at 28°.

───────────────

The displacements are shown in Figure 1.3.

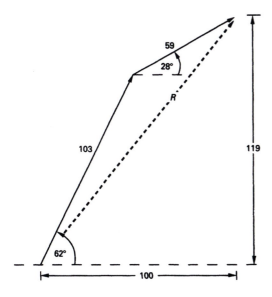

Figure 1.3

The total vertical displacement is equal to

103 sin 62° + 59 sin 28° = 119 m

The total horizontal displacement is equal to

103 cos 62° + 59 cos 28° = 100 m

The magnitude of the resultant R is equal to

$\sqrt{119^2 + 100^2}$ = 155 m

The direction of the resultant R is equal to

$\tan^{-1}$ 119/100 = 50°

Worked Example 1.2

Find the magnitude and direction of the resultant of successive displacements of 25 m at 90°, 30 m at 45° and 20 m at 300°.

The displacements are shown in Figure 1.4.

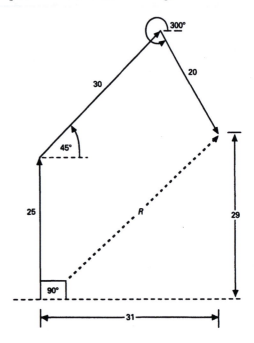

Figure 1.4

The total vertical displacement is equal to

25 sin 90° + 30 sin 45° + 20 sin 300° = 29 m

The total horizontal displacement is equal to

25 cos 90° + 30 cos 45° + 20 cos 300° = 31 m

The magnitude of the resultant R is equal to

$\sqrt{29^2 + 31^2}$ = 42 m

The direction of the resultant R is equal to

$\tan^{-1}$ 29/31 = 43°

Worked Example 1.3

If a boat is being rowed due north at 2 m s^{-1} and there is a current flowing due east at 1.5 m s^{-1}, what is the true velocity of the boat relative to the earth?

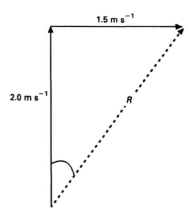

1.5 m s^{-1}

2.0 m s^{-1}

R

Figure 1.5

The velocity diagram is shown in Figure 1.5.
The magnitude of the resultant R is equal to

$\sqrt{2^2 + 1.5^2} = 2.5$ m s^{-1}

The direction of the resultant R is equal to

$\tan^{-1} \dfrac{1.5}{2} = 37°$ east of north

i.e. 53° anticlockwise from the positive x-axis.

Questions

1. Find the magnitude and direction of the resultants of the following pairs of successive displacements:

   (a) 42 m at 31°, 53 m at 72°;
   (b) 117 m at 28°, 67 m at 147°;
   (c) 331 m at 47°, 158 m at 238°;
   (d) 97 m at 72°, 84 m at 163°.

2. A car climbs a hill at a road speed of 10 m s^{-1}. If the slope of the hill is 6° above the horizontal, what are the horizontal and vertical components of the car's velocity?

3. An object travels eastwards at 5 m s^{-1} for 7 s, then northwards at 7.5 m s^{-1} for 12 s, then westwards at 10 m s^{-1} for 15 s. Find how far and in what direction it must travel to return to its starting point.

4. A ship steers north at 5.0 m s^{-1} with a current flowing towards the south-east at 2.0 m s^{-1}. With what velocity must a man cross the deck to maintain a fixed position relative to the seabed?

5. An object travels 52 m in the 1 o'clock direction, then 71 m in the 5 o'clock direction, then 103 m in the 8 o'clock direction, then 43 m in the 11 o'clock direction. What is its resultant displacement?

6. A boat heads due north across a river 300 m wide which

flows from west to east at 1.5 m s^{-1}. If the boat moves at 5.0 m s^{-1} relative to the water:

(a) how long does it take to cross the river?
(b) how far downstream does it land?
(c) in what direction should it have headed to have crossed in the shortest possible distance?
(d) how long would this have taken?

TOPIC 2 FORCES AND MATTER

COVERING:

- mass and gravitational force;
- internal forces and elastic behaviour;
- friction.

Isaac Newton (1642–1727) put forward three propositions concerning the relationship between the motion of a body and the forces acting upon it. These are known as *Newton's laws of motion*. For the moment we shall confine ourselves to the first. (We shall deal with the others in Topic 6.)

Newton's first law tells us that a stationary body remains at rest, or, if in motion, it moves in a straight line at constant speed, unless it is acted upon by a force (or the resultant of a number of forces). In mechanical terms, a force may therefore be regarded as an influence that tends to change a body's state of motion.

Scientists believe that there are only four fundamental types of force that operate in the universe. Two of these need not concern us at this stage because they operate inside the nucleus of the atom (see Topic 15). Of the remaining two, one stems from the gravitational interaction that arises between bodies because of their mass, and the other from the electro-magnetic interaction due to the effects of electric charge.

2.1 MASS AND GRAVITATIONAL FORCE

The mass of a body is the quantity of matter that it contains. The SI unit of mass is the kilogram (kg).

Mass is formally defined in terms of inertia – that is to say, resistance to change of state of motion. We notice the inertia of our bodies when we are in a car that is changing speed or direction. We can feel the car forcing us to accelerate or decelerate against our natural tendency to remain at rest or move at constant speed in a straight line.

In practice, gravitation provides us with a much more convenient way of measuring mass. The law of universal gravitation, also named after Newton (who discovered it), is expressed by the equation

$$F = G \frac{m_1 m_2}{r^2} \qquad (2.1)$$

where F is the gravitational force of attraction between two masses of magnitude m_1 and m_2 that are separated by a distance r. G is known as the *gravitational constant* and has the value 6.7×10^{-11} N m^2 kg^{-2}. Note that this law is an *inverse square law* – that is to say, the force is proportional to the reciprocal of the square of the distance between the masses. If the distance is doubled, the force is reduced to a quarter of its original magnitude; if the distance is halved, the force is quadrupled; and so on. Also note that we can verify the units of G by rearranging Equation (2.1) to give

$$G = \frac{Fr^2}{m_1 m_2}$$

which has the units of

$$\frac{\text{N m}^2}{\text{kg}^2}$$

From a gravitational point of view, we can assume that the earth behaves as though its mass is concentrated at its centre. If we substitute its mass (6.0×10^{24} kg) for m_1 and its radius (6.4×10^6 m) for r in Equation (2.1), and if we give m_2 the value 1.0 kg, then we obtain the gravitational force between the earth and a 1 kg mass at its surface as follows:

$$F = 6.7 \times 10^{-11} \frac{(6.0 \times 10^{24})(1.0)}{(6.4 \times 10^6)^2} = 9.8 \text{ N}$$

This means that there is a gravitational field of influence around the earth which causes any object at its surface to have a *weight* of 9.8 N per kilogram mass. (As a rough guide to magnitude, a medium-sized apple weights about 1 N.)

9.8 N kg^{-1} is the *gravitational field strength* at the earth's surface and, if we give it the symbol g, we can write

$$W = mg \tag{2.2}$$

where W is the weight experienced by a mass m. (Engineers normally tend to interpret g in a different way, as we shall see in Topics 5 and 6.) Note that the weight of a given object will be different in places where the gravitational field strength is different. For example, a 1 kg object would weigh 1.6 N at the surface of the moon.

The gravitational interaction between objects on the human scale is very small. For instance, Equation (2.1) shows us that the force between two 1 kg masses 1 m apart is 6.7×10^{-11} N. Gravitational forces become large where astronomical masses (e.g. the earth) are involved, so that $m_1 \times m_2$ on the top line of Equation (2.1) becomes big enough to offset the tiny value of G.

Note that the centre of gravity of a body is the point at which its entire weight may be considered to act – for instance, the centre of a sphere or the mid-point of a ladder (assuming they are both uniform).

2.2 INTERNAL FORCES

A solid object will tend to resist being compressed or stretched because of opposing internal forces between its constituent atoms. These forces have electrical origins, which we shall discuss in some detail in Topic 16. For the time being, we can think of solid materials as consisting of atoms which are held together by *chemical bonds* that behave like tiny springs. A compressive force acting on a solid tends to push its constituent atoms together and a tensile force tends to pull them apart. In either case the deformation of the bonds results in the generation of an equal and opposite internal force that tends to restore the atoms to their original positions. If we remove the applied force, then the restoring force will return the solid to its original shape. This behaviour, called *elasticity*, is shared by all solids, even materials such as steel or concrete, although the deformation may not be obvious (for example, the deformation of a railway bridge supporting a train or a desk supporting a book). Thus, a force can change the shape as well as the state of motion of a body.

Figure 2.1 represents the elastic behaviour of a steel wire. Let us imagine that the wire is suspended vertically from one end and that we stretch it by hanging a mass from the other. Figure 2.1 shows the relationship between the tensile force in the wire and the extension (increase in length) as the mass is increased. The chemical bonds in the wire stretch just far enough to provide an equal and opposite force to support the mass. The figure shows that if we double the mass so that the bonds have to provide twice the support, then the extension is doubled. If the force is trebled, the extension is trebled, and so on.

This is an example of Hooke's law, which tells us that elastic materials deform in proportion to the force causing their deformation. We shall consider this in more detail in Topic 21 (and we shall see that Hooke's law is not valid for all materials). In the meantime we should note that the argument can be broadened to state that, for the many materials that obey Hooke's law, *strain* is proportional to the *stress* which causes it. Strain ε is the amount of extension or compression per unit length and is given by $\varepsilon = \Delta l / l_0$ where Δl is the change in length and l_0 is the original length. (Δ is a symbol that is used to represent a change in a quantity.) Note that strain is a dimensionless quantity, since it is obtained by dividing a length by a length. It is sometimes expressed as a simple ratio and sometimes as a percentage.

Stress σ is the force per unit cross-sectional area and is given by $\sigma = F/A$, where F is the force (sometimes called the load) and A is the cross-sectional area. The unit of stress, N m^{-2}, is sometimes called the *pascal* (Pa).

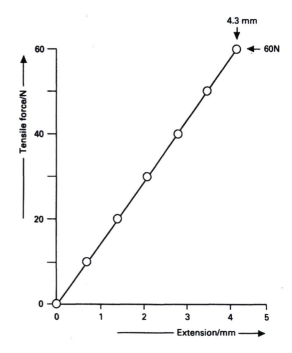

Figure 2.1

The advantage of using stress and strain, as opposed to force and extension, is that we eliminate the effect of the dimensions of the specimen. (Under a given tensile load, the extension of a thick wire would be less than that of a thin one and the extension of a long wire would be greater than that of a short one.) A graph of stress against strain therefore represents the behaviour of the material alone and its slope E (= σ/ε), called the *Young's modulus*, gives a measure of the material's stiffness. We shall deal with stress/strain relationships in Topic 21. In the meantime, the following worked example uses the graph in Figure 2.1 to obtain Young's modulus directly.

Worked Example 2.1

Figure 2.1 (above) is a graph of load against extension for a wire of length 1.72 m and 0.40 mm diameter. Find the Young's modulus of the material.

From the discussion above we obtain the following expression

$$E = \frac{\sigma}{\varepsilon} = \frac{F}{A} \times \frac{l_0}{\Delta l} = \frac{l_0}{A} \times \frac{F}{\Delta l}$$

From the figure we see that the graph is a straight line which passes through both the origin and the point where $F = 60$ N and $\Delta l = 4.3$ mm. Its slope $(F/\Delta l)$ is therefore equal to

$$60/(4.3 \times 10^{-3}) \text{ N m}^{-1}$$

The radius r of the wire is equal to 0.2×10^{-3}m.

Substituting the values for l_0, A $(= \pi r^2)$ and $F/\Delta l$ in the expression for E that we obtained above,

$$E = \frac{1.72}{\pi(0.2 \times 10^{-3})^2} \times \frac{60}{(4.3 \times 10^{-3})} = 1.9 \times 10^{11} \text{ N m}^{-2}$$

therefore,

$$E = 190 \text{ GPa}$$

If we unload the stretched wire represented in Figure 2.1, it will return to its original length. And we can repeat the loading/unloading cycle as many times as we like (within the limits imposed by metal fatigue, which we shall not be considering in this book). It is important to recognise that we must not use too great a load, otherwise the wire will deform *plastically* (i.e. stretch permanently) or even break. We shall discuss this in more detail in Topic 21. In the meantime, we shall confine our general discussion to stresses below the *elastic limit* (i.e. the stress level at which deformation ceases to be entirely elastic).

Note that our suspended mass is ultimately supported by the ground. If the wire is attached to the ceiling, then the mass and the wire are supported by an upward force generated by the slight bending of the joist to which the wire is fastened. (Bending of the joist is, in effect, a combination of tension and compression, where its bottom face tends to lengthen and become convex and its top face tends to shorten and become concave.) The joist is supported by an upward force generated by the compression of the wall which supports it. Finally, the wall is supported by an upward force generated by the deformation of the ground.

2.3 FRICTIONAL FORCE

Frictional forces arise through contact between objects and become apparent when we try to slide one surface over another.

Figure 2.2 shows a body lying on a horizontal plane. The weight W of the body is supported by the *normal* (i.e. perpendicular) force N exerted on it by the plane. Even the flattest surfaces are rough to varying degrees, so there will be high spots in the contact area between the

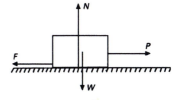

Figure 2.2

body and the plane which tend to interlock when they move relative to one another. The applied horizontal force P, acting towards the right, will therefore be opposed by a frictional resistance which will tend to prevent the body from sliding. In effect, P is opposed by a frictional force F which is the sum of all the tiny horizontal forces generated at the contact points where the surface irregularities push against one another. If we increase P, then, initially, F will increase to match it and the body will not move. Eventually P will reach a value just large enough to break down the interlocking between the surface irregularities. At this stage, when the body is just at the point of sliding, the frictional force F is given by the relationship

$$F = \mu_s N \tag{2.3}$$

The constant μ_s is called the *coefficient of static friction* and depends on the nature of the surfaces. The greater the friction between them then the greater will be the value of μ_s. Depending on conditions, the value for metals on certain plastics can be as low as 0.04, compared with around 1 for rubber on dry concrete.

As Equation (2.3) suggests, if we increase the normal force N by pressing the contact surfaces harder together, then the friction between them becomes proportionally greater.

Once sliding begins, the frictional force F usually falls slightly, then tends to remain at a constant value that is independent of P. We therefore distinguish between the coefficients of static and kinetic (or dynamic) friction, μ_s and μ_k, respectively. μ_k (hence, F) tends to remain constant over a fairly wide range of sliding velocities and in many instances the values of μ_k and μ_s are not very different. The relationship between F and P is summarised in Figure 2.3.

From the figure we can see that:

$F (= P) < \mu_s N$ under static conditions

$F (= P) = \mu_s N$ at the point of sliding

$F = \mu_k N (< P)$ under sliding conditions

Equation (2.3) and its dynamic counterpart ($F = \mu_k N$) are approximate but, within their range of validity for a given system, they indicate that the frictional force is independent of the contact area.

Friction is a very complex phenomenon and it is impossible to devise a single physical model to cover all cases. Nevertheless it is worth making some general observations.

It is believed that, because of roughness, the points of real contact between two solid surfaces pressed together are usually few and far between. This means that the actual contact area is much smaller than the apparent area suggested by the overall dimensions of the surfaces involved. A normal force N will therefore be concentrated into regions of intense localised pressure at the contact points. It is believed that at

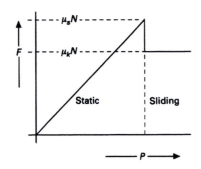

Figure 2.3

these points some materials, particularly metals, tend to flow plastically, thereby causing an increase in the actual contact area (where the contact points are deformed) and even localised adhesion.

It appears that an increase in the value of N causes an increase in the actual contact area and, hence, an increase in frictional resistance. Also, for a given value of N, the sum of the areas of all the contact points appears to remain more or less the same whether they are spread out over a large apparent surface area or concentrated into a relatively small one. Furthermore, under sliding conditions, the areas of contact between the surfaces are being continually broken and remade. The fact that μ_k tends to be less than μ_s seems to suggest that, under sliding conditions, a proportion of the potential contact points across the interface are not connected at any given moment.

Another measure of frictional resistance is the *angle of friction*, which is illustrated in Figure 2.4. If the plane on which a body rests is tilted to an angle θ where the body is on the point of sliding downwards, then θ is the angle of friction, where

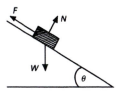

Figure 2.4

$$\mu_s = \tan \theta \tag{2.4}$$

Equation (2.4) is easy to verify. If the body is on the point of slipping, then its weight W is balanced by the normal force N supporting it perpendicular to the plane plus the frictional force F just stopping it from sliding down. N and F, in effect, are combined in an upward resultant force R (that is equal and opposite to W), as shown in Figure 2.5(a).

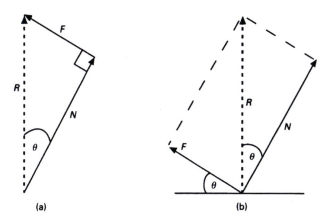

(a)   (b)

Figure 2.5

Note that, since they are vector quantities, we treat the forces N and F in the same way as the displacements in the previous topic. We represent them by arrows indicating their direction, with length proportional to their magnitude. It is sometimes helpful to represent two forces acting at a point as the sides of a parallelogram (a rectangle in this particular case) where the diagonal from the point at which the forces meet represents the resultant (Figure 2.5b).

As Figure 2.5(b) shows, the angle between R and N must be θ. (F and N are mutually perpendicular, so they must make an angle θ with the horizontal and vertical respectively, as shown.) From the right-angled triangle in Figure 2.5(a) we can see that

$$F = R \sin \theta$$

and

$$N = R \cos \theta$$

So, from Equation (2.3),

$$\mu_s = \frac{F}{N} = \frac{R \sin \theta}{R \cos \theta} = \tan \theta$$

as in Equation (2.4).

Rather than combine N and F to find the resultant force R, we could equally well tackle the problem by *resolving* the weight W into components parallel and perpendicular to the plane, as in Figure 2.6. These two components can then be equated to F and N, respectively, and

$$\mu_s = \frac{F}{N} = \frac{W \sin \theta}{W \cos \theta} = \tan \theta$$

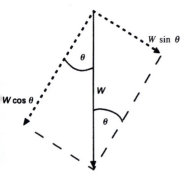

Figure 2.6

Friction is sometimes reduced by lubrication. Putting oil between metal contact surfaces tends to keep them apart, so that frictional forces are reduced. Another approach is to mount things on wheels. This does not entirely solve the problem, because of *rolling friction* between the wheel and the surface over which it is running. The wheel tends to create a slight depression out of which it continuously tries to climb and, at the same time, it tends to flatten where it touches the surface.

Note that, in the present context, the term 'smooth' is sometimes used to describe a surface that is frictionless or that can be considered to be frictionless as an approximation.

Worked Example 2.2

(a) Assuming that $\mu_s = 0.55$, what is the minimum horizontal force needed to start a 50 kg box sliding across a horizontal floor?

(b) What force is required if it is applied to the box by means of a rope inclined at 25° above the horizontal?

(a) The forces involved are shown in Figure 2.2 (page 13). The weight W of the box is supported by the normal force N, so that

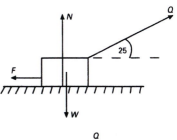

$$N = W = 9.8 \times 50 = 490 \text{ newtons}$$

Therefore, from Figure 2.3 (page 14), to start the box sliding,

$$P = F = \mu_s N = 0.55 \times 490 = 270 \text{ newtons}$$

(b) Figure 2.7 shows that the applied force Q has a horizontal component $Q \cos 25°$ and a vertical component $Q \sin 25°$.
    Considering the horizontal forces, we have

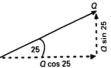

Figure 2.7

$$F = Q \cos 25°$$

Considering the vertical forces, and remembering that $W = 490$ newtons,

$$N + Q \sin 25° = 490$$

Therefore,

$$N = 490 - Q \sin 25°$$

When the box is on the point of sliding,

$$F = \mu_s N = 0.55 \times N$$

and, substituting for F and N,

$$Q \cos 25° = 0.55 (490 - Q \sin 25°)$$

which gives

$$Q = 237 \text{ newtons}$$

Questions

(Where necessary assume that $g = 9.8 \text{ N kg}^{-1}$.)

1. Estimate the mass of the moon, given that its diameter is 3500 km and that a 1.0 kg mass weighs 1.6 N (newtons) at its surface. ($G = 6.7 \times 10^{-11} \text{ N m}^2 \text{ kg}^{-2}$.)

2. A cable, 50 m long, experiences a strain of 0.05%. By how many mm has its original length increased?

3. A 0.4 mm diameter wire supports a mass of 2.5 kg.

(a) What is the tensile stress in the wire?

(b) What is the percentage strain if the value of Young's modulus is 2×10^{11} N m^{-2}?

4. A body of 25 kg mass is resting on a level floor. If $\mu_s = 0.4$, what is the frictional force?

5. A body of 25 kg mass lying on a level floor has a steadily increasing horizontal force applied to it. If $\mu_s = 0.4$, what is the maximum frictional force reached?

6. A body of 25 kg mass is sliding down a 45° slope. If $\mu_k = 0.6$, what is the frictional force?

7. For Figure 2.7, calculate the magnitude of Q if its direction is reversed (so that the box is being pushed by a force inclined at 25° below the horizontal).

8. A 50 kg box is placed on a ramp inclined at 25° to the horizontal.

   (a) Find the angle of friction; hence show that the box will not slide downwards.
   (b) Find the minimum force parallel to the ramp needed to start the box (i) sliding downwards and (ii) sliding upwards. (Assume $\mu_s = 0.55$.)

TOPIC 3 EQUILIBRIUM

COVERING:

- translational and rotational equilibrium;
- the components of forces;
- the moments of forces.

A body is in equilibrium when the forces acting on it balance each other so that it either remains at rest or, if it is moving, remains in a state of uniform motion (i.e. constant speed in a straight line). To understand equilibrium we need to recognise that forces can influence the motion of a body in two ways: they can affect its translational motion from one place to another (with all its parts moving in the same direction) and they can affect its rotation.

Two equal and opposite forces meeting at the same point in the body do not affect its state of motion at all, because they are in both translational and rotational equilibrium. If the two forces are separated so that their lines of action are parallel, as at the opposite ends of bicycle handlebars, then they are still in translational equilibrium, because they cancel each other out and their resultant is zero. However, they are no longer in rotational equilibrium, because they tend to cause the handlebars to turn.

We shall only consider coplanar forces here – that is to say, forces that act in the same plane. First we shall think about translational equilibrium in terms of the components of forces; then we shall move on to rotational equilibrium.

3.1 COMPONENTS OF FORCES

In the previous topic we met simple cases of combining forces into a resultant and of resolving forces into components. We can very easily extend this to more complex systems, as we did with displacements. We can even plot force vectors head to tail on a scale drawing and find the resultant graphically, just as we can with displacements on a map.

If the vectors representing a system of forces form a closed loop, then the resultant is zero and the forces are in translational equilibrium. And if they are concurrent (i.e. if they meet at a point), then they are in rotational equilibrium too. For instance, Figure 3.1(a) shows a weight

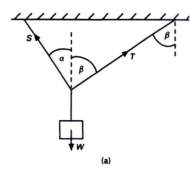

(a)

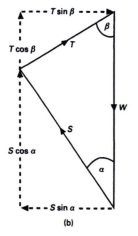

(b)

Figure 3.1

W being supported by two fixed strings which carry tensile forces S and T, respectively. Note that the values of the angles α and β are dictated by the lengths of the strings and the distance between their fixing points on the ceiling. The system is at rest and therefore in equilibrium.

How can the values of S and T be found? We could do it with a ruler and a protractor. In Figure 3.1(b) the length W is drawn to represent the weight. The angles α and β are then marked off at either end to give the directions of S and T. The point where the lines representing S and T intersect can then be found, and, hence, their lengths and the corresponding magnitudes.

Alternatively, since the forces are in equilibrium, the sum of all their horizontal components and of all their vertical components must separately be zero (since there can be no resultant); therefore,

$$T \sin \beta = S \sin \alpha$$

and

$$S \cos \alpha + T \cos \beta = W$$

Knowing the values of W, α and β (and assuming the strings have negligible mass), we can find S and T as in the following worked example.

Worked Example 3.1

Two strings (of negligible mass) support a weight of 80 N, as in Figure 3.1. If $\alpha = 35°$ and $\beta = 60°$, find the tension in each string.

Resolving horizontally,

$$S \sin 35° = T \sin 60°$$

which gives

$$S = 1.5T$$

Resolving vertically,

$$80 = S \cos 35° + T \cos 60°$$

and, since $S = 1.5T$, this gives

$$S = 69 \text{ N and } T = 46 \text{ N}$$

What happens if we disconnect the right-hand string from the ceiling and pull it horizontally, as in Figure 3.2, so that the left-hand string still makes the same angle α with the vertical as before? The triangle of forces is shown in Figure 3.2(b). T no longer has a vertical component. This means that S has to support all of W, which results in an increase in its vertical component and, hence, in its horizontal component. The increase in the latter is supported by T.

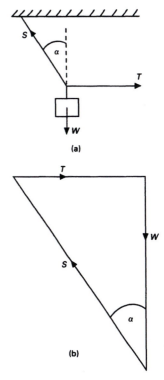

(a)

(b)

Figure 3.2

Worked Example 3.2

If, in Figure 3.2, $\alpha = 35°$ and $W = 80$ N, find the tension in each string.

From the triangle of forces in Figure 3.2(b),

$$S = W/\cos\alpha = 80/\cos 35° = 98 \text{ N}$$

and

$$T = W \tan \alpha = 80 \tan 35° = 56 \text{ N}$$

Finally, if we let go of the right-hand string completely (so that $T = 0$ and $\alpha = 0$), then $S = W$.

Figure 3.3 shows a variation on the same system. In Figure 3.3(a) we can see that the weight W is supported by two counterbalancing forces S and T supplied by masses suspended from the ceiling by pulleys. As long as there is no friction, the tension in each string can be assumed to be constant along its length, provided that its weight can be ignored. If we know S, T and W, then we can find α and β by constructing arcs of radius S and T, respectively from the ends of a line of length W, as in Figure 3.3(b). The point where the arcs intersect completes the triangle and enables us to measure α and β with a protractor. (Alternatively, these angles could be calculated by use of the cosine rule.)

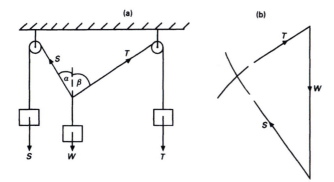

Figure 3.3

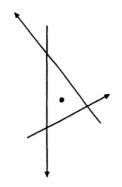

Figure 3.4

Note that in this case the values of S and T are fixed and the strings are free to move, so that the angles adjust themselves to bring the system to equilibrium. (In the fixed-string arrangement the geometry dictates the angles; the values of S and T therefore adjust themselves to bring the system to equilibrium.)

A system of forces will be in translational equilibrium provided that there is no net force in each of two directions (often taken to be mutually perpendicular for convenience). However, if the forces are not concurrent, they will tend to cause rotation. In the cases above, the lines of action of the forces S, T and W pass through the same point (where the strings join) and there is no tendency to rotation. In Figure 3.4 the three forces are not concurrent and the system will tend to rotate in an anticlockwise direction, even though it may be in translational equilibrium.

3.2 MOMENTS OF FORCES

The turning effect, or *moment*, of a force about a point (sometimes called *torque*) is defined as the product obtained by multiplying the force by the perpendicular distance of its line of action from the point. The units are newton metres, N m (not to be confused with joules, which are units of energy). The moment of the force F about the point O in Figure 3.5(a) is therefore $F \times s$ newton metres. This is consistent with practical experience (using a spanner, for example); a turning moment is made greater by increasing either the force F or its perpendicular distance s from the point, or both together. Note that if the line of action of the force passes through the point, then $s = 0$ and its moment is zero, so there will be no rotation.

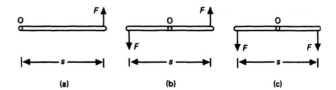

(a) (b) (c)

Figure 3.5

Figure 3.5(b) shows a *couple*, which consists of two equal and opposite parallel forces. This has a moment about O equal to the sum of the two individual moments. Each of these is equal to $F \times s/2$, giving a total moment of $F \times s$. Thus, the moment of a couple is equal to the product of one of the forces and the perpendicular distance between them. A couple clearly has no resultant, so it causes pure rotation, as in turning a key or a tap.

Figure 3.5(c) shows two equal parallel forces – for example, identical weights balanced at equal distances from the pivot of a see-saw. The anticlockwise moment of the left-hand force about O is equal to the clockwise moment of the right-hand force, so they balance each other – in other words, they are in rotational equilibrium. Two forces of different magnitude can be in rotational equilibrium provided that their respective values of $F \times s$ are equal and opposite – for instance, if one person twice the weight of the other is sitting half as far away from the pivot.

In fact we can have as many forces in rotational equilibrium as we like provided that the sum of the clockwise moments equals the sum of the anticlockwise moments. And we can take moments (i.e. find their sum) about any point we choose, although it is best to select one through which a force passes, or several forces in more complex cases; this makes their moments zero about the point and reduces the amount of calculation needed. It also enables us to eliminate an unwanted unknown force from a calculation.

Worked Example 3.3

A 3 m uniform beam of unknown mass, pivoted in the middle, supports a weight of 800 N at one end and another of 400 N at the other. Where must a further weight of 800 N act in order to bring the system to equilibrium?

Figure 3.6 shows the forces involved. The third weight must act between the 400 N weight and the pivot at some distance s from the pivot (which then supports a total weight of 2000 N plus the weight of

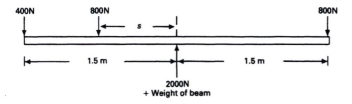

Figure 3.6

the beam). For equilibrium, the sum of the anticlockwise moments must equal the sum of the clockwise moments about any point. Taking moments about the pivot (to eliminate the unknown weight of the beam),

$$(400 \times 1.5) + (800 \times s) = (800 \times 1.5)$$

Therefore,

$$s = \frac{1200 - 600}{800} = 0.75 \text{ m}$$

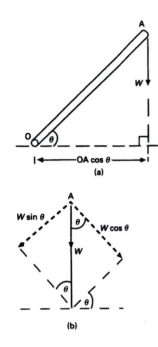

(a)

(b)

Figure 3.7

(In fact it would not have mattered if we had assumed s to be on the wrong side of the pivot – giving the third weight a clockwise moment. In this case we would have found that $s = -0.75$ m, the minus sign indicating a distance of 0.75 m from the pivot on the opposite side from the one we selected.)

Earlier we defined the moment of a force about a point as the product of the force and the perpendicular distance of its line of action from the point. In the examples we have considered so far the distance has been very obviously perpendicular to the line of action. Figure 3.7 shows a different case. In Figure 3.7(a), OA represents a lever arm pivoted at O with a body of weight W suspended from it at A. From our experience we would expect the moment of W about O to be at a minimum value (zero in fact) when OA is vertical and at a maximum when it is horizontal. As we can see from the figure, the perpendicular distance of the line of action of W from O is given by OA cos θ, where θ is the angle that OA makes with the horizontal. The moment of W about O is therefore $W \times$ OA cos θ. If OA is vertical, then $\theta = 90°$ or $270°$ and OA cos $\theta = 0$ (i.e. the line of action passes through O, so W has no turning effect). If OA is horizontal, then $\theta = 0°$ or $180°$, therefore, cos $\theta = 1$ or -1 and the perpendicular distance is OA, which is its maximum value and gives the maximum moment.

Figure 3.7(b) shows an alternative approach where W is resolved into components perpendicular to and along OA, i.e. W cos θ and W sin θ, respectively. The moment of W sin θ about O is zero, since its line of action passes through O, but the moment of the perpendicular component is OA $\times$ W cos θ, which gives the same result as above.

3.3. EQUILIBRIUM CONDITIONS

For a body to be in equilibrium, the forces acting upon it must balance and their clockwise and anticlockwise moments about any point must be equal. For translational equilibrium, we usually resolve the forces into two mutually perpendicular directions, typically vertically and horizontally, or perpendicular and parallel to some convenient plane. And, as noted above, we usually take moments about a point through which at least one of the unknown forces passes.

Worked Example 3.4

A uniform ladder weighing 200 N rests against a smooth (frictionless) vertical wall and makes an angle of 70° with the ground, which is level. Find the force exerted on the bottom end of the ladder by the ground.

Figure 3.8 shows the forces acting on the ladder. A frictionless surface can only exert a force perpendicular to itself; therefore, the only force the ladder experiences where it touches the wall is the normal force P. The ladder is uniform, so its weight acts half-way along its length l. It is at rest, so there is sufficient frictional force F between its bottom end and the ground to prevent if from slipping. N is the upward force exerted on the ladder by the ground. The force we require is the resultant of F and N.

Vertically we find that

$N = 200$ newtons

and horizontally

$F = P$

We can find P by taking moments about the bottom of the ladder (eliminating F and N), as follows

$P \times OB = 200 \times OA$

Therefore,

$P \times l \sin 70 = 200 \times (l/2) \cos 70°$

which, on rearranging, gives

$$p = \frac{200}{2 \tan 70°} = 36 \text{ N}$$

(Note that l cancels out, so we do not need to know the length of the ladder.)

The magnitude of the resultant is equal to

$\sqrt{200^2 + 36^2} = 203$ N

The direction of the resultant is equal to

$\tan^{-1} 200/36 = 80°$

(Be careful not to confuse the direction of the resultant ($\tan^{-1} N/F$) with the angle the ladder makes with the ground.)

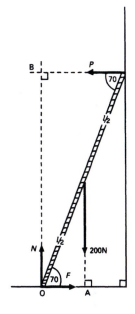

Figure 3.8

Questions

(Where necessary assume that $g = 9.8$ N kg^{-1}.)

1. Determine whether each of the following coplanar systems of concurrent forces is in approximate equilibrium.

If not, then find the magnitude and direction of the counter-balancing force required to bring the system to equilibrium. (Angles are measured anticlockwise from the *x*-axis.)

(a) 5.0 N at 20°, 6.0 N at 60°, 10.3 N at 222°
(b) 28 N at 32°, 34 N at 214°, 56 N at 313°
(c) 125 N at 27°, 240 N at 137°, 530 N at 270°
(d) 35 N at 47°, 43 N at 116°, 61 N at 215°, 54 N at 327°
(e) 63 N at 28°, 47 N at 128°, 38 N at 139°, 34 N at 203°, 85 N at 293°
(f) 119 N at 47°, 116 N at 108°, 124 N at 146°, 196 N at 233°, 207 N at 328°

2. If, in Figure 3.3(a), $S = T = W = 1$ kg, what are the values of α and β?

3. A weightless beam, 10 m long and supported at either end, carries a 50 kg load 2.5 m from one end. Find the forces supporting the beam at either end.

4. A 150 kg steel bar, 2.0 m long and of uniform cross-section, is supported by two vertical wires, one fixed at one end of the bar and the second at a distance of 0.8 m from the other end. Find the tension in each wire.

5. A weightless horizontal cantilever beam projects 5 m from a vertical wall.

   (a) If a 15 kg mass is placed on the end of the beam furthest from the wall, find the moment of its weight about the point where the beam enters the wall.
   (b) If an additional two masses are placed on the beam, 10 kg at 1 m and 5 kg at 3 m from the wall, respectively, find the total moment about the point where the beam enters the wall.
   (c) If the three masses are combined into a single mass, find how far from the wall it must be placed to provide the same moment as in (b).

6. Seven coplanar forces act at the corners of a 1 m square as shown in Figure 3.9.

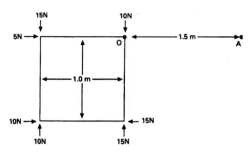

Figure 3.9

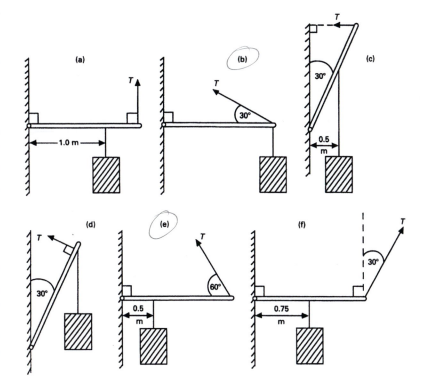

Figure 3.10

(a) Find the magnitude and direction of the resultant translational force.

(b) Find the net moment about O.

(c) Find the net moment about A.

7. Figure 3.10 shows a weightless rod, 1.5 m long, fastened to a vertical wall by a hinge at one end so that it can pivot vertically. A weightless wire is attached to the other end to enable it to support a 5.0 kg mass in the various arrangements shown. In each case (i) find the value of T, the tension in the wire, and (ii) find the vertical and horizontal components, and, hence, the magnitude and direction of the resultant force acting on the rod at the hinge.

8. If, in Worked Example 3.4, the ladder is on the point of slipping, what is the value of μ_s between the ladder and the ground?

9. If, in Worked Example 3.4, $\mu_s = 0.5$ between the ladder and the ground, what is the smallest angle the ladder could make with the ground (i.e. where it is on the point of slipping)?

TOPIC 4 PRESSURE AND UPTHRUST

COVERING:

- density and relative density;
- absolute pressure and gauge pressure;
- transmission of pressure;
- floating and sinking.

This topic is mainly concerned with forces that operate in liquids at rest, although some of it refers to gases as well.

4.1 DENSITY

First we need to recognise that the *density* of a substance is a measure of the mass it contains per unit volume. Expressing this in mathematical terms,

$$\rho = \frac{m}{V} \tag{4.1}$$

where m represents mass (kg), V volume (m^3) and ρ density (kg m^{-3}). Density is sometimes given in grams per cubic centimetre, g cm^{-3}. (Note that 1000 kg m^{-3} is equal to 1 g cm^{-3}.)

Relative density (formerly called *specific gravity*) is a dimensionless quantity which, for a liquid or a solid, is obtained by dividing its density by the density of water. The maximum density of water (1000 kg m^{-3}) is commonly used, in which case the relative density of a substance is obtained by dividing its density in kg m^{-3} by 1000. Although the density of water varies with temperature (its maximum occurring at 4°C), the variation is small over the normal range of room temperatures.

In practice, the relative density of a liquid can be measured by dividing its mass by the mass of an equal volume of water. This can be done by weighing the liquids in a *relative density bottle*, which is a small flask with a special stopper that ensures a reproducible volume each time it is filled.

4.2 PRESSURE

Pressure is a measure of the normal (perpendicular) force acting per unit area on a surface. The SI unit of pressure is called the pascal (Pa) and $1 \text{ Pa} = 1 \text{ N m}^{-2}$. (As we saw in Topic 2, the pascal can also be used as the unit of stress.)

Worked Example 4.1

A metal block, of relative density 8.90, measures $45 \times 70 \times 125$ mm. How much does it weigh? What apparent pressure would each of its faces exert when resting on a level surface? ($g = 9.8 \text{ N kg}^{-1}$.)

Substituting the data into Equation (4.1) ($m = V \times \rho$),

$$m = (0.045 \times 0.070 \times 0.125) \times 8.90 \times 10^3$$

which gives

$$m = 3.50 \text{ kg}$$

The block therefore weighs $3.50 \times 9.8 = 34.3$ N. Since pressure $p = $ force/area, then, when the block stands on its end,

$$p = \frac{34.3}{0.045 \times 0.070} = 1.1 \times 10^4 \text{ Pa}$$

and when it lies on its side,

$$p = \frac{34.3}{0.045 \times 0.125} = 6.1 \times 10^3 \text{ Pa}$$

and when it lies flat,

$$p = \frac{34.3}{0.070 \times 0.125} = 3.9 \times 10^3 \text{ Pa}$$

Pressure exists at any point in the body of a liquid because of the weight of liquid above it. It therefore increases with depth. Figure 4.1 shows an imaginary horizontal area ($= A$) at a depth h below the surface of a liquid. In effect, the area A supports a column of liquid above it which has a volume $h \times A$. Since $m = V \times \rho$ (from Equation 4.1) the mass of liquid in the column will be $h \times A \times \rho$. Multiplying this by g (9.8 N kg^{-1}) gives its weight as $h \times A \times \rho \times g$. Since

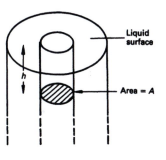

Liquid surface

Area $= A$

Figure 4.1

weight is a force and the area over which it acts is A, the pressure p at A is given by

$$p = \frac{\text{force}}{\text{area}} = \frac{h \times A \times \rho \times g}{A} = \rho g h \qquad (4.2)$$

This means that, since g is constant, the pressure acting at any point in a liquid simply depends on its depth and the density of the liquid above it. In practice, liquids are normally subjected to atmospheric pressure. Taking this into account, the total pressure is given by

$$p = \rho g h + p_{atm}$$

where p_{atm} represents atmospheric pressure. This leads to a distinction between the *absolute pressure* relative to a perfect vacuum and the *gauge pressure* measured relative to atmospheric pressure. (Remember that gauge pressure is zero at atmospheric pressure.)

The *manometer*, which is essentially a U-tube containing a liquid, is a device which measures gauge pressure (Figure 4.2).

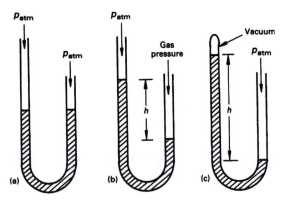

Figure 4.2

In Figure 4.2(a) both ends of the tube are open and the liquid surfaces on each side are level, because they are both subjected to atmospheric pressure. In Figure 4.2(b) the right-hand side is connected to a supply of gas under pressure. The liquid in the left-hand column will rise until the value of h is such that $\rho g h + p_{atm}$ balances the pressure of the gas supply. The difference in the height of the columns is therefore a measure of the gas pressure relative to the atmosphere – that is to say, $\rho g h$ gives the gauge pressure of the gas.

In Figure 4.2(c) the left-hand side has been sealed off with the space at the top under vacuum. The difference in the height of the columns now represents the absolute pressure at the open end compared with the vacuum. This is the basis of the mercury barometer used to measure atmospheric pressure. Atmospheric pressure fluctuates around 760 mmHg – that is to say, h equals 760 mm of mercury. (Hg is the chemical symbol for mercury.)

Worked Example 4.2

If the air pressure is 758 mmHg at the bottom of a mountain and 618 mmHg at the top, find the height of the mountain. (Assume that $\rho_{Hg} = 13.6 \times 10^3$ kg m^{-3} and that the average value of $\rho_{air} = 1.27$ kg m^{-3}.)

A column of air the same height as the mountain is equivalent to a mercury column of height equal to

$$0.758 - 0.618 = 0.140 \text{ m}$$

If $\rho g h$ for air = $\rho g h$ for mercury, then

$$1.27 \times g \times h_{air} = 13.6 \times 10^3 \times g \times 0.140$$

which gives

$$h_{air} = 1500 \text{ m}$$

Although the pascal is the SI unit of pressure, there are a number of others in use. 1 mmHg is a unit called the *torr*, which is often used in measuring low pressures. The *atmosphere* (atm), equivalent to 760 mmHg, is convenient for high pressures. If we take the density of mercury to be 13.6×10^3 kg m^{-3}, then Equation (4.2) tells us that 760 mmHg is equivalent to 101 kPa. (The accurate value for 1 atm is 101.325 kPa.) The *bar* is equal to 100 kPa and equivalent to 750 mmHg. Meteorologists use the *millibar*, which is 100 Pa. The pound per square inch (14.7 lb in^{-2} = 1 atm) was in common use at one time.

Pressure values are occasionally expressed in terms of the height of a column of water. Since the relative density of mercury is 13.6, a water column is much higher than the equivalent mercury column and this makes it easier to measure small pressures.

Liquids lack the rigidity of solids and have the ability to flow. This has important consequences when we think about applying pressure to a liquid at rest. First, the pressure at any point is the same in all directions. (Otherwise the liquid would not remain at rest.) Second, only normal forces can act between a liquid and a surface in contact with it. (Any parallel component would lead to relative movement between the two.) Third, a liquid will transmit an externally applied pressure uniformly throughout its volume, although there will still be the internal variation with depth.

This leads us on to the idea of amplifying forces by means of the

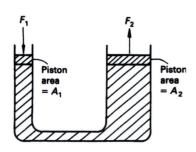

Figure 4.3

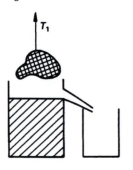

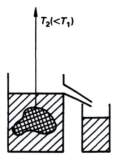

Figure 4.4

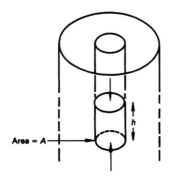

Figure 4.5

pressure in a liquid. Figure 4.3 shows a simplified version of a hydraulic press. It consists of two liquid-filled interconnected cylinders of different diameters, each fitted with a piston. If a downward force F_1 is applied to the smaller piston of area A_1, then the resulting pressure p transmitted by the liquid will be equal to F_1/A_1. The pressure will be converted to an upward force F_2 by the larger piston of area A_2, so that F_2/A_2 equals p. Since

$$\frac{F_2}{A_2} = p = \frac{F_1}{A_1}$$

$$\frac{F_2}{F_1} = \frac{A_2}{A_1}$$

The ratio between the output and the input force is therefore determined by the ratio between the areas of the output to the input piston.

4.3 UPTHRUST

From experience we recognise that an object placed in a liquid seems to get lighter. *Archimedes' principle* tells us that an object that is either totally or partially immersed in a fluid (liquid or gas) experiences an *upthrust* equal to the weight of fluid it displaces.

An object totally immersed in a liquid will displace a volume of liquid equal to its own. Figure 4.4 shows how the displaced liquid can be collected via a side arm fitted to the container from which it is displaced. If the object is suspended from a spring balance, the upthrust it experiences corresponds to the apparent reduction in its weight ($T_1 - T_2$ in the figure) as it is immersed in the liquid. And, from Archimedes' principle, we find that the upthrust is equal to the weight of liquid displaced.

To help us understand this, Figure 4.5 shows a cylindrical solid object, of height h and cross-sectional area A, replacing part of the liquid column we considered earlier. The pressure acting on the top surface of the cylinder will be less than the pressure at the bottom, where the liquid is deeper. Since pressure acts perpendicularly to solid surfaces, the pressure at the top will act downwards and the pressure at the bottom will act upwards. The vertical pressure difference will be ρgh (where ρ is the density of the liquid), so the net upward force or upthrust acting on the cylinder will be $A \times \rho gh$, which is precisely the same as the weight of the liquid ($Ah \times \rho g$) that it replaces. (The horizontal pressure acting on the sides of the cylinder will have no vertical effect.)

If the density of the object is greater than the density of the liquid, its weight cannot be supported by the upthrust and it will sink. If its density is less than that of the liquid, the upthrust will exceed its weight and it will rise. If its density is the same, it will stay where it is.

Archimedes' principle provides us with a very simple method of finding the relative density of a solid by dividing its weight in air by its apparent loss in weight in water.

Worked Example 4.3

A cube of density 2500 kg m^{-3} weighs 24.5 N in air. Find its apparent weight if it is

(a) totally immersed in water ($\rho = 1000$ kg m^{-3});
(b) half immersed in water;
(c) totally immersed in liquid of density 800 kg m^{-3};
(d) totally immersed in mercury ($\rho = 13.6 \times 10^3$ kg m^{-3}).

($g = 9.8$ N kg^{-1}.)

The mass of the cube $= 24.5/9.8 = 2.5$ kg. The volume of the cube $= 2.5/2500 = 1 \times 10^{-3}$ m^3.

(a) From Equation (4.1) ($m = V \times \rho$), the mass of water displaced by the cube is equal to

$$1 \times 10^{-3} \times 1 \times 10^3 = 1 \text{ kg}$$

Therefore, the upthrust (weight of water displaced) is equal to

$$1 \times 9.8 = 9.8 \text{ N}$$

Therefore,

apparent weight of cube $= 24.5 - 9.8 = 14.7$ N

(b) The upthrust on half the cube $= 4.9$ N; therefore,

apparent weight of cube $= 24.5 - 4.9 = 19.6$ N

(c) The mass of liquid displaced by the cube is equal to

$$1 \times 10^{-3} \times 0.8 \times 10^3 = 0.8 \text{ kg}$$

Therefore,

upthrust $= 0.8 \times 9.8 = 7.8$ N

Therefore,

apparent weight of cube $= 24.5 - 7.8 = 16.7$ N

(d) The mass of mercury displaced by the cube is equal to

$$1 \times 10^{-3} \times 13.6 \times 10^3 = 13.6 \text{ kg}$$

Therefore,

$$\text{upthrust} = 13.6 \times 9.8 = 133.3 \text{ N}$$

Therefore,

$$\text{apparent weight of the cube} = 24.5 - 133.3 = -108.8 \text{ N}$$

That is to say, the upthrust exceeds the weight of the cube by 108.8 N, which is therefore the force that would be needed to hold the cube beneath the surface of the mercury.

If an object is floating, so that the upthrust supports its weight exactly, then it must displace its own weight of liquid. If it has the same density as the liquid, then it will be totally immersed. If it is of lower density, then it will only need to displace part of its volume, so only part of it will sink below the surface; and the higher the density of the liquid the less the object has to sink. A steel ship floats because it is hollow and its average density is lower than that of water. But the density of water varies with dissolved salts and with temperature; thus, the ship floats higher in sea-water than in fresh water and lower in warm water than in cold water. The *hydrometer*, used to measure the relative density of liquids (e.g. battery acid), works in a similar way.

Worked Example 4.4

A piece of wood, of density 650 kg m^{-3}, measures 400 × 300 × 50 mm thick. How much of its thickness is immersed when it is floating in water? By how much does it sink if a 1 kg mass is placed centrally on top of it?

$$m = V \times \rho$$

Therefore,

$$m = (0.4 \times 0.3 \times 0.05) \times 650 = 3.9 \text{ Kg}$$

This will displace 3.9 kg of water, which occupies

$$3.9/1000 = 3.9 \times 10^{-3} \text{ m}^3$$

If d is the thickness immersed, then

$$d \times 0.4 \times 0.3 = 3.9 \times 10^{-3} \text{ m}^3$$

which gives

$$d = 33 \text{ mm}$$

The 1 kg mass will displace a further 1.0×10^{-3} m³ of water, so the wood will sink a further distance equal to

$$\frac{1.0 \times 10^{-3}}{0.4 \times 0.3} \text{ m} = 8 \text{ mm}$$

Questions

(Where necessary assume that $g = 9.8$ N kg^{-1} and that $\rho_{\text{water}} = 1000$ kg m^{-3}.)

1. What is the mass of air in a room measuring $7 \times 3 \times 3$ m? (Assume $\rho_{\text{air}} = 1.3$ kg m^{-3}.)

2. What is the volume of a sample of concrete that weighs 28.7 N and is known to have a density of 2.39×10^3 kg m^{-3}?

3. Assuming the earth's mass to be 6.0×10^{24} kg and its radius to be 6.4×10^6 m, what is its average density?

4. A bottle weighs 95.7 g when it is empty and 308.4 g when it is full of water. If 207.3 g of sand is placed in the empty bottle, then 133.6 g of water is required to fill it. Find the density of the sand, giving your answer in kg m^{-3}.

5. The pistons of a hydraulic press have diameters of 16 mm and 32 mm, respectively. What input force should be applied to the smaller piston to produce an output force of 1000 N, and what would be the corresponding pressure?

6. Convert the following to Pa: (a) 3.2 lb in^{-2}, (b) 14 torr, (c) 998 mbar, (d) 765 mmHg, (e) 8.9 atm.

7. An open-ended U-tube of uniform bore is partially filled with mercury. If 38.9 cm³ of liquid of density 800 kg m^{-3} is poured into one side, what volume of water must be poured into the other to keep the mercury levels equal?

8. An irregularly shaped metal object weighs 15.6 N in air, 13.6 N in water and 14.0 N in an unknown liquid.

   (a) Find the density of the metal.
   (b) Find the density of the unknown liquid.

9. A polythene block of mass 138 g is held under water by means of a thread fastened to the bottom of the container. Find the tension in the thread. ($\rho = 920$ kg m^{-3} for the polythene.)

TOPIC 5 DISPLACEMENT, VELOCITY AND ACCELERATION

COVERING:

- accelerated motion in a straight line;
- the equations of motion (kinematic equations);
- acceleration due to gravity;
- projectiles.

So far our discussion has been about systems that are either at rest or in a state of uniform motion. This topic introduces the idea of accelerated motion but, for the moment, without any reference to the forces that cause it.

There are two basic quantities that help us to define motion – namely length and time. The first quantity derived from these is velocity, which is the rate at which displacement changes with time.

Before going any further, it is important to make a careful distinction between speed and velocity. Speed is a scalar quantity measured in terms of magnitude alone – that is to say, it represents the rate of change of distance with time. Velocity is a vector quantity which has direction as well as magnitude. At any particular moment, the speed of a moving body represents the magnitude of its velocity. If a body travels along a curved path at constant speed, its velocity is changing because its direction is changing.

To start at the simplest level, we shall consider motion in a straight line. (In effect, we can then think of any change in displacement in terms of change in distance alone.) Note that because they are both directional, displacement and velocity can have negative values signifying movement opposite to some reference direction already taken as being positive. For example, if we think of an object thrown straight upwards as having a positive velocity, then once it starts to fall it will have a negative velocity.

Acceleration is the rate at which velocity changes. In terms of SI units, velocity is measured in metres per second ($m\ s^{-1}$) and acceleration in metres per second per second ($m\ s^{-2}$). When velocity is decreasing, acceleration has a negative value and is generally called either retardation or deceleration.

These definitions of velocity and acceleration lead to some useful equations that describe the motion of a body undergoing uniform acceleration in a straight line. (Uniform acceleration means that velocity is changing at a constant rate. For example, a body starting from rest with a uniform acceleration of 2 m s^{-2} will have a velocity of 0 m s^{-1} at 0 s, 2 m s^{-1} at 1 s, 4 m s^{-1} at 2 s, 6 m s^{-1} at 3 s, and so on.)

If a body accelerates uniformly from an initial velocity u to a final velocity v in a period of time t, then the acceleration a is given by the change in velocity per unit time, as follows:

$$a = \frac{v - u}{t}$$

which, on rearranging, becomes

$$v = u + at \tag{5.1}$$

Since velocity is defined as displacement per unit time, the displacement s of the body can be obtained by multiplying its average velocity, $(u + v)/2$, by the time t over which the acceleration takes place. Thus,

$$s = \left(\frac{u + v}{2}\right) t \tag{5.2}$$

By substituting $(u + at)$ for the final velocity v (from Equation 5.1), Equation (5.2) becomes

$$s = \left(\frac{u + u + at}{2}\right) t$$

Therefore,

$$s = \frac{2ut + at^2}{2}$$

and

$$s = ut + \frac{1}{2} at^2 \tag{5.3}$$

Finally, by substituting $2s/(u + v)$ for t (from Equation 5.2), Equation (5.1) becomes

$$v = u + a\,\frac{2s}{(u + v)}$$

and, on rearrangement,

$$2as = (v - u)(u + v)$$

Therefore,

$$2as = v^2 - u^2$$

and

$$v^2 = u^2 + 2as \tag{5.4}$$

This gives us four equations involving the five quantities u, v, t, a and s, as summarised in Table 5.1. Careful inspection of the table shows that we can calculate the value of any two quantities provided that we know the values of the other three. For example, given u, a and s, we can calculate v from Equation (5.4) and, quite independently, t from Equation (5.3); given u, v and t, we can calculate a from Equation (5.1) and s from Equation (5.2) – and so on.

Table 5.1

Equation	Involves	Omits
(5.1) $v = u + at$	v, u, a, t	s
(5.2) $s = \dfrac{(u + v)}{2} t$	s, u, v, t	a
(5.3) $s = ut + \dfrac{1}{2} at^2$	s, u, t, a	v
(5.4) $v^2 = u^2 + 2as$	v, u, a, s	t

Worked Example 5.1

An object is uniformly accelerated from rest to 25 m s^{-1} over a period of 10 s. Find:

(a) the acceleration;
(b) the distance travelled;
(c) the extra time needed to reach 50 m s^{-1};
(d) the uniform acceleration needed to bring it to rest from 50 m s^{-1} in a distance of 250 m.

(a) $u = 0$, $v = 25$, $t = 10$, $a = ?$

From Equation (5.1)

$$25 = a \times 10$$

Therefore,

$$a = 2.5 \text{ m s}^{-2}$$

(b) $u = 0$, $v = 25$, $t = 10$, $s = ?$

From Equation (5.2)

$$s = \frac{(0 + 25)}{2} \times 10 = 125 \text{ m}$$

(c) $u = 25$, $v = 50$, $a = 2.5$, $t = ?$

From Equation (5.1)

$$50 = 25 + (2.5 \times t)$$

Therefore,

$$t = 10 \text{ s}$$

(d) $u = 50$, $v = 0$, $s = 250$, $a = ?$

From Equation (5.4) (where a is a retardation)

$$0 = 50^2 + 2 \times (-a) \times 250$$

Therefore,

$$a = 5 \text{ m s}^{-2}$$

Worked Example 5.2

An object at rest experiences a uniform acceleration of 5 m s^{-2} for 6 s. It maintains constant velocity for 14 s and is then brought to rest in 5 s by a uniform retardation. How far has it travelled?

Acceleration stage:

$$u = 0, \quad t = 6, \quad a = 5, \quad s = ?, \quad v = ?$$

From Equation (5.3)

$$s = \frac{1}{2} \times 5 \times 36 = 90 \text{ m}$$

From Equation (5.1)

$$v = 5 \times 6 = 30 \text{ m s}^{-1}$$

Constant velocity stage:

$u = v = 30, \quad t = 14, \quad a = 0, \quad s = ?$

From Equation (5.3)

$s = 30 \times 14 = 420$ m

Retardation stage:

$u = 30, \quad v = 0, \quad t = 5, \quad s = ?$

From Equation (5.2)

$$s = \frac{(30 + 0)}{2} \times 5 = 75 \text{ m}$$

Therefore,

total distance travelled $= 90 + 420 + 75 = 585$ m

Note that all the four equations (5.1–5.4) contain u. If u is one of two unknowns, its value must be calculated first by using the appropriate equation. Then it can be substituted into any of the others to find the value of the second unknown. For example, if u and v are unknown, the value of u must be obtained from Equation (5.3); then the value of v can be obtained from any of the others.

Worked Example 5.3

A moving particle experienced an acceleration of 10 m s^{-2} over a period of 8 s, during which time it travelled 400 m. What were its initial and final velocities?

$t = 8, \quad a = 10, \quad s = 400, \quad u = ?, \quad v = ?$

From Equation (5.3)

$$400 = (u \times 8) + \left(\frac{1}{2} \times 10 \times 64\right)$$

Therefore,

$u = 10$ m s^{-2}

From Equation (5.1)

$v = 10 + (10 \times 8) = 90$ m s^{-1}

Equations (5.1)–(5.4), which are called the *equations of motion* or *kinematic equations*, can be used for calculations that involve bodies moving under the influence of gravity. We shall begin by considering this in terms of motion in a vertical straight line.

The first thing we need to recognise is that all objects allowed to fall freely close to the earth's surface experience the same downward acceleration. This fact may seem surprising and contrary to experience, because it tends to be obscured by the effects of air resistance and, for very light objects, the slight upthrust in air. A leaf or a feather seems to drift downwards at a more or less steady rate, and even a dense object reaches a steady *terminal velocity* when its weight is balanced by the effects of upthrust and air resistance (so that the resultant force is zero). But if an object of any mass or density is allowed to fall *freely* (i.e. with unimpeded motion, as in a vacuum), then it will accelerate uniformly. This *acceleration due to gravity*, sometimes called the *acceleration of free fall*, has the value 9.8 m s^{-2}. (There are slight variations over the earth's surface but we shall ignore them in this book.)

As we shall see in the next topic, it is no coincidence that the strength of the earth's gravitational field has the same numerical value as the acceleration due to gravity. In fact, engineers tend to regard g as an acceleration (9.8 m s^{-2}) rather than a field strength (9.8 N kg^{-1}). So, provided that any effect due to the air is small enough to be ignored, $a = g = 9.8$ m s^{-2} in the equations of motion when they are applied to objects moving vertically under the influence of gravity.

Sometimes the calculations are simplified by either u or v having the value zero. For example, if an object falls from rest (so that $u = 0$), then Equation (5.1) gives the relationship between its velocity and the time it has been falling (i.e. $v = gt$) and we can calculate the velocity after a given time or the time required to reach a given velocity. Similarly, Equations (5.3) and (5.4) give simple relationships between displacement and time and between final velocity and displacement, respectively, if the object falls from rest.

If an object is thrown vertically upwards with a known initial velocity u, then, by setting v to zero, we can calculate the maximum height s it reaches (Equation 5.4) and the time t it takes to get there (Equation 5.1). Note that if we take the upward direction as positive, then u and s have positive values and $a = -g$ because it acts downwards.

Worked Example 5.4

An object is thrown vertically upwards with an initial velocity of 20 m s^{-1}. What height does it reach, and what is the total time it takes to return to its starting point?

$$u = 20, \quad v = 0, \quad a = g = 9.8, \quad s = ?, \quad t = ?$$

From Equation (5.4) (taking upwards as positive)

$$0 = 400 + (2 \times (-9.8) \times s)$$

which gives

$$s = 20.4 \text{ m}$$

From Equation (5.1)

$$0 = 20 + ((-9.8) \times t)$$

which gives

$$t = 2.04 \text{ s}$$

The object will take an equal time to return to its starting point, so the total time is 4.08 s.

Now let us add a second dimension, Figure 5.1 shows the path of an object thrown horizontally from a height s above the ground. Its velocity is resolved into horizontal and vertical components along the x- and y-axes, respectively. Since the object is thrown horizontally, its initial velocity u has no vertical component (i.e. $u_x = u$ and $u_y = 0$) but it immediately starts to fall, so at any subsequent moment its velocity v is the resultant of v_x and v_y. These components can be considered quite independently and the object takes the same length of time to hit the ground as if it had fallen from rest from the same height. Knowing the height s, we can calculate the time of flight t from Equation (5.3) (remembering that $u_y = 0$). If we neglect air resistance, the horizontal component of the velocity will remain constant and equal to u_x throughout the flight. This means that we can calculate the horizontal distance the object travels by multiplying u_x by t.

We can also calculate the magnitude and direction of its velocity by finding the vertical component v_y at any time (Equation 5.1) or at any height (Equation 5.4) and combining it with the horizontal component v_x ($= u_x$).

If the object had been projected from the ground with an equal and opposite velocity to that with which it had landed, then its path would have been the same as in Figure 5.1, but, of course, traversed in the opposite direction. (In this case the figure only shows half the path if the object goes on to complete its trajectory and return to the ground.)

Worked Example 5.5

An object is thrown horizontally at 6.0 m s^{-1} from a height of 3.3 m above level ground. How far does the object travel horizontally, and

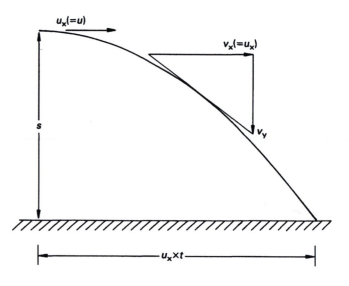

Figure 5.1

what is the magnitude and the direction of its velocity when it hits the ground?

_____

Vertically

$$u_y = 0, \quad a = g = 9.8, \quad s = 3.3, \quad t = ?, \quad v_y = ?$$

From Equation (5.3) (taking downwards as positive)

$$3.3 = \frac{1}{2} \times 9.8 \times t^2$$

which gives

$$t = 0.82 \text{ s}$$

Therefore, the horizontal distance travelled is equal to

$$6 \times 0.82 = 4.9 \text{ m}$$

From Equation (5.4)

$$v_y^2 = 2 \times 9.8 \times 3.3$$

Therefore,

$$v_y = 8.0 \text{ m s}^{-1}$$

The magnitude of the final velocity is equal to

$\sqrt{6.0^2 + 8.0^2} = 10.0 \text{ m s}^{-1}$

The direction of the final velocity is equal to

$\tan^{-1} 8/6 = 53°$ below the horizontal

Worked Example 5.6

An object is projected at 10 m s^{-1} at an angle of 53° above the horizontal. What is the maximum height it reaches, and what horizontal distance does it cover?

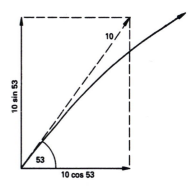

Figure 5.2

The horizontal and vertical components of the initial velocity are shown in Figure 5.2. First, note that 10 sin 53° = 8 and 10 cos 53° = 6. Vertically:

$u_y = 10 \sin 53°, \quad v_y = 0, \quad a = g = 9.8, \quad s = ?, \quad t = ?$

From Equation (5.4) (taking upwards as positive)

$0 = (10 \sin 53°)^2 + (2 \times (-9.8) \times s)$

which gives

$s = 3.3 \text{ m}$

From Equation (5.1)

$0 = 10 \sin 53° + ((-9.8) \times t)$

which gives

$t = 0.82 \text{ s}$

Horizontally:

$u_x = 10 \cos 53°, \; t = 0.82, \; s = ?$

From Equation (5.3) $a = 0$ if we ignore wind resistance, and

$s = 10 \cos 53° \times 0.82 = 4.9 \text{ m}$

The object travels a further 4.9 m in returning to the ground therefore

total horizontal distance $= 2s = 9.8 \text{ m}$

Questions

(Apart from Questions 9 and 10, assume motion in a straight line. Where necessary, assume $g = 9.8$ m s^{-2} and neglect air resistance and upthrust.)

1. An object is travelling at 12 m s^{-1}. For what period of time must it be accelerated at 4 m s^{-2} in order to reach 48 m s^{-1}?

2. What distance is taken for an object to reach 50 m s^{-1} from rest if it is accelerated at 2.5 m s^{-2}?

3. An object is projected vertically upwards from ground level at 49 m s^{-1}. Find its height after 1 s, 3 s, 5 s, 7 s and 9 s. What is its total time of flight?

4. An object is travelling at 12 m s^{-1}. It experiences a uniform acceleration and covers the next 36 m in 6 s. What is its final velocity and its acceleration?

5. An object initially at rest experiences an acceleration of 6 m s^{-2} for 4.5 s. What acceleration is required to return it to rest in 9 s?

6. An object that had fallen from a height of 128 m was found to have penetrated the ground to a depth of 40 mm. Estimate its average deceleration.

7. An object accelerates from 10 m s^{-1} to 50 m s^{-1} over a distance of 120 m. Calculate the acceleration and the time taken.

8. An object thrown vertically upwards reached a height of 28.8 m in 1.75 s. What maximum height did it reach and what time did it take to get there?

9. An object is dropped from an aircraft that is flying horizontally with a velocity of 100 m s^{-1} at a height of 250 m. What horizontal distance does the object cover before hitting the ground and what is the magnitude and direction of its impact velocity?

10. An object is projected with a velocity u at an angle θ above the horizontal. If g is the acceleration due to gravity, derive formulae giving (a) the maximum height to which it rises and (b) the time it takes to get there.

TOPIC 6 FORCE AND MOTION

COVERING:

- Newton's laws of motion;
- force and acceleration;
- action and reaction.

Newton's first law tells us that an object will remain at rest or in a state of uniform motion unless a force acts on it. In effect, this defines force as an influence which tends to change the velocity of an object. (Remember that the force in question might be the resultant of two or more others.)

Newton's second law tells us that, when an object is acted upon by a force, it will accelerate in accordance with the expression

$$F = ma \tag{6.1}$$

where F is the force (measured in N), m is the mass of the object (in kg) and a is its acceleration (in m s^{-2}).

As we might expect, the acceleration lies in the same direction as that of the force. The equation tells us that the larger the mass the larger the force needed to produce a given acceleration. Or, looking at it another way, the magnitude of a force can be found by measuring the acceleration that it gives to a known mass. Furthermore, if there is no force, then there can be no acceleration (which is another way of stating the first law). The equation also tells us that 1 N is the magnitude of the force that gives a 1 kg mass an acceleration of 1 m s^{-2}. (As we shall see in the next topic, the second law can also be stated in terms of momentum rather than acceleration.)

In Topic 2 we saw that the force due to gravity acting on a mass m (i.e. its weight W) provides a measure of the gravitational field strength g. At the earth's surface $g = W/m = 9.8$ N Kg^{-1}. If the mass is allowed to fall freely, then Equation (6.1) tells us that, since $F = W$, W/m also gives its acceleration. So we can regard g either as a gravitational field strength or as an acceleration equal to 9.8 m s^{-2}. As we noted in the last topic, engineers generally regard it as an acceleration, so we shall do the same.

Note that Equation (6.1) does not tell us anything about the displacement or velocity of an object. If we need to know about these, and the force is constant, then we use the equations of motion.

Worked Example 6.1

A steady horizontal force of 12 N is applied to a 6 kg mass that is at rest on a smooth level surface. After 15 s (a) how far has the mass moved and (b) what is the magnitude of its velocity?

———————

(a) $F = ma$

Therefore,

$$a = F/m = 12/6 = 2 \text{ m s}^{-2}$$

Now

$$u = 0, \quad t = 15, \quad a = 2, \quad s = ?$$

so substituting in

$$s = ut + \frac{1}{2} at^2$$

we obtain

$$s = \frac{1}{2} \times 2 \times 15^2 = 225 \text{ m}$$

(b) $u = 0, \quad t = 15, \quad a = 2, \quad v = ?$

Substituting in

$$v = u + at$$

we obtain

$$v = 2 \times 15 = 30 \text{ m s}^{-1}$$

———————

Worked Example 6.2

A 15 kg mass is uniformly accelerated from 10 m s^{-1} to 20 m s^{-1} over a distance of 300 m. What force is being exerted on it?

———————

$$u = 10, \quad v = 20, \quad s = 300, \quad a = ?$$

Substituting in

$$v^2 = u^2 + 2as$$

we obtain

$$400 = 100 + (2 \times a \times 300)$$

which gives

$$a = 0.5 \text{ m s}^{-2}$$

Therefore,

$$F = ma = 15 \times 0.5 = 7.5 \text{ N}$$

Worked Example 6.3

A ball of 100 g mass fell from a height of 10 m and rebounded to a height of 5 m. Assuming that it remained in contact with the ground for 12 ms, what was the average force it exerted on the ground? ($g = 9.8$ m s^{-2}.)

Considering the ball as it fell downwards,

$$u = 0, \quad a = 9.8, \quad s = 10, \quad v = ?$$

Substituting in

$$v^2 = u^2 + 2as$$

we obtain

$$v^2 = 2 \times 9.8 \times 10$$

which gives

$$v = 14.0 \text{ m s}^{-1} \text{ downwards at impact}$$

Considering the ball as it rebounded upwards,

$$v = 0, \quad a = -9.8, \quad s = 5, \quad u = ?$$

Substituting in

$$v^2 = u^2 + 2as$$

we obtain

$$0 = u^2 + (2 \times (-9.8) \times 5)$$

which gives

$u = 9.9$ m s^{-1} upwards after impact

The change in velocity due to the impact was therefore 23.9 m s^{-1} (i.e. from 14.0 m s^{-1} downwards to 9.9 m s^{-1} upwards). This took place over a period of 12×10^{-3} s and corresponds to an average acceleration of

$$\frac{23.9}{12 \times 10^{-3}} = 2.0 \times 10^3 \text{ m s}^{-2}$$

Therefore,

$$F = ma = 0.1 \times 2.0 \times 10^3 = 200 \text{ N}$$

Newton's third law states that action and reaction are equal and opposite. In other words, if one object exerts a force on another, then the second object exerts an equal but opposite force on the first. For instance, viewing Newton's third law in terms of his law of gravitation (Equation 2.1 on page 9), we can say that the earth exerts a gravitational force on an apple above its surface and the apple exerts an equal and opposite gravitational force on the earth. If the apple is released, the second law tells us that

$$\text{force} = m_{apple} \times a_{apple} = m_{earth} \times a_{earth}$$

That is to say, the apple will fall towards the earth and, at the same time, the earth will fall towards the apple though its acceleration ($a_{earth} = \text{force}/m_{earth}$) will be infinitesimal because its mass is so large.

The third law applies equally well to objects in contact. If the apple is lying on the ground, it presses downwards with a force equal to its weight and the ground reacts with an equal and opposite force on the apple. When a motor car tyre pushes backwards on the road, there is an equal and opposite reaction as the road pushes forward on the tyre – and the car moves forwards.

The laws of motion help us where vertical forces and acceleration are involved. Let us imagine an object of mass m resting on the floor of a lift. If the lift is stationary or moving with uniform velocity (i.e. $a = 0$), then there can be no net vertical force acting on the object, because $F = ma = 0$. The object exerts a downward force mg on the floor and there is an equal and opposite reaction as the floor pushes upwards on the object.

If the lift starts to move upwards with a uniform acceleration *a*, then not only does it have to support the object's weight *mg*, it also has to support the additional force $F = ma$ needed to make the object accelerate upwards. The floor of the lift therefore pushes upwards on the object with a total force $(mg + ma)$ and the object experiences an apparent increase in weight.

If the lift starts to move downwards with an acceleration *a*, then subsequent events depend on whether *a* is less than, equal to or greater than *g*, the acceleration due to gravity.

If the lift is propelled downwards so that *a* is greater than *g*, then the object will be left behind and, neglecting the effects of the air, will fall freely with acceleration *g* (until the ceiling of the lift catches up with it). If the cable breaks, then $a = g$ and the lift and the object will both fall freely together.

If *a* is less than *g*, then the resultant force *ma* accelerating the object downwards will be its weight *mg* less the support it receives from the floor (i.e. its apparent weight), so that

$$ma = mg - \text{apparent weight}$$

Therefore,

$$\text{apparent weight} = mg - ma$$

We shall consider further consequences of Newton's laws in the next topic.

Worked Example 6.4

An object of 30 kg mass is being propelled across a horizontal surface with a horizontal force of 202 N. If the coefficient of kinetic friction is 0.55, find the acceleration. (Assume $g = 9.8$ m s^{-2}.)

Frictional force $= \mu_k N = 0.55 \times 30 \times 9.8 = 162$ N

Therefore,

resultant force $= 202 - 162 = 40$ N

and

$$a = F/m = 40/30 = 1.3 \text{ m s}^{-2}$$

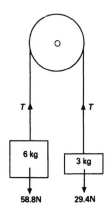

Figure 6.1

Worked Example 6.5

A 6 kg mass is connected to a 3 kg mass by a string that passes over a frictionless pulley. Assuming that the string and pulley have no effect on the motion of the system, find the tension in the string and the acceleration of the masses. (Assume $g = 9.8$ m s^{-2}.)

The weights experienced by the masses are 58.8 N and 29.4 N, as shown in Figure 6.1. Since the pulley is frictionless, the tension T in the string will be uniform along its length. Clearly the 6 kg mass will fall and the 3 kg mass will rise with the same acceleration. Taking this direction (with the pulley rotating anticlockwise) as positive, then the net force acting on the 6 kg mass is $(58.8 - T)$ and, from the second law,

$$a = \frac{F}{m} = \frac{(58.8 - T)}{6}$$

Similarly, for the 3 kg mass

$$a = \frac{F}{m} = \frac{(T - 29.4)}{3}$$

Since a is the same in both cases

$$\frac{(58.8 - T)}{6} = \frac{(T - 29.4)}{3}$$

which gives

$$T = 39.2 \text{ N}$$

and, from above,

$$a = \frac{(58.8 - T)}{6} = \frac{(58.8 - 39.2)}{6} = 3.3 \text{ m s}^{-2}$$

Questions

(Where necessary assume that $g = 9.8$ m s^{-2}.)

1. If an apple weighing 1.0 N falls to the ground, what is the acceleration of the earth towards the apple? (Mass of the earth $= 6.0 \times 10^{24}$ kg.)

2. What is the acceleration of an 18 kg mass upon which a force of 27 N is being exerted? If it starts from rest, how long does it take to cover a distance of 50 m?

3. A mass of 8 kg is being pulled by a force of 34 N in the 3 o'clock direction and by a force of 18 N in the 9 o'clock direction. If its initial speed is 15 m s^{-1}, how long does it take for it to reach 30 m s^{-1}?

4. A horizontal force of 247 N is applied to an object of 25 kg mass resting on a level surface. Assuming $\mu_s = \mu_k = 0.60$, what distance does the object travel in 10 s?

5. A 22 kg mass is subjected to three forces: 12 N at 0°, 18 N at 45° and 24 N at 90°. Find its acceleration.

6. A 6 kg mass on a frictionless horizontal surface is subjected to two forces: 14 N at 30° and 17 N at 60°. If it starts at rest, what point will it reach after 12 s?

7. An object of mass 4.5 kg accelerates at a rate of 2.0 m s^{-2} when acted upon by a force of 35 N parallel to the horizontal surface which supports it. Find the coefficient of kinetic friction between the object and the surface.

8. A catapult is used to throw a 50 g stone horizontally with a velocity of 14 m s^{-1}. If the catapult sling had been drawn back through a distance of 550 mm, estimate the average force it exerted on the stone when it was released.

9. A 10 kg mass hangs from a string which can support a maximum load of 200 N. What is the maximum acceleration that can be used to raise the mass vertically by pulling the string without breaking it?

10. A 10 g projectile exerted an average force of 100 N on its target, which it penetrated to a depth of 0.5 m. Estimate its velocity on impact.

11. A passenger of 60 kg mass stands in a lift which accelerates upwards at 0.5 m s^{-2}. Find (a) the passenger's apparent weight, and (b) the tension in the lift cable. (The mass of the empty lift is 840 kg.)

TOPIC 7 MOMENTUM AND IMPULSE

COVERING:

- linear momentum;
- momentum, force and impulse;
- the principle of conservation of momentum,

Linear momentum is a physical quantity that provides another approach to the behaviour of objects in motion. It is a vector quantity obtained by multiplying the mass of an object by its velocity. It therefore has the unit kg m s^{-1}. Its direction is the same as that of the velocity of the object. The word *linear* is used to distinguish it from the angular momentum of a rotating body, which we shall meet later. The use of the word 'momentum' alone implies linear momentum, and that is the convention we shall adopt here.

Newton's second law of motion is often stated in a form telling us that the rate at which the momentum of an object changes with time is proportional to, and in the same direction as, the net force acting upon it. If the velocity of an object of mass m changes from u to v, then the change of momentum is given by $(mv - mu)$. If this change is brought about by a net force F acting on the object for a period of time t, then, using SI units, the second law can be expressed in the form

$$F = \frac{(mv - mu)}{t} \tag{7.1}$$

Note that, in the case of retardation, u will be greater than v and that, in any case, u and v may have positive or negative values, depending on the reference direction chosen. F may therefore have a negative value, which simply means that it is acting in the opposite direction to that originally chosen as being positive. Also note that $(v - u)/t$ gives the acceleration a, so Equation (7.1) can be reduced to the form we used in the previous topic. Thus,

$$F = m\,\frac{(v - u)}{t} = ma$$

Rearranging Equation (7.1), we find that

$$Ft = (mv - mu)$$

Ft represents the *impulse* given to the body which changes its momentum. Impulse is a vector quantity like momentum and has the unit N s, which is, as the equation suggests, equivalent to momentum. (Remember that, by definition, $1 \text{ N} = 1 \text{ kg m s}^{-2}$, so $1 \text{ N s} = 1$ kg m s^{-1}.) Note that a small force exerted over a long period can provide the same change in momentum as a large force over a short period. (It is better to fall onto a mattress than a concrete floor, because the same change in momentum is spread over a longer period and the average retarding force is correspondingly smaller.)

We now have three ways of viewing Equation (7.1):

force = mass $\times$ acceleration

force = change of momentum per unit time

impulse = force $\times$ time = change of momentum

The last of these leads us to a very important principle. Consider two objects which collide. According to Newton's third law, the force exerted by object A on object B will be equal and opposite to that exerted by B on A. Their respective impulses *Ft* must therefore be equal and opposite (since *t* is the same for both) and so must their respective changes in momentum. The total momentum of the system therefore remains constant. This principle, which is true for any system with any number of interacting objects, is known as the *principle of conservation of momentum*. It can be stated in the form that the total momentum of a system of interacting bodies remains unchanged as long as no external resultant force acts upon it.

This applies to all types of collisions, either where the objects involved move apart afterwards or where they stick together. If two objects of mass m_1 and m_2 collide with initial velocities u_1 and u_2, respectively, and then continue on their separate ways with velocities v_1 and v_2, their total momentum remains constant. Thus, we have

$$m_1 u_1 + m_2 u_2 = m_1 v_1 + m_2 v_2 \tag{7.2}$$

If the objects stick together, then Equation (7.2) becomes

$$m_1 u_1 + m_2 u_2 = (m_1 + m_2) v_c$$

where v_c is the velocity of the combined masses.

The principle can also be applied to a bullet fired from a gun. The momentum of the bullet is equal and opposite to the momentum of the gun as it recoils, so the total momentum remains zero:

$$m_{\text{bullet}} v_{\text{bullet}} - m_{\text{gun}} v_{\text{gun}} = 0$$

Similarly, the thrust of a rocket motor results from the momentum of its exhaust gases.

Worked Example 7.1

An object of mass 9.0 kg moves in a straight line at 5.0 m s^{-1} and collides with another object of mass 5.0 kg moving at 3.0 m s^{-1} in the same direction.

  (a) If the 5.0 kg object moves on at 5.5 m s^{-1}, find the velocity of the 9.0 kg object.
  (b) If the objects stick together, find the velocity of the combination.

(a) $m_1u_1 + m_2u_2 = m_1v_1 + m_2v_2$

Taking the common direction of motion of the objects as positive and substituting the given values,

$$(9.0 \times 5.0) + (5.0 \times 3.0) = (9.0 \times v_1) + (5.0 \times 5.5)$$

which gives

$$v_1 = 3.6 \text{ m s}^{-1} \text{ (in the same direction)}.$$

(b) $m_1u_1 + m_2u_2 = (m_1 + m_2)v_c$

Taking the common direction of motion of the objects as positive and substituting the given values,

$$(9.0 \times 5.0) + (5.0 \times 3.0) = 14.0 \times v_c$$

which gives

$$v_c = 4.3 \text{ m s}^{-1} \text{ (in the same direction)}$$

Worked Example 7.2

An object of 5.0 kg mass moving in the 9 o'clock direction at 10.0 m s^{-1} collides with an object of 2.0 kg mass moving in the 3 o'clock direction at 7.5 m s^{-1}. What is their final velocity if they stick together?

$m_1u_1 + m_2u_2 = (m_1 + m_2)v_c$

Taking the 3 o'clock direction as positive and substituting the given values,

$$(5.0 \times (-10.0)) + (2.0 \times 7.5) = 7.0 \times v_c$$

which gives

$$v_c = -5.0 \text{ m s}^{-1}$$

the minus sign indicating that v_c is in the opposite direction to that chosen as positive.

Worked Example 7.3

A ball of 100 g mass fell vertically to the ground with an impact velocity of 14 m s^{-1} and rebounded at 10 m s^{-1}. Assuming that it remained in contact with the ground for 12 ms, what was the average force it exerted on the ground?

$$F = m \frac{(v - u)}{t}$$

Taking the upward direction as positive and substituting the given values,

$$F = 0.1 \frac{(10 - (-14))}{12 \times 10^{-3}} = 200 \text{ N}$$

Worked Example 7.4

If, in Worked Example 7.2, the 2.0 kg mass had been moving in the 12 o'clock direction at 7.5 m s^{-1}, what would the combined final velocity have been?

Since the momenta of the objects are vector quantities, their resultant can be found from the momentum diagram in Figure 7.1. From the diagram

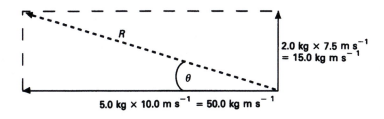

5.0 kg × 10.0 m s^{-1} = 50.0 kg m s^{-1}

2.0 kg × 7.5 m s^{-1} = 15.0 kg m s^{-1}

Figure 7.1

$$R = \sqrt{50.0^2 + 15.0^2} = 52.2 \text{ kg m s}^{-1}$$

Therefore,

$$\text{velocity} = \text{momentum/mass} = 52.2/7.0 = 7.5 \text{ m s}^{-1}$$

and

$$\theta = \tan^{-1} 15.0/50.0 = 17°$$

i.e. 163° anticlockwise from the positive x-axis.

Questions

(Where necessary assume that $g = 9.8 \text{ m s}^{-2}$.)

1. What force is required to uniformly accelerate a 1000 kg vehicle from 40 km per hour to 60 km per hour in 10 s?

2. A projectile of mass 10 g exerts an average force of 20 N on its target as it is brought to a halt 0.05 s after its initial impact. Estimate the magnitude of its impact velocity.

3. A bullet of 15 g mass leaves the muzzle of a 5 kg rifle at 450 m s^{-1}. Find the recoil velocity of the rifle.

4. An object of 6000 kg mass, initially at rest, is subjected to a steady force of 1000 N for a period of 2 min. Assuming linear motion under frictionless conditions, calculate the object's final velocity. Compare its momentum with that of a 2000 kg mass after they have both been subjected to the same 1000 N force for a period of 3 min.

5. An object of 10 kg mass falls from a height of 10 m. Estimate the average retarding force acting upon it if it falls onto (a) a concrete floor which stops it in 0.01 s, and (b) a mattress which stops it in 0.06 s.

6. Water emerging from a 12 mm diameter horizontal pipe at 4.0 m s^{-1} strikes a vertical wall normal to its surface. What is the force exerted on the wall, assuming that all the momentum of the water is lost on impact. ($\rho_{\text{water}} = 1 \times 10^3 \text{ kg m}^{-3}$.)

7. A 3.00 kg object, at rest on a smooth frictionless horizontal surface, is struck by a bullet travelling horizontally at 370 m s^{-1}. The bullet is embedded in

the object and they move off together at 4.27 m s^{-1}. Assuming motion in a straight line, find the mass of the bullet.

8. A 10 kg object falls vertically onto a 25 kg object travelling at 20 m s^{-1} in the 3 o'clock direction on a smooth horizontal frictionless surface. Find their combined velocity.

9. A 5 kg object is moving at 5 m s^{-1} in the 3 o'clock direction on a smooth horizontal surface. A 50 g bullet fired from the 6 o'clock direction strikes the object at 400 m s^{-1} and becomes embedded in it. What is their joint final velocity?

10. A rocket motor producing a thrust of 5.4 kN burns fuel at a rate of 3 kg s^{-1}. At what speed are the exhaust gases being ejected?

TOPIC 8 WORK, ENERGY AND POWER

COVERING:

- energy as the capacity to do work;
- potential energy (including strain energy);
- kinetic energy;
- the principle of conservation of energy;
- power and efficiency.

8.1 WORK

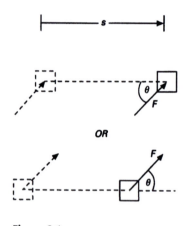

Figure 8.1

When an object moves under the influence of a force, then work is done according to the equation

$$W = F \times s \qquad (8.1)$$

where W is the work done, F is the force (N) and s is the displacement (m) in the direction of the force. The SI unit of work is the *joule* (J), which is a scalar quantity defined as the work done when the point of application of a force of 1 N moves 1 m in the direction along which it is being applied. If the force is applied at an angle to the displacement, as in Figure 8.1, then we must use the magnitude of its component in the displacement direction, in which case Equation (8.1) becomes

$$W = F \cos \theta \times s \qquad (8.2)$$

where θ is the angle which the force makes with the displacement. If $\theta = 0°$, then $\cos \theta = 1$ and we have Equation (8.1). If $\theta = 90°$, $\cos \theta = 0$, so the force has no component and can do no work in the displacement direction.

Worked Example 8.1

A box is pulled a distance of 15 m across a level floor by a force of 237 N applied to it via a rope inclined at 25° above the horizontal (as in Figure 2.7 on page 17). How much work is done?

$$W = F \times \cos \theta \times s$$

and substituting the values given,

$$W = 237 \times \cos 25° \times 15 = 3.2 \text{ kJ}$$

Note that there must be both force and displacement for work to be done. If an object is moving with uniform velocity with no net force acting upon it, then there is no work being done on it or by it. Nor is any work being done if the object remains stationary, no matter how large a force there might be acting on it.

8.2 ENERGY

Energy, which is also a scalar quantity measured in joules, is the capacity to do work. It exists in many forms (electrical energy, mechanical energy, thermal energy, and so on) and is transformed from one form to another when work is done. (There is also energy associated with mass in accordance with Einstein's theory of relativity but we shall not consider it here.)

In effect, we can regard energy as being stored by a system when work is done on it, or as being changed into another form, or forms, when work is done by a system. In the present context we shall consider potential energy and kinetic energy as representing mechanical work stored by an object by virtue of its position and motion, respectively.

8.3 POTENTIAL ENERGY

In the broadest sense, the potential energy of a system is derived from the relative position of its components. An apple hanging from a tree has potential energy. It has the capacity to do work by virtue of its position above the ground; it could, for example, be connected to a generator so that its potential energy is converted to electrical energy as it falls. In fact, the potential energy is possessed by the earth/apple system because of the gravitational force between them but, in view of the infinitesimal influence of the apple on the earth, it is more sensible to think in terms of the apple's potential energy in the earth's gravitational field.

The potential energy stored by an object by virtue of its height above the ground (or any other reference level) can be regarded as the work done in raising it against its weight mg through a vertical distance h. That is to say,

$$\text{potential energy} (= W = Fs) = mg \times h \tag{8.3}$$

The potential energy of the object remains constant anywhere on an *equipotential surface* at any fixed height above a given reference level. It follows that, since horizontal movement has no effect on the potential energy, the route taken by an object to reach a given height, no matter how circuitous, has no effect on its final potential energy.

It is often helpful to be able to view work and energy in pictorial terms. The plot of force against displacement in Figure 8.2(a) gives a graphical description of how potential energy is stored in raising an object of weight mg vertically to a height h. Since the force required $(= mg)$ is constant, the plot is a horizontal straight line of length h at a distance mg above the x-axis. The rectangle under the line therefore has an area $mg \times h$ which, as Equation (8.3) shows, represents the work done and, hence, the final potential energy of the object.

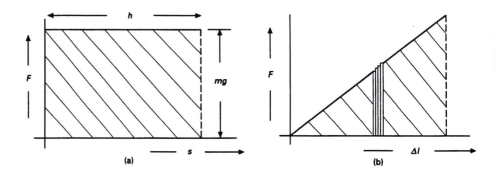

Figure 8.2

Strain energy, stored in a stretched wire, for example, can be treated in a similar way. Figure 8.2(b) represents the linear (i.e. proportional) relationship, following Hooke's law, between the stretching force F and the extension Δl (see Topic 2). (In effect, the extension is the displacement of the point of application of the force at the free end of the wire.) Like the potential energy example in Figure 8.2(a), the work done in stretching the wire is represented by the area under the graph.

We can think of this area as the sum of a very large number of extremely narrow strips (a few of which are shown in Figure 8.2b). Each strip is so narrow that it can be regarded as a rectangle where F is virtually constant. F, however, varies from one rectangle to the next, so we can regard the area under any force/displacement graph, no matter what its shape, as being made up of these narrow rectangles each representing a tiny amount of work done. In the case of a wire to which Hooke's law applies, the heights of the rectangles increase uniformly to give a triangle under the graph with an area equal to $\frac{1}{2}(F \times \Delta l)$ representing the work done and, hence, the strain energy stored in the wire.

Worked Example 8.2

From Figure 2.1 (page 12) find the energy stored in the wire when it is stretched by a force of 60 N.

Taking the extension to be 4.3 mm at 60 N,

strain energy $= \frac{1}{2} \times F \times \Delta l = \frac{1}{2} \times 60 \times 0.0043 = 0.13$ J

8.4 KINETIC ENERGY

From Equation (5.4) (page 39) we know that if an object is accelerated from rest (i.e. $u = 0$) to a speed v, then $v^2 = 2as$, so the acceleration it experiences is given by $a = v^2/2s$. From Newton's second law of motion, the force required to produce this acceleration in a body of mass m is

$$F = ma = m \times \frac{v^2}{2s}$$

and, rearranging,

$$Fs = \frac{1}{2} mv^2 \tag{8.4}$$

where Fs is the work done in accelerating the object to velocity v. $\frac{1}{2}mv^2$ therefore represents the *kinetic energy* stored by an object by virtue of its motion. Since the kinetic energy of an object depends on v^2, then doubling v will increase its kinetic energy by a factor of 4 and trebling v will increase it by a factor of 9, and so on.

If the object had had an initial velocity u and, therefore, an initial kinetic energy $\frac{1}{2}mu^2$, then the work done to bring about the change in kinetic energy in accelerating or retarding it to velocity v would have been

$$Fs = \frac{1}{2} mv^2 - \frac{1}{2} mu^2 \tag{8.5}$$

Worked Example 8.3

An object of mass 5.0 kg falls to the ground from a height of 10.0 m. Find its total energy (a) initially, (b) half-way down and (c) immediately before impact. (Assume $g = 9.8$ m s^{-1} and that the air has no effect.)

(a) Initially, at a height of 10.0 m, the object possesses only potential energy and

$$mgh = 5.0 \times 9.8 \times 10.0 = 490 \text{ J}$$

(b) After falling 5.0 m, the velocity of the object (from Equation 5.4) is given by

$$v^2 = 2gs = 2 \times 9.8 \times 5.0 = 98$$

so its kinetic energy is

$$\frac{1}{2} mv^2 = \frac{1}{2} \times 5.0 \times 98 = 245 \text{ J}$$

Its potential energy (5.0 m above the ground) is given by

$$mgh = 5.0 \times 9.8 \times 5.0 = 245 \text{ J}$$

The total energy of the object is the sum of its kinetic energy and its potential energy, which is equal to

$$245 + 245 = 490 \text{ J}$$

(c) Immediately before impact (after falling 10.0 m) the potential energy of the object is zero and its velocity is given by

$$v^2 = 2gs = 2 \times 9.8 \times 10.0 = 196$$

so its total energy is entirely kinetic energy, given by

$$\frac{1}{2} mv^2 = \frac{1}{2} \times 5.0 \times 196 = 490 \text{ J}$$

(Note that the total energy of the object remains constant as it falls.)

Worked Example 8.4

If, in Worked Example 8.3, the object penetrates the ground to a depth of 50 mm after impact, estimate the average retarding force.

$v = 0$, $s = 0.05$ and, from Worked Example 8.3, $u = \sqrt{196}$.

$$Fs = \frac{1}{2} mv^2 - \frac{1}{2} mu^2$$

Therefore, taking downwards as the positive direction,

$$F = -\frac{mu^2}{2s} = -\frac{5.0 \times 196}{2 \times 0.05} = -9.8 \text{ kN}$$

(The minus sign indicates that the force acts upwards.)

8.5 CONSERVATION OF ENERGY

Although energy can be changed from one form to another, the *principle of conservation of energy* tells us that it cannot be created or destroyed. Worked Example 8.3 illustrates how the total energy of a falling object remains constant as its gravitational potential energy is traded for kinetic energy. Car engines change chemical energy (in petrol) into mechanical energy, electric generators change mechanical energy into electrical energy, and so on.

In practice, energy transformation always involves wastage. Much of the chemical energy stored in petrol is wasted in the form of heat as a by-product of providing a motor car with kinetic energy. Even a falling mass will not normally reach its theoretical velocity (hence, its theoretical kinetic energy) because of the air; and lifting the mass in the first place involves losses due to friction and raising moving parts of the lifting gear. Nevertheless none of the energy is truly lost, only changed into unwanted by-products.

Some energy changes are not so obvious. In the last topic we saw that momentum is conserved in collisions. But some, or even all, of the kinetic energy may be transformed. A collision between steel ball-bearings will be almost perfectly elastic; any kinetic energy converted to strain energy around the contact points will be recovered as the balls spring apart, although there might be slight losses in the form of heat or sound energy. At the other extreme, inelastic collisions can involve total kinetic energy loss. Two identical lumps of putty colliding with equal and opposite velocities (and momenta) along the same line will stick to each other and come to rest, so that all their kinetic energy is lost as heat and sound.

8.6 POWER

Power is a scalar quantity which gives the rate at which energy is transformed or work is done. The unit of power is the *watt* (W), which is equivalent to one joule per second, so

$$W \text{ (watts)} = \text{J s}^{-1} = \text{N m s}^{-1} = \text{N} \times \text{m s}^{-1}$$

$\text{N} \times \text{m s}^{-1}$ are the units of force times velocity. That is to say, if an object moves at a velocity v under the influence of a force F, then the mechanical power delivered to the object is $F \cos \theta \times v$, where θ is the angle between the line of action of the force and the direction of motion.

(Note that 1 horsepower is a unit of power that is equivalent to 746 W.)

8.7 EFFICIENCY

The efficiency of a machine or a process is given by the ratio between the useful energy output and the energy input, commonly expressed as a percentage. A motor that requires 1.0 kW of electrical power to provide 0.75 kW of mechanical power has an efficiency of

$$\frac{0.75}{1.00} \times 100 = 75\%$$

Questions

(Where necessary assume that $g = 9.8$ m s^{-2}.)

1. 735 J is available to raise an object off the ground. What is the largest mass that can be lifted to a height of 15 m?

2. An object of 12.5 kg mass rests on a smooth (frictionless) horizontal surface. Calculate the work done if a 50 N force is applied to the object for a period of 10 s (a) parallel to the surface, (b) vertically upwards.

3. An object of 3 kg mass is travelling with a uniform velocity of 12 m s^{-1} across a smooth horizontal surface when a force of 38 N is suddenly applied in the displacement direction. What is the velocity of the object after it has travelled a further 30 m?

4. An object of 15 kg mass travels at 40 m s^{-1} across a smooth horizontal surface. Find the retarding force needed to bring it to a halt in a distance of 50 m.

5. An object 15 m above the ground has a potential energy of 735 J and a kinetic energy of 12.25 kJ. What is its speed?

6. What is the efficiency of a pump which uses 1.5 kW of electrical power in raising 85 litres of water through a height of 12 m in 10 s? (1 litre = 1000 cm^3 and ρ_{water} = 1000 kg m^{-3}.)

7. A ball-bearing suspended from a weightless thread swings to and fro so that its highest point is 94 mm and its lowest point is 32 mm above the table top. Find its maximum velocity.

8. An 80 kg man climbs a staircase 3 m high in 9.8 s. If he worked just as hard to drive a pedal-operated generator with an efficiency of 75%, how much electrical power would he produce?

9. An object of mass 2.5 kg, resting on a horizontal surface, is subjected to a horizontal force of 20 N. If $\mu_s = \mu_k = 0.49$, find the velocity of the object after it has travelled 10 m.

10. What is the power consumption when an object is pushed 16 m up a 26° slope in 12 s by a horizontal force of 50 N?

---

efficiency = useful work done / actual work done

TOPIC 9 MOTION IN A CIRCLE

COVERING:

- angular displacement, velocity and acceleration;
- angular equations of motion;
- centripetal force and acceleration.

So far we have tended to think about objects that are either at rest or moving in straight lines. Now we need to consider circular motion and find angular equivalents of the linear parameters that we have already met.

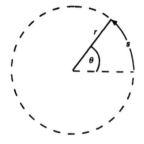

Figure 9.1

9.1 ANGULAR DISPLACEMENT, VELOCITY AND ACCELERATION

Let us begin by assuming that an object is moving round a circular track at a constant speed.

There are two easy ways of describing its displacement over a given period. As Figure 9.1 suggests, we can use the distance s it has moved round the circumference from its starting point. Alternatively we can use the *angular displacement* θ, i.e. the angle through which the radius has moved.

The SI unit for the measurement of angles is the radian (rad). In terms of Figure 9.1, the angle θ in radians is given by the length of the arc s divided by the radius r. Thus,

$$\theta = \frac{s}{r} \text{ rad} \tag{9.1}$$

It follows that 1 rad is the angle where $s = r$ and that a linear (as opposed to angular) displacement round the circumstance is given by $s = r\theta$. Note that for one complete lap of the circle $s = 2\pi r$ and

$$\theta = \frac{2\pi r}{r} = 2\pi = 6.28 \text{ rad}$$

which is equivalent to 360°; therefore, 1 rad = 57.3°. *Angular velocity* ω is the rate of angular displacement with time t and

$$\omega = \frac{\theta}{t} \text{ rad s}^{-1} \tag{9.2}$$

This can also be written in terms of the object's linear speed v (= s/t) around the circumference of the circle, since by combining Equations (9.1) and (9.2) to eliminate θ we have

$$\omega = \frac{s}{r \times t}$$

and because $v = s/t$

$$\omega = \frac{v}{r} \tag{9.3}$$

Note that the object's linear speed is related to its *period T* (the time for one complete revolution) by the expression $v = 2\pi r/T$, i.e. the circumference divided by the time taken to travel round it. Substituting ωr for v (Equation 9.3),

$$\omega r = \frac{2\pi r}{T}$$

Therefore,

$$T = \frac{2\pi}{\omega} \tag{9.4}$$

Now let us imagine that the angular velocity of the object is varying uniformly with time. If ω changes from ω_1 to ω_2 in time t, then the *angular acceleration* α is given by

$$\alpha = \frac{(\omega_2 - \omega_1)}{t} \text{ rad s}^{-2} \tag{9.5}$$

From Equation (9.3), $(\omega_2 - \omega_1) = (v_2 - v_1)/r$, so that

$$\alpha = \frac{(v_2 - v_1)}{rt}$$

and, since the linear acceleration a round the circumference of the circle is given by $(v_2 - v_1)/t$, then

$$\alpha = \frac{a}{r} \tag{9.6}$$

Thus, angular displacement, velocity and acceleration can all be obtained by dividing their linear counterparts by the radius r.

9.2 ANGULAR EQUATIONS OF MOTION

In Topic 5 we obtained four equations (5.1–5.4) giving relationships between displacement (s), velocity (u and v), uniform linear acceleration (a) and time (t). By using similar arguments (or by dividing each linear quantity by r) we can obtain four equivalent equations for motion in a circle where θ replaces s, ω_1 and ω_2 replace u and v, respectively, and α replaces a. Thus, we have

$$\omega_2 = \omega_1 + \alpha t \tag{9.7}$$

$$\theta = \frac{(\omega_1 + \omega_2)}{2} t \tag{9.8}$$

$$\theta = \omega_1 t + \frac{1}{2} \alpha t^2 \tag{9.9}$$

$$\omega_2^2 = \omega_1^2 + 2\alpha\theta \tag{9.10}$$

Worked Example 9.1

A car with 500 mm diameter wheels moves in a straight line at 85 km per hour. How fast do the wheels turn (a) in rpm (revolutions per minute), and (b) in rad s^{-1}?

Converting km per hour to m s^{-1}, the linear speed of any point on the wheels' circumference is given by

$$85 \times \frac{1000}{3600} = 23.6 \text{ m s}^{-1}$$

and

$$\text{circumference} = 2\pi \times 0.25 = 1.57 \text{ m}$$

Therefore,

(a) the number of revolutions per minute is given by

$$\frac{\text{linear speed}}{\text{circumference}} \times 60 = \frac{23.6}{1.57} \times 60 = 900$$

and

(b) from Equation (9.3)

$$\omega = \frac{v}{r} = \frac{23.6}{0.25} = 94 \text{ rad s}^{-1}$$

Worked Example 9.2

A wheel initially at rest experiences a uniform angular acceleration of 5 rad s^{-2} for 6 s. It maintains a constant angular velocity for 14 s and is then brought to rest in 5 s by a uniform angular retardation. Find the total angular displacement.

Acceleration stage:

$$\omega_1 = 0, \quad t = 6, \quad \alpha = 5, \quad \theta = ?, \quad \omega_2 = ?$$

From Equation (9.9)

$$\theta = \omega_1 t + \frac{1}{2} \alpha t^2 = \frac{1}{2} \times 5 \times 36 = 90 \text{ rad}$$

From Equation (9.7)

$$\omega_2 = \omega_1 + \alpha t = 5 \times 6 = 30 \text{ rad s}^{-1}$$

Constant velocity stage:

$$\omega_1 = \omega_2 = 30, \quad t = 14, \quad \alpha = 0, \quad \theta = ?$$

From Equation (9.9)

$$\theta = 30 \times 14 = 420 \text{ rad}$$

Retardation stage:

$$\omega_1 = 30, \quad \omega_2 = 0, \quad t = 5, \quad \theta = ?$$

From Equation (9.8)

$$\theta = \frac{(\omega_1 + \omega_2)}{2} t = \frac{(30 + 0)}{2} \times 5 = 75 \text{ rad}$$

Therefore,

total angular displacement $= 90 + 420 + 75 = 585$ rad.

(Now compare this with Worked Example 5.2 on page 40).

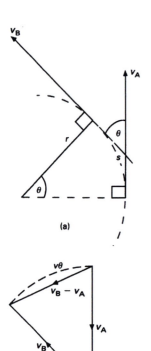

(a)

(b)

Figure 9.2

9.3 CENTRIPETAL ACCELERATION AND FORCE

For an object to move in a circle, there must be an inward force acting on it to overcome its natural tendency to follow a straight line. This *centripetal* (i.e. centre-seeking) force can be provided in many ways. For example, by gravitation, as in keeping the moon orbiting round the earth, or by the tension in a piece of string used to swing an object round in a circle. At any instant the object is moving in the direction given by the tangent to the circle at that point, so that, in the case of the second example, if the string breaks, the object will fly off along the tangent. The object is suddenly liberated from the centripetal force provided by the tension in the string, so it will continue in a straight line (or it would do in the absence of gravity). Centripetal force is also provided by friction between the tyres and the road when a car turns a corner.

An object moving in a circle experiences a change of velocity, even though its linear speed might be constant, because of the continuous change in its direction. Change of velocity means acceleration, and for acceleration we need a force such as the tension in a string continuously pulling an object away from a straight path. This centripetal acceleration must be distinguished from the angular acceleration α and its linear counterpart (which in any case would both be zero if ω is constant). Let us consider how it can be quantified.

Figure 9.2(a) shows the path of an object moving round a circle of radius r at constant linear speed. In moving through an angle θ (equivalent to a linear distance s) its velocity changes from v_A to v_B in time t. Its change in velocity $(v_B - v_A)$ is represented vectorially in Figure 9.2(b); note that the direction of v_A is reversed to make it negative, so that

$$v_B + (-v_A) = v_B - v_A$$

(The angle between v_A and v_B is θ, since this is the angle between the respective radii which they meet perpendicularly in Figure 9.2a.)

θ is shown as a large angle for clarity. If we make it very small, then the straight line representing the change of velocity $(v_B - v_A)$ in Figure 9.2(b) will merge with the broken curve which represents the arc of the circle of radius v (where v is the magnitude of v_A and v_B). Thus, if θ is in radians, then, to a very close approximation, Equation (9.1) gives us the change of velocity $(v_B - v_A)$ as $v\theta$, which, since $\theta = s/r$ from Figure 9.2(a), is equal to vs/r. Since $s = vt$, change of velocity $= \dfrac{v^2 t}{r}$

and the magnitude of the associated centripetal acceleration a_c is therefore given by

$$a_c = \frac{\text{change of velocity}}{t} = \frac{v^2 t}{r} \times \frac{1}{t}$$

Therefore,

$$a_c = \frac{v^2}{r} \tag{9.11}$$

Furthermore, since $v^2 = \omega^2 r^2$ (from Equation 9.3),

$$a_c = \frac{v^2}{r} = \frac{\omega^2 r^2}{r} = \omega^2 r \tag{9.12}$$

If θ is made extremely small, so that v_A and v_B virtually overlap, then $(v_B - v_A)$ is perpendicular to the tangent at that point and is therefore directed towards the centre of the circle along the radius. The instantaneous acceleration therefore acts towards the centre of the circle.

If the mass of the object is m, then the centripetal force acting on it is given by Newton's second law as follows:

$$F \ (= ma_c) = m \times \frac{v^2}{r} \tag{9.13}$$

or

$$F = m\omega^2 r \tag{9.14}$$

where a_c is obtained from Equations (9.11) and (9.12), respectively.

As noted earlier, centripetal force can be provided in a number of ways. Figure 9.3 shows an object of mass m swinging round in a

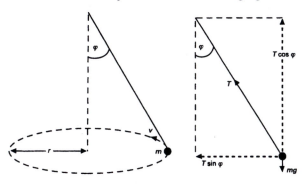

Figure 9.3

horizontal circle of radius r on the end of a string, fixed at the top, which makes an angle of φ with the vertical. At a given linear speed v the horizontal component of the tension T in the string provides the centripetal force, so that

$$T \sin \varphi = \frac{mv^2}{r}$$

The vertical component of T supports the weight of the object, so that

$$T \cos \varphi = mg$$

Putting these together,

$$\tan \varphi \left(= \frac{T \sin \varphi}{T \cos \varphi} \right) = \frac{mv^2}{r} \times \frac{1}{mg}$$

Therefore,

$$\tan \varphi = \frac{v^2}{gr} \tag{9.15}$$

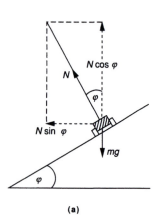

(a)

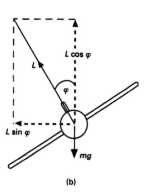

(b)

Figure 9.4

We can use a similar approach to banking a curved road to reduce the risk of skidding. Figure 9.4(a) shows the end view of a car of mass m travelling at a linear velocity v round a curve of radius r banked at an angle φ. The horizontal component of the normal force N acting on the car provides the centripetal force, the vertical component supports the car's weight, and putting them together as above again gives $\tan \varphi = v^2/gr$. So, for a given curve (where φ and r are fixed), there is a particular linear speed v, independent of the car's mass, where friction makes no contribution to the centripetal force. (Note that friction is still needed to prevent the car from slipping inwards at lower speeds and outwards at higher speeds.)

As Figure 9.4(b) suggests, the same equation applies to an aircraft making a banked turn; the vertical component of the lift force L experienced by the wings supports the aircraft's weight, while the horizontal component provides the centripetal force.

Going back to the car, if the road is horizontal ($\varphi = 0°$), then friction has to supply all the centripetal force. If the tyres are not slipping, then the contact area can be assumed to be instantaneously at rest. Therefore, the maximum possible centripetal force F corresponding to the maximum possible linear speed v is given by

$$F = \mu_s N = \mu_s mg = \frac{mv^2}{r}$$

so

$$\mu_s = \frac{mv^2}{r} \times \frac{1}{mg} = \frac{v^2}{gr}$$

and the maximum possible linear speed to negotiate the curve of radius r without skidding is given by $\sqrt{\mu_s gr}$.

Finally, let us briefly consider what happens when the object in Figure 9.3 moves through a vertical circle rather than a horizontal one. In this case the object, the string and its fixing point at the centre all lie within the plane of the circle. At the top of the circle the centripetal force is provided by the tension T in the string plus the object's weight, so that

$$\frac{mv^2}{r} = T + mg$$

and

$$T = \frac{mv^2}{r} - mg \tag{9.16}$$

Obviously there is a critical value of v where mg equals mv^2/r and the tension in the string is zero, so that the object only just manages to complete the circle. If v is even slightly below this value, the object will start to fall before it reaches the top. The same basic argument applies to swinging a bucket of water around vertically without spilling it.

At the bottom of the circle the weight of the object acts away from the centre, so T must provide the centripetal force and support the weight of the object, so that

$$T = \frac{mv^2}{r} + mg \tag{9.17}$$

In this case v must not be too large; otherwise the string may break.

Worked Example 9.3

Find the period of rotation of an object swinging round in a horizontal circle of 500 mm radius on the end of a string 1300 mm long (as in Figure 9.3). ($g = 9.8$ m s^{-2}.)

The vertical distance of the plane of rotation below the suspension point is given by

$$\sqrt{1.3^2 - 0.5^2} = 1.2$$

Therefore,

$$\tan \varphi = \frac{0.5}{1.2} = \frac{v^2}{gr} = \frac{v^2}{9.8 \times 0.5}$$

which gives

$$v = 1.43 \text{ m s}^{-1}$$

but, since $v = 2\pi r/T$,

$$T = \frac{2\pi r}{v} = \frac{2\pi \times 0.5}{1.43} = 2.2 \text{ s}$$

Worked Example 9.4

(a) Find the minimum linear velocity for the object in Worked Example 9.3 to maintain continuous circular motion in a vertical plane. (b) If the mass of the object is 250 g, and if its linear velocity remains the same as in (a), then find its apparent weight at the bottom of the circle. ($g = 9.8 \text{ m s}^{-2}$.)

(a) The minimum value of v is where the tension in the string is zero, so that, from Equation (9.16),

$$\frac{mv^2}{r} = mg$$

Therefore,

$$v = \sqrt{rg} = \sqrt{1.3 \times 9.8} = 3.6 \text{ m s}^{-1}$$

(b) Substituting $\sqrt{rg}$ for v in Equation (9.17),

$$T = \frac{mrg}{r} + mg = 2mg = 2 \times 0.25 \times 9.8 = 4.9 \text{ N}$$

which is the apparent weight of the object exerted on the string (i.e. twice its actual weight).

Questions

(Where necessary assume that $g = 9.8$ m s^{-2}.)

1. If the minute hand of a clock is 115 mm long, find the linear speed at its tip.

2. A turntable takes 1.8 s to reach $33\frac{1}{3}$ rpm from rest. How many revolutions does this take?

3. The moon's mass is 7.3×10^{22} kg and its orbit round the earth has an average radius of 3.8×10^8 m and a period of 27 days. Estimate the magnitude of the gravitational force between the earth and the moon.

4. A truck travels round a circular track of 80 m radius at a uniform linear speed of 10 m s^{-1}. Find the angle to which the track would need to be banked to eliminate the need for radial friction.

5. An object is turning through a vertical circle of 0.69 m radius on the end of a string. Find the minimum angular velocity required to maintain circular motion.

6. If the mass of the object in the previous question is 0.1 kg, estimate the minimum possible breaking strength of the string.

7. Estimate the angle at which an aircraft should be banked to make a horizontal turn of 3.5 km radius at a speed of 450 km per hour.

8. What is the tension in the string in Worked Example 9.3 if the mass of the object is 190 g?

9. A rotating shaft experienced an angular acceleration of 10 rad s^{-2} over a period of 8 s, during which time its angular displacement was 400 rad. What were its initial and final angular velocities?

10. An aircraft loops the loop with a radius of 750 m at a constant linear speed of 360 km h^{-1}. What is the magnitude of the force exerted by the seat on a 70 kg pilot (a) at the top, and (b) at the bottom of the loop?

TOPIC 10 ROTATION OF SOLIDS

COVERING:

- moment of inertia;
- angular momentum;
- rotational kinetic energy;
- torque, work and power.

In the previous topic we considered the translational motion of an object, treating it as a particle moving round a circular path. Now we shall consider a solid object, such as a shaft or a flywheel, rotating about an axis without necessarily moving from one place to another.

We have already met the idea that an object resists change in its state of translational motion because of its inertia. In an analogous way, an object resists change in its rotational state because of its *moment of inertia*.

10.1 MOMENT OF INERTIA

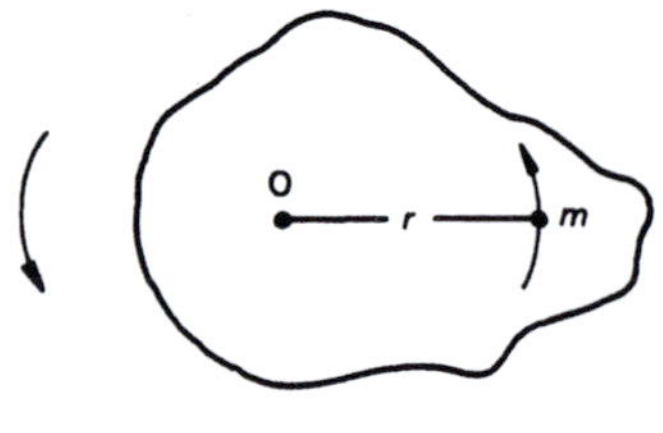

Figure 10.1

Figure 10.1 represents a solid object rotating about a fixed axis O perpendicular to the page. Let us focus on a single component particle of mass m rotating about O at a distance r. If we want to change the speed of the particle, then, considering it separately from all the others, we would have to apply a tangential force to it (in the same direction as its motion) in accordance with Newton's second law

$$F = ma$$

but, since $a = \alpha r$ (Equation 9.6 on page 69), then

$$F = m\alpha r$$

where α is the angular acceleration.

In Topic 3 (Section 3.2) we saw that a *torque* about a point is found by multiplying the force producing it by the perpendicular distance of the line of action of the force from the point. The torque T needed to change the angular velocity of our particle is therefore given by

$$T = F \times r = m\alpha r \times r = mr^2\alpha$$

But the object consists of a large number of component particles, each with its own particular value of mr^2, so the total torque needed to change the angular velocity of the object as a whole is given by

$$T = (\textstyle\sum mr^2)\,\alpha$$

where $\sum mr^2$ is the sum of the individual mr^2 values of all the component particles about the axis. (Note that Σ is simply a mathematical symbol meaning 'the sum of'.) In fact, the quantity $\sum mr^2$, which is given the symbol I and has units of kg m^2, represents the moment of inertia of the object, so

$$T \;(= (\textstyle\sum mr^2)\,\alpha) = I\alpha \qquad\qquad (10.1)$$

This equation is the rotational version of Newton's second law where moment of inertia is analogous to mass. The greater the moment of inertia of a body the greater the torque needed to provide a given angular acceleration.

Since the value of mr^2 for each component particle in the object is the product of its mass and the square of its distance from the axis, it follows that the moment of inertia of a body depends on the way in which the total mass is distributed about the axis. Figure 10.2 shows some examples. If the total mass M is distributed in the form of a thin

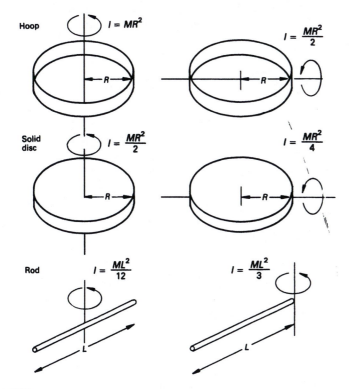

Figure 10.2

hoop of radius R, and the axis of rotation passes through the centre of the hoop normal to its plane, then $I = MR^2$, because all the component particles are at a distance R from the axis. For a given mass M, I increases as R^2; thus, the moments of inertia of a series of hoops of different sizes, but identical mass, increase rapidly with radius. For a solid disc rotating about the corresponding axis $I = MR^2/2$, which is less than for the hoop, because the mass is distributed closer to the axis. For this reason flywheels tend to have their mass concentrated at the rim, because this increases their moment of inertia. If the axis of rotation is changed, as on the right-hand side of Figure 10.2, then I changes, because the mass is distributed differently about it. For example, $I = MR^2/4$ for a thin solid disc with its axis of rotation along a diameter.

10.2 ANGULAR MOMENTUM

Angular momentum is the rotational counterpart of linear momentum (mv), which we met in Topic 7. Moment of inertia and angular velocity are the rotational counterparts of mass and linear velocity and

angular momentum $= I\omega$

In Topic 7 we noted that Newton's second law can be expressed in terms of change of linear momentum. The same is true of angular momentum, since, from Equation (10.1),

$$T = I\alpha = I\frac{(\omega_2 - \omega_1)}{t} = \frac{I\omega_2 - I\omega_1}{t}$$

And, as with linear momentum, there is the *principle of conservation of angular momentum*, which states that the total angular momentum of a system remains constant if there is no net torque acting on it. Skaters make use of this principle to control the rate at which they spin. If they tuck their arms in close to their bodies, then their moment of inertia decreases, so their angular velocity increases in order to keep the value of $I\omega$ constant. Similarly, if they stretch their arms out, then their angular velocity decreases.

10.3 ENERGY, WORK AND POWER

A rotating object must possess kinetic energy, since its component parts are moving. As we might expect from its translational counterpart ($\frac{1}{2}mv^2$),

rotational kinetic energy $= \frac{1}{2} I\omega^2$

But this energy was originally provided in accelerating the object from an initial angular velocity of zero. Let us see how we can quantify the work done.

Figure 10.3 shows a tangential force F that, in moving through the distance s, has produced rotation through the angle θ. In this case the work done W is given by

$$W = F \times s$$

and, since $s = r\theta$,

$$W = Fr\theta$$

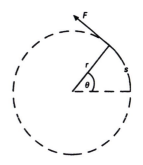

Figure 10.3

Therefore, since $Fr = T$ (torque), we have

$$W = T\theta$$

If this is the work done in accelerating the object from rest to an angular velocity ω, then

$$T\theta = \frac{1}{2} I\omega^2$$

Alternatively, $T\theta$ might represent the energy transmitted over a period of time t through a rotating shaft, say from a turbine to an electric generator, in which case the power P transmitted is given by

$$P = \frac{\text{work}}{\text{time}} = \frac{T\theta}{t}$$

and, since $\omega = \theta/t$,

$$P = T\omega$$

which is equivalent, in linear terms, to force times velocity.

Finally, we should note that an object may possess both rotational and translational kinetic energy, as in the case of a round object rolling along the ground.

10.4 SUMMARY

The angular quantities that we have met in this topic are listed below (with their linear counterparts in parentheses):

- moment of inertia I (mass m);
- torque $T = Ia$ (force $F = ma$);
- angular momentum $I\omega$ (mv);

- kinetic energy $\frac{1}{2}I\omega^2$ ($\frac{1}{2}mv^2$);
- work $T\theta$ (Fs);
- power $T\omega$ (Fv).

Worked Example 10.1

Find the total kinetic energy of a 50 kg disc, 400 mm in diameter, rolling across a level surface at 70 rpm.

Considering translational plus rotational motion, the total kinetic energy E_k is given by

$$E_k = \frac{1}{2} mv^2 + \frac{1}{2} I\omega^2$$

but since $v = r\omega$ (Equation 9.3 on page 69) and $I = \frac{1}{2}mr^2$ in this case

$$E_k = \frac{1}{2} mr^2\omega^2 + \frac{1}{2}\left(\frac{mr^2}{2}\right)\omega^2 = \frac{3}{4} mr^2\omega^2$$

then, on substituting,

$$E_k = \frac{3}{4} \times 50 \times (0.2)^2 \times \left(\frac{70}{60} \times 2\pi\right)^2 = 81 \text{ J}$$

Worked Example 10.2

If the disc in the previous example is mounted as a flywheel, find the torque required to raise it from rest to 200 rpm in 5 revolutions.

In this case $I = \frac{1}{2}mr^2$ and, on substituting,

$$I = \frac{50 \times (0.2)^2}{2} = 1 \text{ kg m}^2$$

$$200 \text{ rpm} = \frac{200}{60} \times 2\pi = 21 \text{ rad s}^{-1}$$

and

$$\omega_1 = 0, \quad \omega_2 = 21, \quad \theta = 5 \times 2\pi, \quad \alpha = ?$$

so, from Equation (9.10) (page 70),

$(21)^2 = 0 + (2 \times \alpha \times 5 \times 2\pi)$

which gives

$\alpha = 7$ rad s^{-2}

Therefore,

$T (= I\alpha) = 1 \times 7 = 7$ N m.

Worked Example 10.3

The flywheel in the previous example, rotating freely (i.e. no external torque) at 21 rad s^{-1}, is connected via a friction clutch of zero I to a second flywheel, with $I = 2$ kg m^2, that is at rest. Find (a) the final combined angular velocity, and (b) the heat energy dissipated in the clutch.

(a) Angular momentum is conserved; therefore,

$I_A\omega_A + I_B\omega_B = (I_A + I_B)\, \omega_{A+B}$

and, substituting,

$(1 \times 21) + 0 = (1 + 2)\, \omega_{A+B}$

which gives

$\omega_{A+B} = 7$ rad s^{-1}

(b) Kinetic energy before engaging clutch is equal to

$\frac{1}{2} I_A\omega_A^2 + \frac{1}{2} I_B\omega_B^2 = \left(\frac{1}{2} \times 1 \times (21)^2\right) + 0 = 220.5$ J

Kinetic energy after engaging clutch is equal to

$\frac{1}{2} I_{A+B}\omega_{A+B}^2 = \frac{1}{2} \times 3 \times 7^2 = 73.5$ J

Therefore, the heat dissipated is equal to

$220.5 - 73.5 = 147$ J

Worked Example 10.4

A shaft is being driven at a constant 200 rpm by a steady torque of 7 N m. Find the power consumption.

$$\text{power} = T\omega = 7 \times \frac{200 \times 2\pi}{60} = 147 \text{ W}$$

Questions

1. A 5 kg disc, 200 mm in diameter, is mounted as a flywheel and rotates at 300 rpm. Find (a) its moment of inertia, (b) its kinetic energy, (c) its total kinetic energy, if it is allowed to roll across a level surface with the same angular velocity.

2. A 12 g coin 30 mm in diameter is tossed so that it spins at 300 rpm about an axis along its diameter. Estimate its rotational kinetic energy.

3. A steady torque of 12 N m is applied to a flywheel at rest that has a moment of inertia of 6 kg m^2. After 15 s what is (a) the angular displacement, (b) the angular velocity, (c) the work done, (d) the power consumption and (e) the kinetic energy of the flywheel?

4. Find the moment of inertia about the axis of rotation of an object, initially at rest, which is accelerated to an angular velocity of 180 rad s^{-1} in 15 s by a torque of 15 N m.

5. What is the angular acceleration of a flywheel with a moment of inertia of 18 kg m^2 upon which a torque of 27 N m is acting? If the flywheel starts from rest, how long does it take to turn through 50 rad?

6. A 200 mm diameter flywheel rotating at 200 rad s^{-1} is brought to rest in 125 revolutions by a braking torque of 0.26 N m. Find the mass of the flywheel, assuming that it is a disc.

7. A shaft, with $I = 5$ kg m^2, freely rotating clockwise at 50 rad s^{-1}, is connected via a friction clutch to a shaft with $I = 10$ kg m^2 that is freely rotating anticlockwise at 100 rad s^{-1}. Find (a) their combined angular velocity and (b) the kinetic energy lost.

8. A rod of negligible mass rotates horizontally about its centre with two identical masses attached, one at each

side, at equal distances from the axis of rotation. Assuming free rotation with no external torque applied, and an angular velocity of 40 rad s^{-1} with the masses 500 mm apart, find the new angular velocity if the masses move to a position 550 mm apart as they rotate.

9. Shaft A, driven at a constant 150 rad s^{-1}, is connected via a friction clutch to shaft B, which has a moment of inertia of 0.04 kg m^2. Shaft B accelerates uniformly to the same angular velocity as shaft A over a period of 0.75 s. By consideration of the kinetic energy acquired by shaft B and the torque acting upon it, find the heat energy dissipated in the clutch.

10. A round object of 2 kg mass and 750 mm diameter, starting from rest, takes 3.8 s to roll 15 m down a 25° slope. Find (a) the object's final total kinetic energy and (b) its moment of inertia about the axis of rotation; hence (c) establish whether it is a disc or a hoop.

TOPIC 11 SIMPLE HARMONIC MOTION

COVERING:

- a mathematical model;
- the simple pendulum;
- vertical oscillation of a mass hanging from a spring;
- damping.

Having considered linear, circular and rotational motion, we now move on to vibrational motion, or oscillation, such as that of a pendulum, where an object is displaced from some central equilibrium position, then released so that it oscillates backwards and forwards about it. Such behaviour can often be described in terms of *simple harmonic motion*, which is characterised by an acceleration towards the equilibrium position that has a magnitude proportional to the displacement from it.

11.1 A MATHEMATICAL MODEL

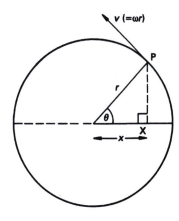

Figure 11.1

Figure 11.1 provides the basis of a mathematical model of simple harmonic motion.

The figure shows a point P moving round a circle of radius r with a constant linear speed v ($= \omega r$). As P makes successive revolutions, the point X, i.e. the vertical projection from P onto the horizontal diameter, moves to and fro along it with simple harmonic motion about the centre.

Before examining this idea more closely, note that the positive direction runs along the positive x-axis from the centre of the circle. The displacement x of the point X from the centre is given by $r \cos \theta$. When X is to the left of centre, then the value of x is negative, since $\cos \theta$ is always negative on that side. Since $x = r \cos \theta$ and $\theta = \omega t$ (Equation 9.2 on page 69),

$$x = r \cos \omega t \tag{11.1}$$

We have implied here that $\theta = 0$ at $t = 0$, which is not necessarily true. If there is already an angular displacement, say φ, at time $t = 0$, then $\theta = (\omega t + \varphi)$. Also note that the maximum displacement, $x = r$ or $-r$, is called the *amplitude*.

Figure 11.2 illustrates how displacement varies with time. T, the unit of time used in the figure, is the period $T = 2\pi/\omega$ that corresponds to one complete oscillation (Equation 9.4 on page 69). As Figures 11.1 and 11.2 both imply, X is moving at its fastest through the central position, then it slows down until it stops and reverses direction at the maximum displacement, then it accelerates towards the centre – and so on.

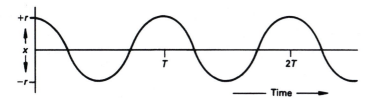

Figure 11.2

From Topic 9 we know that P must experience a centripetal acceleration ($= \omega^2 r$) inwards along the radius (Equation 9.12 on page 73). The acceleration a of X is given by the horizontal component of the centripetal acceleration as follows:

$$a = -\omega^2 r \cos \theta$$

and, since $x = r \cos \theta$,

$$a = -\omega^2 x \qquad\qquad (11.2)$$

This means that, since ω is constant, the acceleration of X is proportional to its displacement x. The negative sign indicates that the acceleration is directed towards the centre (i.e. in the negative direction when X is to the right of centre and x is positive, and in the positive direction when X is to the left of centre and x is negative). X therefore executes simple harmonic motion in accordance with the characteristics noted at the beginning of the topic.

Now let us consider the force involved where a mass is moving with simple harmonic motion. If an object of mass m experiences an acceleration $-\omega^2 x$, then, from Newton's second law, it will experience a force F, given by

$$F (= ma) = -m\omega^2 x$$

Since m and ω are both constants,

$$F = -kx \qquad\qquad (11.3)$$

where the constant $k (= m\omega^2)$ represents the force per unit displacement.

This tells us that, in simple harmonic motion, the object experiences a restoring force acting towards the central equilibrium point, which, like the acceleration, is proportional to the object's displacement from it. Note that any oscillating mechanical system with a proportional relationship between the restoring force and the displacement will execute simple harmonic motion. Now since $\omega = \sqrt{k/m}$ (from above), the period of oscillation is given by

$$T = \frac{2\pi}{\omega} = \frac{2\pi}{\sqrt{k/m}}$$

and therefore

$$T = 2\pi \sqrt{m/k} \tag{11.4}$$

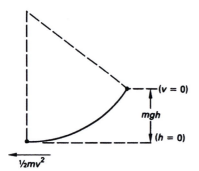

Figure 11.3

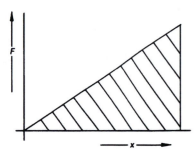

Figure 11.4

The energy of the object is the sum of its potential and kinetic energies, which are continuously interchanging from 100% potential energy at the extreme positions to 100% kinetic energy at the centre. For example, Figure 11.3 shows how the total energy of a mass swinging to and fro at the end of a weightless string is its potential energy (mgh) at its extreme positions, where it is stationary, and its kinetic energy ($\frac{1}{2}mv^2$) at the central position, where it is at the lowest point of its travel.

First let us consider the general case. The total energy of any object undergoing simple harmonic motion can be viewed in terms of the initial work done in displacing it from its central equilibrium position to one of the extreme positions prior to releasing it in the first instance. The work is done against the restoring force F, which, as we saw earlier, must increase proportionally to the displacement x. The potential energy stored by this process is therefore given by the area $\frac{1}{2}Fx$ under the plot of displacing force (of magnitude F) against x in Figure 11.4. When the object is at its maximum displacement, where $x = r$, then its total energy E is given by

$$E = \frac{1}{2} Fr$$

and since the magnitude of F is given by $kr = m\omega^2 r$ (see Equation 11.3), then

$$E = \frac{1}{2} m\omega^2 r^2 \tag{11.5}$$

When the object is released and starts to move, its potential energy is traded for kinetic energy. Its potential energy at any given displacement x is then given by $\frac{1}{2}m\omega^2 x^2$, since this would have been the work needed to displace the object from rest at equilibrium to that position. The kinetic energy of the object at that point will be the difference between its total energy and its potential energy.

$$\frac{1}{2} mv^2 = \frac{1}{2} m\omega^2 r^2 - \frac{1}{2} m\omega^2 x^2$$

Therefore,

$$v^2 = \omega^2(r^2 - x^2)$$

so the magnitude of the velocity is given by

$$v = \omega \sqrt{r^2 - x^2} \tag{11.6}$$

Equation (11.6) confirms our earlier observations about the velocity, i.e. that v is zero where $x = r$ at the maximum displacement, and v has its greatest magnitude (ωr) at the centre. (In general, we shall try to avoid any possible confusion with the sign convention – positive to the right for our model – by using the magnitudes of the quantities involved and specifying their directions where necessary.)

Note that the rate at which a system oscillates is often expressed in terms of the frequency f, which is the reciprocal of the period T:

$$f = \frac{1}{T} \tag{11.7}$$

Frequencies are expressed in hertz (Hz), where 1 Hz is 1 cycle per second. Also note that, since $\omega = 2\pi/T$ (Equation 9.4 on page 69), then

$$\omega = 2\pi f \tag{11.8}$$

Worked Example 11.1

A 950 g mass moves in simple harmonic motion with a frequency of 20 Hz and an amplitude of 100 mm. Find

(a) its maximum and minimum speeds and where these occur;
(b) its maximum and minimum accelerations and where these occur;
(c) its speed and acceleration 30 mm from the extreme positions;
(d) the maximum restoring force acting upon it.

(a) Since $v = \omega \sqrt{r^2 - x^2}$ (Equation 11.6), the maximum speed occurs where $x = 0$ (at the central position); hence, $v = \omega r$ and, since $\omega = 2\pi f$ (Equation 11.8),

$$v = 2\pi f r$$

Substituting,

$$v = 2\pi \times 20 \times 0.1 = 12.6 \text{ m s}^{-1}$$

The minimum speed, $v = 0$, occurs where $x = r$ (at the extreme positions).

(b) Since $a = \omega^2 x$ towards the central point (Equation 11.2) and $\omega = 2\pi f$ (Equation 11.8), the maximum acceleration occurs where $x = r$ (at the extreme positions) and is given by

$$a = 4\pi^2 f^2 r$$

Substituting,

$$a = 4\pi^2 \times (20)^2 \times 0.1 = 1580 \text{ m s}^{-2} \text{ (inwards)}$$

At the central position $x = 0$, therefore, $a = 0$.

(c) From Equations (11.6) and (11.8)

$$v = 2\pi f \sqrt{r^2 - x^2}$$

Therefore, 30 mm from the extreme positions, where $x = 0.07$ m,

$$v = 2\pi \times 20 \times \sqrt{0.1^2 - 0.07^2} = 9 \text{ m s}^{-1}$$

Furthermore,

$$a = 4\pi^2 f^2 x$$

and, substituting,

$$a = 4\pi^2 \times (20)^2 \times 0.07 = 1105 \text{ m s}^{-2} \text{ (inwards)}$$

(d) Since the maximum acceleration occurs at the extreme positions, then so does the maximum restoring force, which is given by

$$F = ma = m \times 4\pi^2 f^2 r$$

and, substituting,

$$F = 0.95 \times 4\pi^2 \times (20)^2 \times 0.1 = 1500 \text{ N}$$

Now let us consider some practical examples of simple harmonic motion.

11.2 THE SIMPLE PENDULUM

The simple pendulum consists of a mass m that swings through a small angle on the end of a string of negligible mass hanging from a fixed point (Figure 11.5).

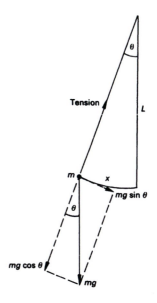

The figure shows the pendulum at a moment when the string makes an angle θ with the vertical and the mass is displaced a distance x along an arc of radius L (where L is the length of the string). The weight mg is resolved into two components: $mg \cos \theta$, supported by the tension in the string, and $mg \sin \theta$, which is the magnitude of the force F restoring the mass to the central position. If θ is small, then $\sin \theta$ is very close to θ in radians (for example, $10° = 0.1745$ rad and $\sin 10° = 0.1736$). Thus,

$$F = mg \sin \theta$$

and, if θ rad is small,

$$F = mg\theta$$

Figure 11.5

and since, from the figure, $\theta = x/L$ rad,

$$F = \frac{mgx}{L} = \frac{mg}{L} \times x$$

Therefore,

$$F = kx, \text{ where } k = \text{constant} = mg/L$$

So, provided that the amplitude is small, the restoring force is proportional to the displacement and the pendulum will execute simple harmonic motion. Furthermore, since $k = mg/L$ and therefore $m/k = L/g$ then, from Equation (11.4).

$$T (= 2\pi \sqrt{m/k}) = 2\pi \sqrt{L/g} \qquad (11.9)$$

which means that the period of oscillation of the pendulum depends only on its length, assuming that g is constant.

Worked Example 11.2

A 20 g bullet is fired at a 20 kg stationary target suspended by a rope. The bullet becomes embedded in the target, which subsequently swings to and fro in simple harmonic motion with period 4 s and amplitude 255 mm. Assuming the mass of the rope may be ignored, estimate the impact velocity of the bullet.

Equation (11.6) tells us that the magnitude of the velocity of the swinging target is at its maximum value $v = \omega r$ at the central position. Since $\omega = 2\pi f$ (Equation 11.8) and $f \ (= 1/T) = 1/4 = 0.25$ Hz, and $r = 0.255$ m, then, at the central position,

$$v = 2\pi f r = 2\pi \times 0.25 \times 0.255 = 0.40 \text{ m s}^{-1}$$

Assuming the system is undamped (see below), this value of v is the combined velocity following the impact of the bullet on the target. The impact velocity of the bullet may therefore be obtained by consideration of the conservation of momentum. Where the subscripts b, t and c refer to the bullet, the target and their combination, respectively,

$$m_b v_b + m_t v_t = m_c v_c$$

and, substituting,

$$(0.02 \times v_b) + 0 = (20 + 0.02) \times 0.40$$

which gives

$$v_b = 400 \text{ m s}^{-1}$$

11.3 A MASS HANGING FROM A SPRING

Let us assume that the spring in Figure 11.6(a) obeys Hooke's law and is of negligible mass. If an object of mass m is suspended from the lower end and extends the spring by a distance d, then, at the equilibrium position in Figure 11.6(b), the tension in the spring is given by

$$mg = kd$$

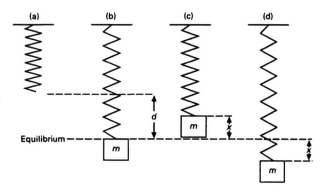

Figure 11.6

where k ($= mg/d$) is the *stiffness* of the spring (i.e. force per unit extension).

If the object is raised slightly, then released so that it oscillates vertically, then at any subsequent displacement x the restoring force towards the equilibrium position will be the resultant of the object's weight pulling it downwards and the tension in the spring pulling it upwards.

If the object is above the equilibrium position, as in Figure 11.6(c), then the tension in the spring is $k(d - x)$ and the downward restoring force R_d is given by

R_d = weight of object − tension in spring

Therefore,

$$R_d = mg - k(d - x)$$

and, since $mg = kd$ (from above),

$$R_d = kd - kd + kx = kx$$

If the object is below the equilibrium position, as in Figure 11.6(d), the tension in the spring is $k(d + x)$ and the upward restoring force R_u is given by

R_u = tension in the spring − weight of object

Therefore,

$$R_u = k(d + x) - mg$$

and, since $mg = kd$,

$$R_u = kd + kx - kd = kx$$

Thus, the restoring force acts towards the equilibrium position and is proportional to the displacement, so the object will execute simple harmonic motion.

Furthermore, since $m/k = d/g$ (from above), then Equation (11.4) gives

$$T \ (= 2\pi \sqrt{m/k}) = 2\pi \sqrt{d/g} \qquad\qquad (11.10)$$

Thus, the period of the oscillation depends on d, the extension at equilibrium, which can be varied by changing m.

11.4 DAMPING

There are many other examples of simple harmonic motion: for example, the oscillation of a vertical float of uniform cross-section in a liquid or of a liquid in a U-tube of uniform cross-section, or the rotational oscillation of a torsion pendulum (e.g. an object twisting about the vertical axis of a wire from which it is suspended).

Many practical systems are more complex than our discussion might seem to suggest. For instance, restoring forces may not be proportional to displacement, and *damping* may be an important factor.

So far we have assumed that the energy of the oscillating system remains constant and that we are dealing with *free* oscillations that continue indefinitely with constant amplitude. Damping causes the amplitude to decay, as in Figure 11.7, for example, because energy is lost to the surroundings. Even a simple pendulum will come to rest eventually because of the damping loss due to the frictional effect of air resistance. Mechanical energy is converted to heat energy and the temperature of the surrounding air increases. The pendulum in a clock loses energy by friction in the mechanism as well as to the air, and the damping losses are topped up by mechanical energy stored in a spring or a raised weight.

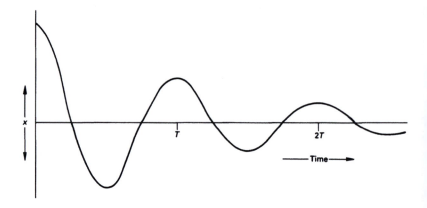

Figure 11.7

Damping is often deliberately introduced into mechanical systems. For instance, shock absorbers are used in a motor car to damp the suspension and minimise oscillation.

Questions

(Where necessary assume that $g = 9.8$ m s^{-2}. Assume free (undamped) oscillations.)

1. Find the length of the simple pendulum with a period of (a) 1 s, (b) 2 s and (c) 4 s.

2. A spring is extended by 25 mm because of a load of 500 g suspended from it. If the load is increased to 1 kg and allowed to oscillate vertically, find the period of its oscillation.

3. If the length of a simple pendulum is reduced by 1.2 m with the result that its frequency is doubled, what was its original length?

4. A spring is extended by 20 mm because of a mass suspended from it. Find the frequency if the mass is allowed to oscillate vertically.

5. Find (a) the maximum speed and (b) the maximum acceleration of an object moving in simple harmonic motion with a period of 2 s and an amplitude of 95 mm.

6. If the period of vertical oscillation of a 0.80 kg mass suspended from a spring is 0.75 s, find the stiffness of the spring.

7. An object is placed on a horizontal surface which oscillates vertically with simple harmonic motion. When the frequency is increased to 7 Hz, the object is on the point of losing contact with the surface. At what point of the cycle does this occur and what is the amplitude of the oscillation?

8. A point that is moving with simple harmonic motion has a velocity of 5 m s^{-1} at a distance of 12 m from its central position, and 12 m s^{-1} at a distance of 5 m from it. Find the frequency of its oscillation.

9. A 2.5 kg mass moves with simple harmonic motion at a frequency of 15 Hz and with an amplitude of 50 mm. Find (a) its total energy, (b) its maximum speed and (c) the maximum restoring force.

10. The period of a torsion pendulum is given by the rotational analogy of Equation (11.4), where m is replaced by the moment of inertia and k by the torsional stiffness (i.e. torque per unit angular displacement, N m rad^{-1}).

    A rod, 1200 mm long, is suspended from its centre by a wire. A torque of 0.175 N m is required to turn the rod 10° about its suspension point. If the period of oscillation is 1.5 s, then, with reference to Figure 10.2 (page 79), estimate the mass of the rod.

TOPIC 12 MECHANICAL WAVES

COVERING:

- the description of mechanical waves;
- reflection, refraction, diffraction and interference;
- wave speed;
- standing waves;
- resonance.

In the previous topic we considered the continuous interchange of potential and kinetic energy in oscillating systems where the total energy remains trapped or would remain trapped in the absence of damping. Figure 11.2 (page 87) shows a wave-like relationship between displacement and time for an isolated oscillating system of this kind.

In this topic we shall consider wave motion via oscillations in a continuous medium which enables energy to be carried from one place to another.

12.1 THE NATURE OF WAVE MOTION

Figure 12.1 shows a rope, stretched horizontally, that is being forced to oscillate vertically at one end. The figure represents snapshots taken at intervals of a quarter of a period. The energy fed into one end of the rope is transferred from one part to the next in a *progressive* (i.e. travelling) wave that moves towards the other end.

The *amplitude* is half the total wave height from trough to crest. The *wavelength* is the distance between any two adjacent points along the wave train that are *in phase*, i.e. exactly 360° apart, such as from crest to crest or from trough to trough. If the frequency of the oscillation initiating the wave motion is f hertz, then the number of waves passing any particular point along their path is f per second. If the wavelength is λ metres, then the *wave speed* v (m s^{-1}) is given by

$$v = f\lambda \tag{12.1}$$

This is sometimes expressed in the form $v = \lambda/T$, where T is the period.

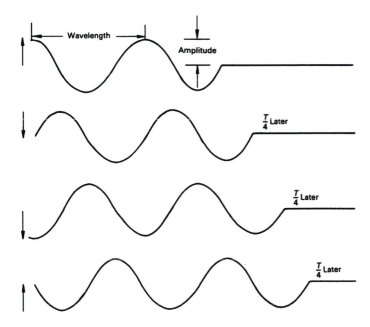

Figure 12.1

The type of wave in Figure 12.1 is described as *transverse*, because the oscillations are perpendicular to the direction in which the wave is travelling. By contrast, *longitudinal* waves involve oscillations which are parallel to the direction of travel, as shown in Figure 12.2. This figure might represent, say, a snapshot of successive pulses of tension and compression passing down the coils of a stretched spring as one end is forced to oscillate along its longitudinal axis.

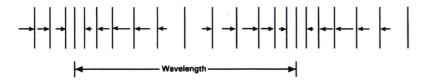

Figure 12.2

It is important to note that Figure 11.2 (page 87) represents displacement against time for a particular point, whereas Figure 12.1 and 12.2 represent displacement against distance at a particular time.

We shall use sound as one illustration of mechanical wave motion. Sound travels through air as longitudinal waves consisting of successive compressions and rarefactions moving at about 340 m s^{-1} at frequencies in the range of 20–20 000 Hz. Corresponding wavelengths therefore range from about 17 m down to 17 mm. (Note that the pitch of musical notes is related to their frequency, based on 440 Hz for the A above middle C.)

Forgetting the sensitivity of a listener's hearing, the loudness of a sound is determined by its *intensity*. This is measured in terms of the amount of energy that sound waves would carry in 1 second through a 1 m^2 aperture perpendicular to their direction of propagation. The unit of intensity is therefore W m^{-2}. (As a rough guide to magnitude, the threshold of hearing is about 10^{-12} W m^{-2}, conversation is about 10^{-6} W m^{-2} and the threshold of pain about 1 W m^{-2}.)

Sound waves normally tend to spread out in all directions from their source. A point source in a uniform medium lies at the centre of a series of expanding spherical wavefronts (i.e. surfaces of constant phase) rather like the ripples spreading out when a stone is dropped into a pond. This means that the intensity progressively decreases as the surface area of the wavefronts increases with their expansion. At a distance r from a sound source of power P watts the energy is distributed over the area of a sphere of radius r, which is equal to $4\pi r^2$, so the intensity I at that distance is given by the inverse square relationship

$$I = \frac{P}{4\pi r^2} \tag{12.2}$$

Surface waves on water depend on a combination of transverse and longitudinal motion. Figure 12.3 shows how individual water molecules move in circular orbits, completing one lap for each wave that passes. The figure represents a snapshot of one complete wavelength.

Figure 12.3

As well as moving vertically, each molecule moves forwards with the wave on the crest and backwards in the trough. Similar orbits occur below the surface, rapidly becoming smaller with depth. In shallow water these orbits interact with the bottom and the wave speed decreases as the depth decreases. If the waves continue to arrive with the same frequency, then the wavelength decreases in accordance with Equation (12.1); that is to say, if $f = v/\lambda$ and f remains constant, then λ must decrease if v decreases.

12.2 WAVE BEHAVIOUR

Wave behaviour is conveniently demonstrated in a device called a *ripple tank*, which has a transparent bottom and contains a shallow layer of water. Waves, or ripples, travelling across the surface of the water are viewed on a screen by passing light through the bottom of the tank. Figure 12.4(a) represents an overhead view of straight, parallel waves being generated by means of a horizontal bar, just dipping into the water, which is made to oscillate vertically. If a small sphere is used instead of the bar, as in Figure 12.4(b), then circular waves diverge from it like those from a stone dropped into a pond. The movement of the waves can be frozen with a stroboscope.

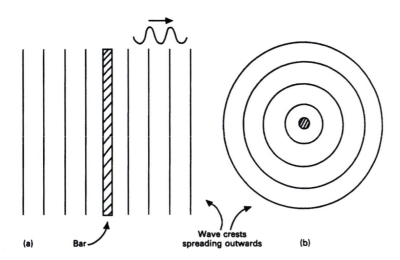

(a)   Bar   Wave crests spreading outwards   (b)

Figure 12.4

Reflection, refraction, diffraction and interference are important aspects of wave behaviour which can be demonstrated with the ripple tank.

Figure 12.5 shows *reflection* demonstrated by placing a barrier diagonally across the path of a train of straight, parallel waves.

Echoes are simply reflections of sound waves. Sound is reflected from hard, flat surfaces in a similar way to light from a mirror (see Figure 13.3 on page 109); the angle of reflection is equal to the angle of incidence.

The reflection of ultrasonic waves (i.e. those above the frequency range of the human ear) is used at sea for echo-sounding and in engineering for non-destructive testing, e.g. for locating cracks. It is also used in medicine for forming images of the inside of the body.

Figure 12.6 shows *refraction*, which is the change of direction a wavefront experiences when it passes obliquely through the interface between two media in which it has different speeds. This can be demonstrated in the ripple tank by placing a flat sheet of glass on the bottom so that the depth of water (hence, the wave speed and the wavelength) changes abruptly. It is important to remember that the frequency remains constant as long as the bar continues to oscillate at the same rate.

Refraction explains why waves tend to reach the beach parallel to the shoreline. When a wave approaches at an angle from deep water,

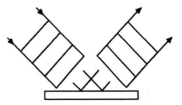

Figure 12.5

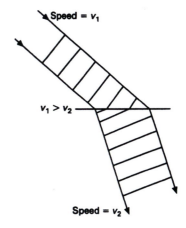

Speed = v_1

$v_1 > v_2$

Speed = v_2

Figure 12.6

then the part of the wavefront closest to the beach slows first as it encounters shallow water, while the parts further out are unaffected and tend to catch up. As these start to slow, the part closest to the shore slows even further and the process continues as progressive refraction along the wavefront tends to turn it parallel to the shore.

As we shall see in the next topic, refraction is an important aspect of the behaviour of light.

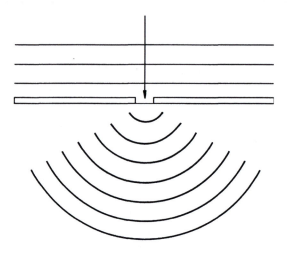

Figure 12.7

Figure 12.7 shows an example of *diffraction* in the ripple tank, where a straight, parallel wavefront passes through an aperture with a width that is similar to the wavelength. Waves passing through the aperture are diffracted – that is to say, they spread out from it with circular wavefronts just as though they had originated there. If the aperture is large compared with the wavelength, then the wavefronts pass through, straight and parallel, more or less unaltered apart from bending at the ends, where they are diffracted round the edges of the aperture. Because of their wavelength, sound waves are diffracted round the corners of buildings and by apertures such as doorways. The diffraction of light waves, which have wavelengths of less than a thousandth of a millimetre, operates on a much smaller scale, as we shall see in the next topic.

Before moving on to interference, we need to take note of the *principle of superposition*, which tells us that, where two or more waves meet, the total displacement at any given point is the sum of their individual displacements. This general idea applies to any waves of the same type but not necessarily of the same shape or wavelength. Figure 12.8(a) shows two waves, identical in this case, which are exactly in phase and superpose constructively to give increased displacement where they meet. On the other hand, Figure 12.8(b) shows that if the same two waves are exactly out of phase, they superpose destructively and cancel each other. Note that if two waves pass

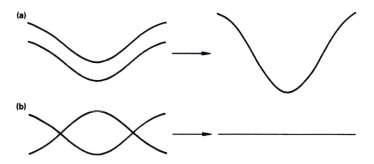

Figure 12.8

through each other, they will continue on their separate ways unaffected by their temporary superposition where they met.

Interference occurs as a result of the superposition of two or more coherent wave motions of comparable amplitude, such as those in Figure 12.8 (*Coherent* waves are those that have a constant phase relationship.)

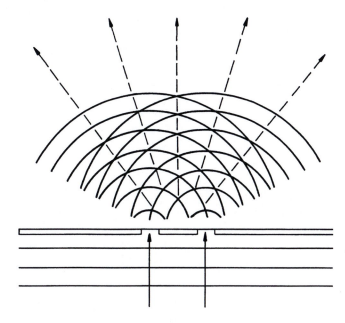

Figure 12.9

Figure 12.9 shows waves originating from two small apertures so that their wavefronts overlap.

Constructive interference occurs along the broken lines where the two wave trains are exactly in phase; crest meets crest where the wavefronts cross and trough meets trough between the crests, to give correspondingly higher crests and deeper troughs, as in Figure 12.8(a).

Destructive interference occurs at those angles in between the broken lines, where the waves are exactly out of phase, i.e. where troughs meet crests and the waves cancel each other out, as in Figure 12.8(b).

12.3 WAVE SPEED

Now we need to consider the speed of mechanical waves passing through matter. In the last topic we met the idea that mechanical oscillations involve the interconversion of potential and kinetic energy and depend on mass and on some kind of restoring force. These factors are also involved in the propagation of mechanical waves. For instance, the speed of a wave passing down a taut string is given by

$$v = \sqrt{\frac{F}{m/l}} \qquad (12.3)$$

where v (m s^{-1}) is the wave speed, F(N) is the tension in the string and m/l (kg m^{-1}) is the mass per unit length of string. If the tension is increased, then the force restoring any displacement in the string will be greater and will tend to return the string to its equilibrium position more quickly. A heavier string, with a greater mass per unit length, has greater inertia and will therefore be returned more slowly.

Similar arguments apply to sound waves passing through matter. The speed of sound can generally be expressed in terms of the elastic behaviour and density of the medium as follows:

$$v = \sqrt{\frac{\text{elastic modulus}}{\text{density}}} \qquad (12.4)$$

Young's modulus (Topic 2) is used for the elastic modulus where longitudinal waves pass down a solid rod of small diameter relative to the wavelength. The stiffer the chemical bonds in the material the more rapid its response to the displacement of its constituent atoms. The less dense the material the less its inertia and the faster its response to restoring forces.

In fluids, sound is transmitted by pressure changes, so the *bulk modulus* is used. Bulk modulus is a measure of the volume elasticity of a substance, as we shall see in Topic 21.

The speed of sound in air at 1 atm is about 331 m s^{-1} at 0 °C and 344 m s^{-1} at 20 °C and, by comparison, about 1500 m s^{-1} in water and about 5000 m s^{-1} in steel.

Interesting things happen when a sound source moves at a speed that is significant relative to the speed of sound. For example, the pitch of a car horn appears to drop as it passes a stationary observer.

This is an example of the Doppler effect. Figure 12.10(a) shows that

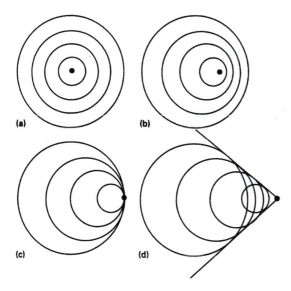

Figure 12.10

a stationary sound source remains at the centre of the spherical wavefronts that it produces, but in Figure 12.10(b), where the source is moving from left to right, it tends to catch up with the wavefronts ahead and leave behind those at the rear. Owing to the effective squashing and stretching of the wavelengths, a stationary observer hears a frequency that appears to be higher than the actual frequency as the source approaches and lower as it moves away. The Doppler effect also occurs if the source is stationary and the observer is moving.

As the velocity is further increased, the wavefronts crowd closer together until at the speed of sound (Figure 12.10c) they can no longer outrun the source and so form a barrier which the source must penetrate to achieve supersonic speeds. Above the speed of sound, as in Figure 12.10(d), the source outstrips the wavefronts, leaving a conical shock wave behind it that is defined by the tangential envelope enclosing the wavefronts. This is the source of the sonic boom from supersonic aircraft.

12.4 STANDING (STATIONARY) WAVES

Standing or stationary waves, as opposed to progressive or travelling waves, are so called because they do not appear to move. They occur, for example, when a progressive wave is reflected straight back along its own path, so that it interacts with the wave moving forwards in the opposite direction. Under the right conditions the superposition of two identical progressive waves moving in opposite directions will cause a standing wave to be set up.

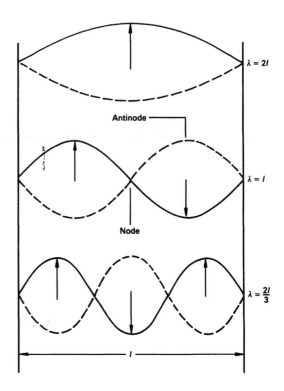

Figure 12.11

Standing waves are formed in taut strings where progressive transverse wave trains move in opposite directions and are continuously reflected at either end to give patterns such as those illustrated in Figure 12.11. The figure represents the times when the displacements are at their maxima. There are points of zero displacement called *nodes*, which never vibrate, and points of maximum amplitude called *antinodes*. The solid and broken lines indicate how the amplitude varies in between. The simplest mode of vibration, called the *fundamental* mode, has one loop corresponding to half a wavelength, so that $\lambda = 2l$, where l is the length of the string. For the *first overtone*, with two loops, $\lambda = l$; for the *second overtone* $\lambda = 2l/3$; and so on. We can see that, in general, $\lambda = 2l/n$, where n is the number of loops. The corresponding frequencies are given by $f (= v/\lambda) = vn/2l$, where v is the speed of the transverse waves in the string.

These various frequencies, sometimes referred to as *harmonics*, are therefore simple multiples of the fundamental frequency; thus, the fundamental frequency is referred to as the *first harmonic* ($n = 1$), the first overtone ($n = 2$) as the *second harmonic*, the second overtone ($n = 3$) as the *third harmonic*, and so on.

Musical instruments produce mechanical standing waves (transverse in strings and longitudinal in air columns) which cause progressive longitudinal waves to spread through the surrounding air to the

listener's ears. Although the predominant mode is usually the fundamental, overtones are produced as well – and the combination of different
amplitudes of the various overtones give different instruments their
characteristic sound quality.

From above, the fundamental frequency of a taut string is given by
$f = v/2l$ (since $n = 1$), so, substituting for v from Equation (12.3),

$$f = \frac{1}{2l} \sqrt{\frac{F}{m/l}} \tag{12.5}$$

Thus, a string may be tuned to a particular fundamental frequency by
varying its tension. The equation also suggests why long, heavy strings
are used for low notes, and short, light strings for high ones.

12.5 RESONANCE

If an object is subjected to vibrations of a frequency that coincides
with one of its own natural frequencies, then *resonance* occurs. Thus,
a note sung near a piano will cause some of the strings to resonate or
vibrate in sympathy.

Resonance is of great interest to engineers. Sometimes it merely causes
irritating vibrations, but sometimes it can be destructive, particularly
when the energy supplied by the source exceeds the damping losses,
so that the amplitude builds up. (As we know from experience, the
amplitude of a pendulum or a swing can be greatly increased by pushing
it quite gently in time with its own natural frequency.) Powerful singers
are said to be able to break wine glasses by hitting a resonant frequency,
and marching soldiers break step crossing bridges to avoid the same
basic type of problem. Machinery can cause resonance in the floor
that supports it, and some readers will have seen the famous film of
the Tacoma Narrows suspension bridge collapsing because of resonance effects due to the wind.

Questions

1. A thin card produces a musical note when it is held
   lightly against the spokes of a rotating wheel. If the
   wheel has 32 spokes, how quickly must it rotate, in
   revolutions per minute, in order to produce the A above
   middle C (i.e. 440 Hz)?

2. Assume the speed of sound in air to be 340 m s^{-1}.
   (a) A clap of thunder arrives 5 s after the lightning
       flash. Assuming that light travels at infinite speed,
       how close is the storm?
   (b) Find the wavelength in air of the musical note C
       with the frequency 262 Hz.

(c) How long does it take for the echo to reach a person who fires a gun while standing 68 m from the base of a cliff?

(d) The wavelength of sound ranges from 17 mm up to 17 m in air. Find the corresponding frequency range.

3. A person stands between two parallel cliffs and fires a gun. The first echo arrives after 1 s and the second after 2 s. How far apart are the cliffs? (Assume the speed of sound is 340 m s^{-1}.)

4. Two people are standing 85 m from the base of a straight cliff. One fires a gun and the other hears two reports, 0.75 s and 0.90 s after seeing the smoke. Find (a) the speed of sound and (b) the distance between the two people.

5. Find the power output of a sound source which produces a sound intensity of 1.4×10^{-4} W m^{-2} at a distance of 25 m.

6. An underwater sound source is operating at 256 Hz. Find the wavelength of the sound (a) under the water and (b) after it has passed through the surface into the air above. (Assume the speed of sound is 1460 m s^{-1} in the water and 340 m s^{-1} in the air.)

7. If the speed of transverse waves is 384 m s^{-1} along a taut string 0.75 m long, find the fundamental frequency of the string and the frequency of its second and third harmonics.

8. The length l of a wire under constant tension was adjusted by varying the distance between its two supports in order to tune it to various frequencies, as follows:

f/Hz	250	300	350	400	450	500
l/m	0.877	0.730	0.629	0.549	0.490	0.440

Manipulate the data to give a straight line relationship and, from a plot, read off the frequency corresponding to $l = 0.500$ m.

9. (a) If it takes 1.9 ms for a sound pulse to travel down a steel rod of length 9.5 m and density 7.8×10^3 kg m^{-3}, estimate the Young's modulus of the steel.

(b) If E represents Young's modulus and ρ represents density, show that the base units of $\sqrt{E/\rho}$ can be reduced to those of velocity.

10. A sound pulse enters one end of a lead rod and of an aluminium rod at the same moment. What must be the relative lengths of the rods if the pulse is to emerge from both simultaneously? (For lead $E = 1.6 \times 10^{10}$ N m^{-2} and $\rho = 1.1 \times 10^4$ kg m^{-3}, and for aluminium $E = 7.0 \times 10^{10}$ N m^{-2} and $\rho = 2.7 \times 10^3$ kg m^{-3}.)

TOPIC 13 ELECTRO-MAGNETIC WAVES

COVERING:

- the nature of electromagnetic waves;
- reflection and refraction;
- total internal reflection;
- diffraction and interference;
- polarisation.

An electromagnetic wave can be considered as a progressive transverse wave that consists of a fluctuating electric field coupled with a fluctuating magnetic field at right angles to it, as shown in Figure 13.1. Don't worry if this seems a difficult idea at this stage; it will become clearer when we discuss electric and magnetic fields in later topics. For the moment the important thing to remember is that, unlike mechanical waves, electromagnetic waves do not necessarily require a medium for their propagation. They travel through empty space (vacuum) at a speed of very nearly 3×10^8 m s^{-1}, commonly called the speed of light (symbol c), and their frequency can be obtained from their wavelength via Equation (12.1) ($v = f\lambda$). The speed of light in air is very slightly less than in vacuum but considerably less in some other materials, as we shall see later.

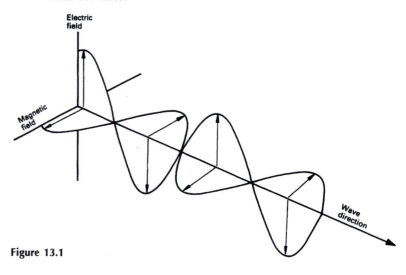

Figure 13.1

The *electromagnetic spectrum* (Figure 13.2) is divided into various types of waves (X-rays, light, radio waves, etc.) which we recognise from everyday experience. These have different names, because they are produced in different ways and have different effects, but the essential difference between them is their wavelength. The boundaries in the figure are approximate, since there are regions of overlap.

In this topic we shall use light to illustrate the nature of electromagnetic waves. Like mechanical waves, they exhibit reflection, refraction, diffraction and interference. And when they fall on the surface of an object they may be reflected by it, transmitted through it, absorbed by it, or some combination of the three. (They may also be scattered, but this is beyond the scope of our discussion.)

Light occupies a narrow band of wavelengths between infrared and ultraviolet. Note that, in a similar way to pitch and sound frequency, colour is a sensation attributable to different wavelengths of light ranging across the colours of the rainbow from about 700 nm (1 nm = 1 × 10^{-9} m) at the red end down to about 400 nm at the violet end. White light is a mixture of all these wavelengths. In white light a red object appears red because it reflects red light and absorbs the other colours; similarly, a red filter transmits red light. In white light white objects appear white because they reflect all the wavelengths and black objects appear black because they absorb them all. (The absorbed energy serves to increase the internal energy of the object, i.e. the kinetic and potential energy of its component particles (Topic 17), and this usually results in a temperature rise.)

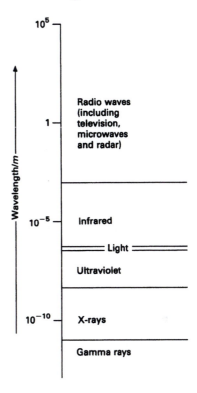

Figure 13.2

13.1 REFLECTION

Objects are made visible by light reflected from their surface. If the surface is rough, then it will appear dull because the reflected light is diffuse, i.e. reflected in all directions by surface irregularities. On the other hand, a smooth, flat, shiny surface like a mirror will reflect a beam of light more or less intact.

Figure 13.3 shows a *ray* representing the direction in which the waves are travelling in a beam of light that is being reflected by a mirror. (Note that the direction of a ray is normal to the wavefronts which it represents.) There are two laws governing reflection. One tells us that the incident ray and the reflected ray lie in the same plane with the normal to the reflecting surface where they meet. The other tells us that the angle of reflection is equal to the angle of incidence relative to the normal.

Figure 13.4 illustrates how light rays diverging from an object form an image in a mirror which gives the impression that the object lies behind the reflecting surface. The direction of the rays changes where they are reflected by the mirror, but the observer interprets them as travelling in straight lines.

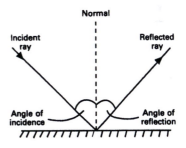

Figure 13.3

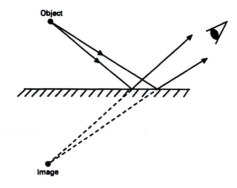

Figure 13.4

13.2 REFRACTION

Figure 13.5 shows how light is refracted when it passes obliquely through the interface between two media in which it has different speeds. The first thing to note is that the incident ray, the refracted ray and the normal all lie in the same plane but, unlike reflection, the angle of refraction r is different from the angle of incidence i.

The figure illustrates the case where the speed of light c is greater in medium 1 than in medium 2, i.e. $c_1 > c_2$. Figure 13.5(b) enables us to find the relationship between i, r, c_1 and c_2. This is less complicated than it might seem. Just follow the argument very carefully, a step at a time.

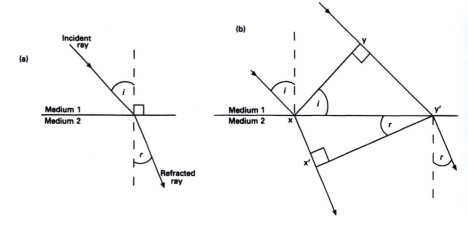

Figure 13.5

Two parallel rays through X and Y approach the interface between the media at an angle of incidence i. Since $c_1 > c_2$, any point on the wavefront XY slows down immediately it crosses the interface. If the wavefront travels from Y to Y' in time t, then the distance YY' equals

$c_1 t$ (i.e. speed × time). During the same interval the other end of the wavefront travels the correspondingly shorter distance XX', equal to $c_2 t$, through medium 2. The wavefront at X'Y' has therefore changed direction and the refracted rays are bent inwards towards the normal so that the angle of refraction r is less than the angle of incidence i. The greater the difference between c_1 and c_2 the greater the change of direction.

Since the ray approaching X is perpendicular to the wavefront XY, and since the normal at X is perpendicular to the interface, then $\text{YXY}' = i$. By a similar argument $\text{XY}'\text{X}' = r$. It follows that

$$\text{YY}' = \text{XY}' \sin i = c_1 t$$

and

$$\text{XX}' = \text{XY}' \sin r = c_2 t$$

Therefore, dividing one equation by the other,

$$\frac{\sin i}{\sin r} = \frac{c_1}{c_2} \tag{13.1}$$

Note that the rays in the figure are reversible and that light will travel along the same path in the opposite direction. In other words, Equation (13.1) still applies where $c_2 > c_1$, but in such cases $r > i$ and refracted rays are bent away from the normal.

13.3 REFRACTIVE INDEX

Equation (13.1) is the basis of *Snell's law*, which tells us that, for two given media, sin i/sin r is a constant. This constant ($= c_1/c_2$) is called the *relative refractive index* $_1 n_2$ for waves passing from medium 1 to medium 2.

The *absolute refractive index* n of a particular medium is given by the ratio between the velocity of light in vacuum, c, and the velocity of light in the medium, c_m, so that $n = c/c_m$. Therefore, $c_1 = c/n_1$ and $c_2 = c/n_2$, so

$$\frac{c_1}{c_2} = \frac{c}{n_1} \times \frac{n_2}{c} = \frac{n_2}{n_1}$$

and, substituting n_2/n_1 for c_1/c_2 in Equation (13.1),

$$\frac{\sin i}{\sin r} = \frac{n_2}{n_1} \tag{13.2}$$

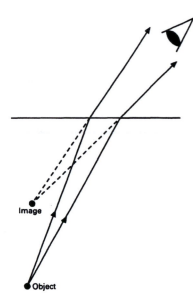

Figure 13.6

At ordinary temperature and pressure the absolute refractive index of air has a value of 1.0003, so if medium 1 is air, then, to a very close approximation, Equation (13.2) reduces to $\sin i / \sin r = n_2$.

To take a few examples, the absolute refractive index of water is 1.33 and that of glass is typically about 1.5–1.7, depending on its type, while that of diamond is about 2.44.

Refractive index varies slightly with wavelength, so values are often quoted for monochromatic light (i.e. a single wavelength), commonly $\lambda = 589.3$ nm, which is in the yellow part of the spectrum.

Figure 13.6 illustrates how refraction accounts for the fact that objects submerged in water appear to be closer to the surface than they really are. Light rays from the object bend away from the normal where they leave the water but the observer interprets them as travelling in straight lines. This is also the reason why straight objects appear to bend upwards where they are partly under water.

When light passing through one medium meets the interface with another of lower refractive index, then *total internal reflection* may occur. Figure 13.7 shows that at zero or relatively low angles of incidence ((a) and (b)) a light ray will pass through the interface and, except at $i = 0°$, will be refracted away from the normal. When the angle of incidence is increased to the *critical angle* i_c, as in (c), the refracted ray passes along the interface and $r = 90°$. When the angle of incidence is greater than i_c, as in (d), total internal reflection occurs, i.e. the ray is reflected from the interface. (Weak internal reflections may be seen for angles of incidence less than i_c; hence the term *total* internal reflection for greater angles.) Substituting i_c for i and $90°$ for r in Equation (13.2), we have

$$\sin i_c = \frac{n_2}{n_1} \tag{13.3}$$

If medium 2 is air, then, to a close approximation, the absolute refractive index of medium 1 is given by $1/\sin i_c$.

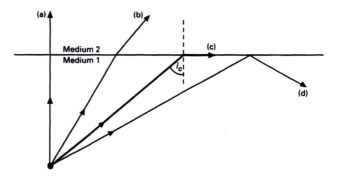

Figure 13.7

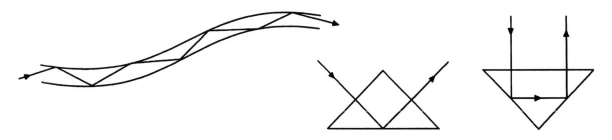

Figure 13.8

Figure 13.8 illustrates how total internal reflection traps light inside glass fibres so that it can be piped from one place to another. It also shows how prisms can be used for reflecting light in optical equipment such as periscopes and binoculars.

13.4 PRISMS AND LENSES

Figure 13.9 illustrates the deviation of light by refraction through a prism. Figure 13.9(a) shows a ray passing through a rectangular block of material. The path by which the light emerges from the block is parallel to the path by which it enters, because the deviation it experiences on entry is cancelled by the equal and opposite deviation on leaving. In the case of the prism in Figure 13.9(b), the angles are such that the ray is bent twice in the same direction and the deviation on leaving the prism is added to the deviation on entering it.

Figure 13.10 shows how the behaviour of a lens can be viewed in terms of a series of prisms. Convex lenses (thicker at the centre) are called *converging* lenses because they bend a parallel beam of light inwards so that it converges to a point called the *principal focus* (at F in Figure 13.10a). Concave lenses (thinner at the centre) are called *diverging* lenses because they bend a parallel beam of light outwards so that it seems to diverge from a principal focus (at F in Figure 13.10b).

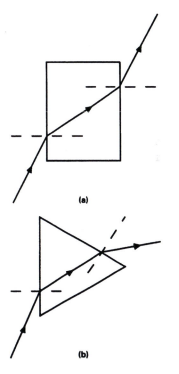

Figure 13.9

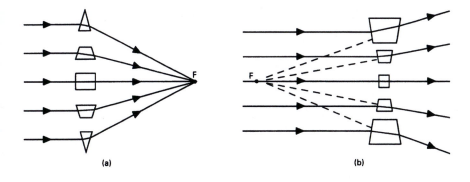

Figure 13.10

A prism will disperse a beam of light into a spectrum consisting of its component wavelengths. *Dispersion* is a consequence of the variation of refractive index with wavelength mentioned earlier. For many substances, refractive index increases with decreasing wavelength; thus, the violet end of the spectrum will be refracted through the greatest angle when white light is dispersed by a glass prism (Figure 13.11). Raindrops disperse sunlight to form rainbows, and simple lenses produce images with coloured fringes because they focus different wavelengths at slightly different positions.

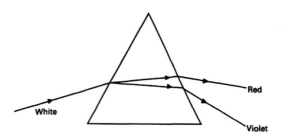

Figure 13.11

13.5 DIFFRACTION AND INTERFERENCE

If monochromatic light passes through a fine slit to illuminate two other fine slits, closely spaced and parallel to the first, then the double slit behaves as a pair of coherent sources which produce interference patterns, as represented in Figure 12.9 (page 101). The effect of constructive and destructive interference may be seen as alternate light and dark bands, parallel to the slits, that are known as *Young's fringes* (after Thomas Young, who demonstrated them in 1801).

A *diffraction grating* consists of many uniformly spaced parallel slits, normally up to about 1000 per mm. These are made by ruling extremely fine lines on suitable materials such as glass, or more commonly (and more cheaply) by casting plastic replicas. (*Reflection gratings* are opaque and work with reflected rather than transmitted light.)

Figure 13.12(a) represents light waves passing through a transmission grating with the diffracted wavefronts emerging from the slits. (Only some of the diffracted wavefronts are shown.) The dotted diagonal line represents the tangential envelope to a series of diffracted wavefronts in which each is exactly one wavelength out of step with its neighbours. All the wavelengths along this line are therefore exactly in phase and reinforce one another. This produces a beam of diffracted light which deviates from the direction of the original beam by an angle θ.

In Figure 13.12(b) d represents the distance between centres of adjacent slits. θ is the angle of deviation of the beam, which is found when the path difference between adjacent wavefronts is equal to λ,

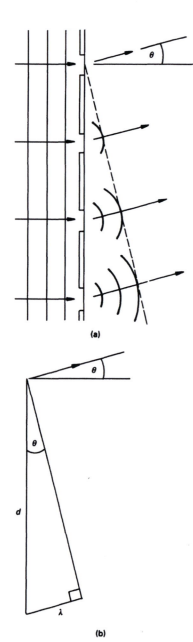

(a)

(b)

Figure 13.12

the wavelength of the light. Simple trigonometry gives $\lambda = d \sin \theta$. Hence, by knowing d and measuring θ, the wavelength can be calculated. If white light is used instead of monochromatic, then each component wavelength deviates to the angle given by the equation, with the result that a spectrum is obtained. As the equation tells us, the deviation will be greatest for the longer wavelengths towards the red end of the spectrum (as opposed to the case of the prism, where the shorter wavelengths suffer the greatest deviation).

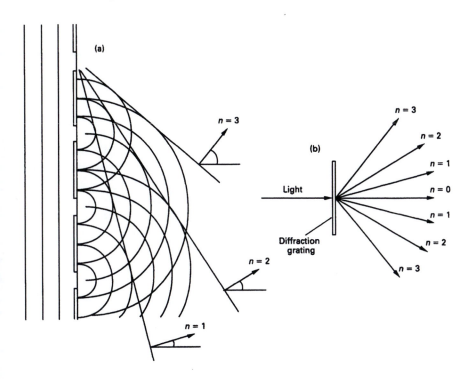

Figure 13.13

As Figure 13.13(a) indicates, there are other tangential envelopes corresponding to path differences of 2λ, 3λ, and so on. In general, therefore, we have

$$n\lambda = d \sin \theta \qquad (13.4)$$

where n is the integral number of wavelengths that constitutes the path difference. This gives rise to first-, second- and third-order angles of diffraction where n equals 1, 2, 3, and so on, as indicated in Figure 13.13(b). Note that there will be a zero-order beam along the centre line where the path difference is zero and θ is therefore zero. On the other side of this there will be a second series of diffraction angles, symmetrical with the first, arising in precisely the same way but with the tangential envelopes facing the other direction, as shown. If white light is used, then corresponding orders of spectra will be seen (apart from a zero-order band of white light along the centre line, where $n = 0$ and $\theta = 0$).

Note that, since $\sin \theta$ cannot exceed 1, $n\lambda$ cannot exceed d, which places a limit on the number of orders which can be obtained.

13.6 POLARISATION

The polarisation of light is best explained by a mechanical analogy using the rope in Figure 12.1 (page 97). Normally transverse waves of any orientation would travel along the rope, but if it is threaded through a vertical slot, then only vertical waves will be able to pass through. Similarly, a horizontal slot will only transmit horizontal waves. A vertical slot followed by a horizontal slot will stop any transverse waves but would have no effect on longitudinal waves (for example, those passing along a stretched spiral spring threaded through them).

Polaroid, the material used in certain types of sunglasses, acts as the optical equivalent of a slot, and two pieces of Polaroid crossed at 90° will stop light passing through. In a beam of unpolarised light the electric field vibrates at all angles within the plane perpendicular to the direction in which the light is travelling. In polarised light these vibrations are confined to one particular direction. As the nature of longitudinal waves suggests, they cannot be polarised.

Polarisation occurs on reflection from certain surfaces and Polaroid sunglasses can be used to cut down glare from the reflected light. Some transparent plastics will polarise light when they are under stress; these can be used to make models of engineering components which enable internal stress patterns to be made visible.

Questions

(Assume that $c = 3.00 \times 10^8$ m s^{-1}.)

1. If an electromagnetic wave has a period of 1.96×10^{-15} s, in what part of the spectrum does it lie?

2. Find the wavelengths transmitted by radio stations broadcasting on (a) 97.6 MHz and (b) 1215 kHz.

3. A source produces light of frequency 4.57×10^{14} Hz. What is the wavelength of the light (a) in air and (b) in water? (Assume $n_{water} = 1.33$.)

4. (a) If a light beam enters the surface of a smooth pond at 55° to the normal, by how much will it be deflected from its original path?
(b) If a light beam emerges from under the surface of a smooth pond at 55° to the normal, by how much

has it been deflected from its original path? (Assume $n_{water} = 1.33$.)

5. What is the refractive index of a substance for which the critical angle in air is 49°?

6. The refractive index of diamond was found to be 2.44, using monochromatic light of frequency 5.09×10^{14} Hz. Find (a) the speed of light in diamond, (b) the wavelength of the monochromatic light in diamond, and the critical angle for internal reflection of diamond (c) in air and (d) in water. (Assume $n_{water} = 1.33$.)

7. A fish's eye view concentrates everything above the water surface into a circle of light which subtends an angle of 98° at the eye. (a) Explain this by assuming $n_{water} = 1.33$. (b) What happens outside the circle?

8. A source of monochromatic light gave a third-order angle of deviation of 48.6° with a particular diffraction grating. Find the first-order angle of deviation.

9. Find the wavelength of monochromatic light giving a first-order angle of deviation of 17.1° using a diffraction grating with 500 lines per mm.

10. What is the angular spread of (a) the first-order spectrum and (b) the third-order spectrum of visible light, from 390 nm to 740 nm, using a diffraction grating with 250 lines per mm?

Part 2: Structure and Properties of Matter

TOPIC 14 ATOMIC STRUCTURE AND THE ELEMENTS

COVERING:

- the electron, the proton and the neutron;
- the nucleus;
- the electronic structure of atoms;
- the elements and the periodic table;
- atomic mass.

So far our everyday general knowledge of gases, liquids and solids has provided us with sufficient background for our discussion. Now we have reached the point where we need to concern ourselves with the internal structure of matter. We shall start with atoms and see how differences in atomic structure lead to the various chemical elements such as hydrogen, carbon, oxygen, and so on.

14.1 THE CONSTITUENT PARTICLES OF ATOMS

Atoms are extremely small, with radii of about 10^{-10} m. We can view their structure in terms of three constituent fundamental particles – the *electron*, the *proton* and the *neutron*. These can be distinguished from one another by their mass and their charge (Table 14.1).

Table 14.1

Particle	Mass/kg	Mass/u	Charge/C
Electron	9.11×10^{-31}	5.5×10^{-4}	-1.60×10^{-19}
Proton	1.67×10^{-27}	1.0	$+1.60 \times 10^{-19}$
Neutron	1.67×10^{-27}	1.0	0

Because these particles are so small, it is sometimes convenient to express their mass in terms of *atomic mass units* (symbol u), where $1 \text{ u} = 1.66 \times 10^{-27}$ kg. (We shall see how we arrive at this value later in the topic.) Note that the mass of the electron can often be considered to be negligible compared with that of the proton and the neutron.

Most of us are aware of the fact that we can electrically charge certain objects such as plastic combs and pens by rubbing them on cloth so that they attract scraps of paper and hairs or even a thin stream of water running from a tap. There are two types of charge, positive and negative, and we need to remember that like charges repel each other, whereas opposite charges attract. The force between two charges obeys an inverse square law, called Coulomb's law, that is analogous to Newton's law of gravitation, which we met in Topic 2. In Topic 26 we shall see that this force, whether attractive or repulsive, is proportional to Q_1Q_2/r^2, where Q_1 and Q_2 represent the magnitudes of the charges and r the distance between them. The unit of charge, which we shall define in Topic 29, is the coulomb (symbol C).

As Table 14.1 shows, the charge on the electron is opposite but equal in magnitude to that on the proton. Atoms contain equal numbers of each and are therefore electrically neutral. (Neutrons are neutral, so they make no contribution to the balance of charge in the atom.)

Now we need to give some thought to how these particles are arranged.

14.2 THE NUCLEUS

At the centre of the atom lies the *nucleus*, which contains all the protons and neutrons and therefore all the positive charge and most of the mass. The number of protons characterises the nucleus as being that of a particular element and is called the *atomic number* (symbol Z). Obviously there must also be Z electrons in a neutral atom. To take a few examples, hydrogen has an atomic number of 1, carbon 6, chlorine 17 and iron 26. The protons and neutrons are held together by an extremely powerful nuclear force, which we shall consider when we deal with the nucleus in more detail in the next topic.

Light elements tend to have about equal numbers of neutrons and protons, but the neutron/proton ratio increases with atomic number to about one and a half for heavy elements. However, the number of neutrons varies for nuclei of the same element. This gives rise to *isotopes*, which are atoms with the same atomic number but which differ in the number of neutrons their nuclei contain. Thus, chlorine-37 has 17 protons and 20 neutrons, whereas chlorine-35 has 17 protons but only 18 neutrons. Note that the *mass number* (symbol A) is the total number of protons and neutrons in the nucleus (37 and 35 in the case of these two examples). It follows that the number of neutrons in the nucleus is given by $(A - Z)$. The convention for representing a particular atom, say of element *X*, is ${}_{Z}^{4}X$. For example, the carbon-12 isotope is represented by ${}_{6}^{12}C$, where C is the chemical symbol for carbon (see Table 14.2 on page 122).

Nuclei have radii of approximately 10^{-14}–10^{-15} m and are therefore tiny compared with the overall size of the atoms which contain them. Since they contain most of the mass, they are of extremely high density.

14.3 THE ELECTRONIC STRUCTURE

The relative size of the nucleus means that the atom is mostly empty space. Early models of the atom suggested that the electrons revolve in orbits around the nucleus like the planets round the sun, the attractive force between the positive nucleus and the negative electrons providing the centripetal force. The total energy of the electron was viewed in terms of its potential energy due to its distance from the nucleus and its kinetic energy due to its motion. More recent developments have led to the idea of three-dimensional *orbitals* representing regions within the atom where the probability of finding an electron is high.

Rather than try to picture the atom, we shall simply think of the electrons as being arranged in energy levels. These are traditionally called *shells* (numbered 1, 2, 3, etc.) and are divided into *subshells*. The first and lowest energy shell has one subshell (labelled 1*s*), the second shell has two subshells (2*s* and 2*p*), the third has three (3*s*, 3*p* and 3*d*) and the fourth has four (4*s*, 4*p*, 4*d* and 4*f*). (The use of the letters *s*, *p*, *d* and *f* has an historical basis.) Figure 14.1 shows these and higher subshells arranged in columns according to which main

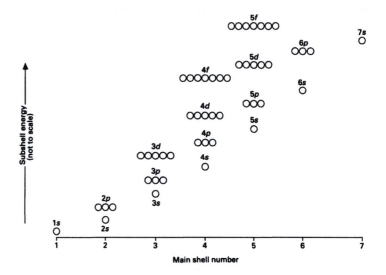

Figure 14.1

shell they belong. They are plotted vertically in order of increasing energy. Note that, from the 4s upwards, the subshell energy levels overlap between the main shells.

Each of the circles in the figure corresponds to one of the orbitals mentioned above and, in effect, represents an 'address' that can accommodate two electrons. Each s subshell can accommodate two electrons in one orbital, each p subshell can accommodate six in three orbitals, each d can accommodate ten in five orbitals and each f fourteen in seven. Note that two electrons occupying the same orbital must be of opposite *spin*. (In some respects an electron behaves as if it is spinning on its axis like a top; in terms of our simple model, it is as though the two electrons must spin in opposite directions, clockwise and anticlockwise, in order to coexist in the same orbital).

We shall now make use of Figure 14.1 as the basis of a paper exercise where we picture the chemical elements as a series, built up by successively adding electrons to the subshells. As we might expect, the lowest energy levels are filled first. (We should bear in mind that there must be an equal number of protons and the appropriate number of neutrons added to the nucleus at the same time.)

Table 14.2 shows the first thirty-six elements. Hydrogen has a single electron in the 1s subshell. The completion of this subshell with a second electron gives helium, whereupon the first main shell has its full complement of two electrons. The third electron enters the next lowest subshell, the 2s, to give lithium, and the fourth electron completes it, to give beryllium. The 2p subshell fills next to give the six elements from boron to neon. With neon the second main shell has its full complement of eight electrons.

Complications begin in the third shell. The 3s and 3p subshells fill first to give the elements from sodium to argon. But, as Figure 14.1 shows, the 4s subshell lies at a lower energy level than the 3d and therefore fills next to give potassium and calcium. The 3d follows then, at gallium, the 4p. This is as far as we need to go to get the general idea of viewing the elements in terms of filling subshells in order of increasing energy.

A very important consequence of the overlap between energy levels is that the outermost main shell of any atom cannot contain more than eight electrons; as Figure 14.1 shows, d and f subshells do not begin to fill until there are electrons present in a higher main shell.

The actual number of electrons in the outermost main shell is of enormous importance in determining the properties of an atom. For example, Figure 14.2 shows the *ionisation energy* of the first twenty elements in order of increasing atomic number. The ionisation energy is simply the amount of energy that would be required to completely remove an outermost electron from an atom against the attractive force due to the positive charge on the nucleus.

First, we can see that helium (He), neon (Ne) and argon (Ar) have high ionisation energies. This means that atoms of these elements are resistant to the removal of an outer electron and therefore have

Table 14.2

Element	Chemical Symbol	Atomic Number	1s	2s	2p	3s	3p	3d	4s	4p
Hydrogen	H	1	1							
Helium	He	2	2							
Lithium	Li	3	2	1						
Beryllium	Be	4	2	2						
Boron	B	5	2	2	1					
Carbon	C	6	2	2	2					
Nitrogen	N	7	2	2	3					
Oxygen	O	8	2	2	4					
Fluorine	F	9	2	2	5					
Neon	Ne	10	2	2	6					
Sodium	Na	11	2	2	6	1				
Magnesium	Mg	12	2	2	6	2				
Aluminium	Al	13	2	2	6	2	1			
Silicon	Si	14	2	2	6	2	2			
Phosphorus	P	15	2	2	6	2	3			
Sulphur	S	16	2	2	6	2	4			
Chlorine	Cl	17	2	2	6	2	5			
Argon	Ar	18	2	2	6	2	6			
Potassium	K	19	2	2	6	2	6		1	
Calcium	Ca	20	2	2	6	2	6		2	
Scandium	Sc	21	2	2	6	2	6	1	2	
Titanium	Ti	22	2	2	6	2	6	2	2	
Vanadium	V	23	2	2	6	2	6	3	2	
Chromium	Cr	24	2	2	6	2	6	5	1	
Manganese	Mn	25	2	2	6	2	6	5	2	
Iron	Fe	26	2	2	6	2	6	6	2	
Cobalt	Co	27	2	2	6	2	6	7	2	
Nickel	Ni	28	2	2	6	2	6	8	2	
Copper	Cu	29	2	2	6	2	6	10	1	
Zinc	Zn	30	2	2	6	2	6	10	2	
Gallium	Ga	31	2	2	6	2	6	10	2	1
Germanium	Ge	32	2	2	6	2	6	10	2	2
Arsenic	As	33	2	2	6	2	6	10	2	3
Selenium	Se	34	2	2	6	2	6	10	2	4
Bromine	Br	35	2	2	6	2	6	10	2	5
Krypton	Kr	36	2	2	6	2	6	10	2	6

particularly stable electronic structures which make them extremely reluctant to combine chemically with other elements. This is reflected in their name, the *inert gases*, sometimes called the noble gases. Table 14.2 shows that the outermost main shells of neon and argon contain the maximum number of eight electrons. Later we shall see that there

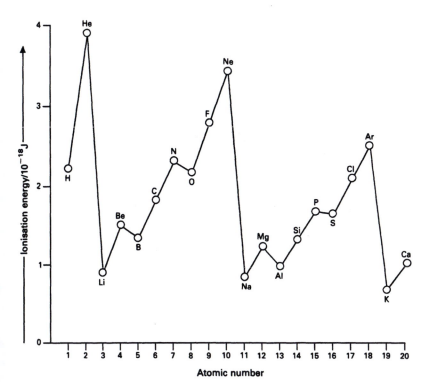

Figure 14.2

are heavier elements with this outer octet which are members of the same family. Note that helium is also an inert gas but can only possess two outer electrons, because the first main shell has no *p* subshell.

Lithium (Li), sodium (Na) and potassium (K) have low ionisation energies, which means that their outermost electrons are loosely held. Each of these elements has one more electron than the preceding inert gas, and Table 14.2 reminds us that the extra electron starts a new main shell. This makes these elements particularly susceptible to chemical combination with others. This family, of which these are the lighter members, is called the *alkali metals*.

Figure 14.2 shows that there is an increase in ionisation energy in building up from one alkali metal to the next inert gas. For example, it increases fourfold from lithium to neon, corresponding to atomic numbers from 3 to 10. This is hardly surprising, since the outer electron in lithium is attracted to the nucleus by the positive charge due to only three protons; the outer electrons in neon are attracted by ten protons, so we would expect them to be much more strongly held.

Note that there is an overall decrease in ionisation energy from the first 'octave' of elements (lithium to neon) to the second (sodium to argon). This is due to a larger 'screen' of inner electrons shielding the outer electrons from the attraction of the nucleus and making them easier to remove. We can regard the electrons in hydrogen and helium as being in direct sight of the nucleus (and helium has the highest ionisation energy of all the elements). From lithium to neon there are two screening electrons in the first main shell. From sodium to argon there are a total of ten, two in the first main shell and eight in the second, and there is a general reduction in ionisation energy.

The abrupt decrease in ionisation energy from an inert gas to the succeeding alkali metal is due to the sudden increase in the number of screening electrons accompanying an increase of only one proton in the nucleus.

14.4 THE PERIODIC TABLE

The cyclical, or periodic, variation in ionisation energy that we see in Figure 14.2 is reflected in the *periodic table*, which is used to classify the elements. Table 14.3 shows the first 36 elements arranged in *groups*, corresponding to the vertical columns, according to the number of outer electrons they possess. The group I elements have one outer electron, the group II elements have two, and so on. By convention, the inert gases in column 8 are given the group number 0.

The *s* and *p* blocks dividing the first four horizontal rows correspond to filling the *s* and *p* subshells in the first four main shells. Note

Table 14.3

GROUP NUMBER	I	II		III	IV	V	VI	VII	0
	s-BLOCK			*p*-BLOCK					

H 1							He 2
Li 3	Be 4	B 5	C 6	N 7	O 8	F 9	Ne 10
Na 11	Mg 12	Al 13	Si 14	P 15	S 16	Cl 17	Ar 18
K 19	Ca 20	Ga 31	Ge 32	As 33	Se 34	Br 35	Kr 36

TRANSITION ELEMENTS									
d-BLOCK									
Sc 21	Ti 22	V 23	Cr 24	Mn 25	Fe 26	Co 27	Ni 28	Cu 29	Zn 30

that there are no *p* subshells in the first main shell and, although helium can only have two electrons, it is properly regarded as an inert gas and therefore placed in column 8.

As we have already seen, following calcium (Z = 20), the 4*p* subshell does not fill until the 3*d* is complete. So, although the *d*-block elements from scandium to zinc (21 to 30) belong between calcium and gallium, they do not readily fit into the fourth row as it stands. Instead they form the first row of an inner series, called the transition elements, which fits in between groups II and III.

Table 14.4 shows the full periodic table, which includes the subshells above the 4*p*. Further transition elements occur where the higher *d* subshells are being filled, and more complications begin with the lanthanide series, where the 4*f* subshell fills before the 5*d*.

We needn't concern ourselves with the detailed structure of the full periodic table here other than to note that we can still view it in terms of filling subshells in order of increasing energy.

The elements below and to the left of the heavy line in Table 14.4 are metals. Be careful to note that this division is not precise, because some elements close to the line show behaviour that is partly metallic and partly non-metallic. Before we can appreciate the distinction, we need some basic understanding of the chemical bonds that occur between atoms. But first let us take a closer look at atomic mass.

Table 14.4

14.5 ATOMIC MASS

As we noted earlier, mass on the atomic scale is measured in atomic mass units. By convention 1 u is taken to be equal to one-twelfth of the mass of the carbon-12 atom, which gives it the value 1.66×10^{-27} kg. (For carbon-12, $A = 12$ and $Z = 6$, so this atom contains six neutrons and six protons in the nucleus and six electrons to balance the protons.)

The *relative atomic mass* (atomic weight) of an element is the average mass per atom expressed in atomic mass units; it is relative to one-twelfth of the mass of the carbon-12 atom and it is an average value, since elements occur naturally as mixtures of their isotopes. Table 14.5 gives some examples.

Table 14.5

Element	Symbol	Atomic no.	Relative atomic mass
Hydrogen	H	1	1.0
Carbon	C	6	12.0
Nitrogen	N	7	14.0
Oxygen	O	8	16.0
Sodium	Na	11	23.0
Aluminium	Al	13	27.0
Chlorine	Cl	17	35.5
Calcium	Ca	20	40.1
Iron	Fe	26	55.8
Copper	Cu	29	63.5
Zinc	Zn	30	65.4
Silver	Ag	47	107.9
Gold	Au	79	197.0

Although mass is measured in kg, the SI base unit for *amount of substance* is the mole (symbol mol), which is the amount containing a fixed number of the 'entities' that constitute the substance under consideration. This number, called the Avogadro constant, is 6.02×10^{23} mol^{-1} whatever the substance happens to be. The entities might be atoms, or ions or molecules (which we shall meet in the next topic). Thus, 1 mol of water contains 6.02×10^{23} water molecules and 1 mol of carbon contains 6.02×10^{23} carbon atoms. The mole is useful where the relative number of entities is important – for example, in studying a chemical reaction.

It is a simple matter to convert moles to mass. For example, carbon has a relative atomic mass of 12.0 and carbon atoms therefore have an average mass of $(12.0 \times 1.66 \times 10^{-27})$ kg each. The mass of 1 mol of carbon, i.e. 6.02×10^{23} carbon atoms, is therefore given by

$$6.02 \times 10^{23} \times (12.0 \times 1.66 \times 10^{-27}) \text{ kg}$$

which is equal to

0.012 kg = 12 g

1 mol is actually defined as the amount of substance that contains as many of the specified entities as there are atoms in 0.012 kg of carbon-12. For our purposes we can say that the mass of 1 mol of carbon or of any other element is equal to its relative atomic mass expressed in grams. Looking at this another way, 6.02×10^{23} atomic mass units have a mass of 1 g, because

$6.02 \times 10^{23} \times 1.66 \times 10^{-27} = 0.001$ kg = 1 g

Note that, although the SI unit of mass is the kilogram, scientists often work in grams (g). Remember that 1 g $= 1 \times 10^{-3}$ kg.

Questions

(Where necessary assume that 1 u $= 1.66 \times 10^{-27}$ kg *or* that the Avogadro constant $= 6.02 \times 10^{23}$ mol^{-1}.)

Making use of the information tabulated in this topic:

1. Identify by name the elements with the following atomic numbers and state which are metals: 3, 7, 10, 16, 19, 21, 25 and 35.

2. Find the number of neutrons contained in each of the following atoms: helium-4, nitrogen-14, carbon-14, oxygen-16, argon-40, potassium-40 and calcium-40.

3. Find the mass in kg of the following atoms: carbon, oxygen and iron.

4. Find the mass of silver that contains the same number of atoms as (a) 63.5 g of copper, (b) 3 g of carbon and (c) 4925 kg of gold.

5. Find the number of atoms in 1 kg of each of the following: hydrogen, copper and silver.

6. An 18 carat gold ring weighs 8.72 g. How many gold atoms does it contain? (18 carat gold contains 75% gold by weight.)

7. Find the number of atoms in a 1 mm cube of copper. ($\rho_{copper} = 8900$ kg m^{-3}.)

8. The smallest entity identifiable as water is the water molecule, which contains two atoms of hydrogen and one of oxygen. If 180 g of water is completely decom-

posed into hydrogen and oxygen, what mass of hydrogen is produced?

9. What is the volume in mm^3 of a piece of copper containing 5×10^{22} atoms? ($\rho_{copper} = 8900$ kg m^{-3}.)

10. In a certain silver/copper alloy 12.1% of the total number of atoms is copper. Find the percentage of silver by weight.

11. Assuming the value of the Avogadro constant and using the data in Table 14.5, (a) suggest what element it is that has atoms with an average mass of 9.27×10^{-26} kg; (b) find the mass of the oxygen atom; and (c) estimate the volume of the copper atom if $\rho_{copper} = 8900$ kg m^{-3}.

See also Further Questions on page 252.

TOPIC 15 THE NUCLEUS

COVERING:

- stability of the nucleus;
- mass defect and binding energy;
- radioactivity;
- α-, β- and γ-radiation;
- radioactive decay;
- half-life;
- nuclear reactions;
- nuclear fission and thermonuclear fusion.

In the previous topic we treated the nucleus as a tiny speck at the centre of the atom that determines its identity, contains most of its mass and keeps its electrons under control. In this topic we shall examine the nucleus in more detail, briefly considering its stability, together with *radioactivity* and *transmutation* (transformation of one element into another) by *nuclear reactions*.

Let us begin by reviewing some of the terms that we shall be using. The protons and neutrons in the nucleus are jointly called *nucleons*, and an atom in which the number of protons and neutrons is specified is called a *nuclide*. The *mass number* or *nucleon number* (symbol A) of a particular nuclide is the number of nucleons that it contains; for example, A equals 35 and 37 respectively for chlorine-35 and chlorine-37. The number of neutrons, sometimes called the *neutron number* N, is equal to $(A - Z)$ where Z (the atomic number) gives the number of protons.

As we saw in the previous topic, the convention for representing a particular nuclide is $^A_Z X$ where X is the chemical symbol for that particular element. The chemical symbol for chlorine is Cl; therefore, chlorine-35 and chlorine-37 are represented by $^{35}_{17}Cl$ and $^{37}_{17}Cl$ respectively.

15.1 STABILITY OF THE NUCLEUS

Nucleons are held together by an extremely powerful nuclear force that overcomes the strong repulsion between the positively charged protons confined within the tiny volume of the nucleus. (There is also a weak nuclear force, which we shall not consider.) The repulsive force

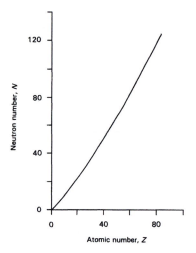

Figure 15.1

is a long-range force that acts between all the protons in the nucleus, whereas the attractive nuclear force has a very short range and is only effective between nucleons that are near neighbours. As we shall see shortly, this limits the size of nuclei that are stable.

Nuclides with unstable nuclei are *radioactive*, that is to say they *decay* with the emission of radiation to form *daughter* nuclides which may or may not be radioactive. If a daughter is radioactive then, in its turn, it becomes a *parent* nuclide which decays – and the process is repeated until a stable nuclide is reached which terminates the radioactive series.

The factors that govern the stability of nuclides are beyond the scope of our discussion; nevertheless it is helpful to make a few relevant observations. Stable (non-radioactive) nuclides lie in a narrow band along the curved line in Figure 15.1, which shows the neutron number N plotted against the number of protons Z. The figure shows that stable nuclides with relatively low atomic numbers have neutron/proton ratios of about 1/1. As the atomic number increases, the curve becomes steeper and the neutron/proton ratio increases up to about 1.5/1 for the heavy elements; in effect, the extra neutrons provide additional nuclear force to help offset the progressively increasing repulsion as the number of protons increases. Eventually, there is a point where the nucleus is so large that short-range nuclear forces are scarcely able to offset the long-range repulsive forces between the protons. The band of stable nuclei ends with $^{209}_{83}\mathrm{Bi}$.

There are certain *magic numbers* of neutrons and protons (2, 8, 20, 28, 50, 82 and so on) that correspond to particularly stable nuclei, suggesting a shell model of the nucleus that is analogous to the shell model of the electronic structure of the atom. Furthermore, the majority of stable nuclides contain either even numbers of protons or neutrons, or, in most cases, even numbers of both. The two protons and two neutrons in the helium nucleus ($^4_2\mathrm{He}$) are a particularly stable combination, and there are certain nuclei with some degree of extra stability that can be regarded as multiples of the helium nucleus (for example $^{12}_6\mathrm{C}$ and $^{16}_8\mathrm{O}$).

15.2 BINDING ENERGY

The mass of any nucleus is very slightly less than the total mass of the separate neutrons and protons from which it is formed. For example, the mass of the helium nucleus is about 6.65×10^{-27} kg, whereas the total mass of 2 protons and 2 neutrons is about 6.70×10^{-27} kg. The difference, which amounts to 0.05×10^{-27} kg, is called the *mass defect*. According to Einstein's famous equation $E = mc^2$, which gives the energy E equivalent to mass m (where c is the speed of light in a vacuum), the mass defect of the helium nucleus is equivalent to

$$E = mc^2 = (0.05 \times 10^{-27}) \times (3.00 \times 10^8)^2 = 4.5 \times 10^{-12} \text{ J}$$

This is the *binding energy* of the helium nucleus. The binding energy of any nucleus reflects its cohesion and is equal to the amount of energy that would be required to separate it into its component neutrons and protons (if that were possible). The *binding energy per nucleon* is the binding energy divided by the number of nucleons in the nucleus. In the case of helium this is equal to about 1.1×10^{-12} J. Figure 15.2

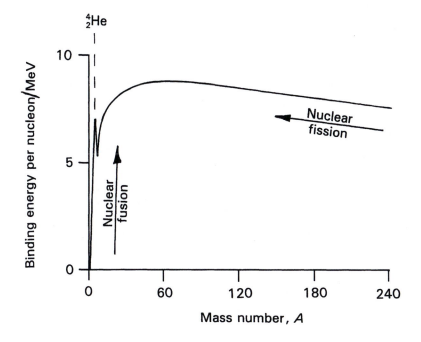

Figure 15.2

shows how binding energy per nucleon varies with mass number up to about $A = 240$. It is important to recognise that the highest binding energies per nucleon are associated with mass numbers in the region of 60; this means that it would take more energy per nucleon to separate these nuclei into their component neutrons and protons than it would for the lighter and the heavier nuclides on either side. *Nuclear fusion* is the fusing together of light nuclei and *nuclear fission* is the splitting of heavy nuclei. Both processes lead to an increase in binding energy per nucleon and, as we shall see shortly, the release of an equivalent amount of *nuclear energy*. In the meantime, note the anomalous peak corresponding to the helium nucleus $^4_2$He, which indicates that the cohesion between its nucleons is particularly great. As we shall now see, helium nuclei can exist on their own in the form of *α-particles* (alpha particles), which have a very important role in radioactive decay.

Note that energy in Figure 15.2 is expressed in terms of the *electronvolt* (eV), which is a non-SI unit that is very widely used for dealing with events on an atomic and nuclear scale. As we shall see in Topic 27, it can be defined as the kinetic energy an electron acquires as it undergoes free acceleration through a potential difference of 1 volt, and

$$1 \text{ eV} = 1.602 \times 10^{-19} \text{ J}$$

15.3 RADIOACTIVITY

There are various modes of radioactive decay that enable unstable nuclides to adjust the number of neutrons and protons that they contain.

α-decay is accompanied by the emission of *α-radiation*, which consists of *α-particles* that are generally travelling at speeds of the order of a few percent of the speed of light. As we have just seen, α-particles are simply helium nuclei, which contain 2 neutrons and 2 protons. They can be represented by the symbol $^4_2\alpha$, where the superscript 4 and the subscript 2 remind us that they have a mass of about 4 u (i.e. 4 atomic mass units) and contain 2 protons which give them a double positive charge. This means that, when a nucleus emits an α-particle, its mass number (A) decreases by 4 and its atomic number (Z) decreases by 2. The daughter resulting from α-decay is therefore the element that lies two places in front of the parent in order of atomic number. In general terms we can write

$$^A_Z P \rightarrow {}^{(A-4)}_{(Z-2)} D + {}^4_2\alpha$$

where P and D represent the parent and daughter elements respectively. To take an example, radium-226 decays to radon-222 as follows:

$$^{226}_{88}\text{Ra} \rightarrow {}^{222}_{86}\text{Rn} + {}^4_2\alpha$$

The arrow indicates that this is an irreversible (i.e. one-way) process. It is important to recognise that A and Z must balance on either side of the arrow so that, for A,

$$226 = 222 + 4$$

and for Z,

$$88 = 86 + 2$$

α-emitters tend to be heavy nuclides, with nuclei that are too large, where the short-range nuclear force of attraction is unable to offset the long-range repulsion between the protons. Since an α-particle contains 2 neutrons and 2 protons then α-decay reduces both N and Z by 2 and, as Figure 15.3 shows, the daughter lies below and to the left of the parent nuclide on a plot of N against Z. An unstable nuclide in the area to the right of the upper end of the stability curve in Figure 15.1, where the slope is steep, would therefore be brought closer to it by α-decay.

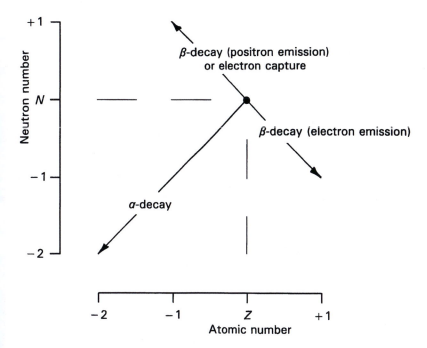

Figure 15.3

β-decay (beta decay) occurs when a neutron in the nucleus is transformed into a proton, as follows, with the emission of an electron that, in some cases, may be travelling at speeds close to the speed of light:

$$_0^1n \rightarrow {}_1^1p + {}_{-1}^0e$$

$_0^1n$, $_1^1p$ and $_{-1}^0e$ represent the neutron, the proton and the electron respectively. Note that, as before, the total mass and the total charge must balance on either side of the arrow. (The mass of the electron is taken to be negligible here.) Since electron emission accompanies the conversion of a neutron to a proton then, as Figure 15.3 shows, N decreases by 1 and Z increases by 1, therefore A remains unaltered, and we can write

$$_Z^A P \rightarrow {}_{(Z+1)}^A D + {}_{-1}^0e$$

A nuclide with excess neutrons therefore moves closer to the stability curve from a position above it; and the daughter element is the element immediately following the parent in order of atomic number. An important example of $β$-decay is electron emission by carbon-14 ($_6^{14}C$), as follows

$$_6^{14}C \rightarrow {}_7^{14}N + {}_{-1}^0e$$

which is used as the basis of radiocarbon dating.

In many cases, α- and β-decay leave the nucleus in an excited state with a surplus of energy which it then emits as a photon of γ-radiation (gamma radiation). As Figure 13.2 shows (page 109), γ-radiation is electromagnetic radiation of shorter wavelength than X-radiation. (This means that γ-rays are of higher frequency and greater energy than X-rays.)

There are two further modes of decay that are of lesser importance in our discussion but which need to be included for completeness. Firstly, the term β-decay also covers the transformation of a proton to a neutron, as follows, by the emission of a *positron* (symbol $_{+1}^{0}e$):

$$_{1}^{1}p \rightarrow {}_{0}^{1}n + {}_{+1}^{0}e$$

Positrons can be regarded as positive electrons, therefore, as a result of positron emission, N increases by 1 and Z decreases by 1 (see Figure 15.3). A remains unaltered and we can write

$$_{Z}^{A}P \rightarrow {}_{(Z-1)}^{A}D + {}_{+1}^{0}e$$

A nuclide with a deficiency of neutrons therefore moves closer to the stability curve from a position below it; and the daughter element is the element that lies immediately in front of the parent in order of atomic number.

Secondly, *electron capture* is a process that results in the same transformation as positron emission, but it occurs when an electron from an inner orbital is captured by the nucleus where it transforms a proton into a neutron, as follows

$$_{1}^{1}p + {}_{-1}^{0}e \rightarrow {}_{0}^{1}n$$

and we can write

$$_{Z}^{A}P + {}_{-1}^{0}e \rightarrow {}_{(Z-1)}^{A}D$$

This is accompanied by the emission of a photon of X-radiation as an outer electron falls into the vacancy left by the captured electron.

To summarise these processes:

α-decay	reduces the size of the nucleus so that $$^A_Z P \rightarrow \, ^{(A-4)}_{(Z-2)} D + \, ^4_2 \alpha$$
β-decay	occurs by (a) *electron emission* $\quad ^1_0 n \rightarrow \, ^1_1 p + \, ^0_{-1} e$ which reduces the neutron/proton ratio so that $$^A_Z P \rightarrow \, ^A_{(Z+1)} D + \, ^0_{-1} e$$ (b) *positron emission* $\quad ^1_1 p \rightarrow \, ^1_0 n + \, ^0_{+1} e$ which increases the neutron/proton ratio so that $$^A_Z P \rightarrow \, ^A_{(Z-1)} D + \, ^0_{+1} e$$
electron capture	$\quad ^1_1 p + \, ^0_{-1} e \rightarrow \, ^1_0 n$ which increases the neutron/proton ratio so that $$^A_Z P + \, ^0_{-1} e \rightarrow \, ^A_{(Z-1)} D$$
γ-radiation	enables the nucleus to shed surplus energy

Depending on its source, radioactivity can release considerable amounts of energy in the form of the kinetic energy of α- and β-particles and electromagnetic energy of γ-radiation. These and certain other types of radiation (including X-rays and high-energy ultraviolet rays) are classified as *ionising radiation* because they cause ionisation in the medium through which they pass. Because of the damage it can cause, particularly to living tissue, there are very stringent safety precautions that must be taken when working with ionising radiation and with radioactive substances.

The relatively high mass and charge of α-particles means that they have a very strong ionising effect. This rapidly absorbs their energy as they pass through matter and limits their powers of penetration. β-particles generally have less ionising effect but greater penetration than α-particles. γ-radiation has still less ionising effect but much greater powers of penetration. Because they consist of charged particles, α- and β-rays are deflected by electrical and magnetic fields (see Part 3, starting on page 256).

Ionising radiation can be detected in various ways by means of the ionisation that it produces. For example, the tracks of α- and β-particles can be made visible in a *cloud chamber* containing supersaturated vapour (e.g. water) in which they leave trails of droplets where they cause ionisation or, similarly, in a *bubble chamber* where they leave trails of bubbles in a liquid (e.g. hydrogen). The *ionisation chamber* is a gas-filled chamber containing two oppositely charged electrodes; ionising radiation entering the chamber ionises the gas, and the resulting electrons and positive ions travel towards the positive and negative electrodes respectively, creating an ionisation current that can be measured. The *Geiger counter* is based on the Geiger–Müller tube, which is a type of ionisation chamber that detects individual ionising events; these can be counted electronically (as counts per minute for example) or registered as a series of clicks from a loudspeaker. Other methods of detection include *solid-state* devices based on semiconductors and *scintillation counters* based on scintillators, which emit minute flashes of light when ionising radiation falls onto them. Photographic emulsion is also sensitive to ionising radiation.

Radioisotopes (radioactive isotopes) have a wide range of uses including medical diagnosis and treatment, thickness and density measurement, non-destructive testing, leak detection, wear measurement, radiotracer studies, radiocarbon dating – and many more.

15.4 HALF-LIFE

Radioactive decay is unaffected by chemical combination of the radionuclide (radioactive nuclide) with other elements or by normal changes in physical conditions. It is a random process, which makes it impossible to predict when a particular nucleus will decay or, in a collection of identical nuclei, which one will decay first. However, where there is a large number of nuclei, the time taken for half of them to decay has a characteristic value that, for any given nuclide, is called its *half-life*. If the initial number of nuclei of a particular radionuclide is N_0, then after 1 half-life there will be $N_0/2$ (or $0.5N_0$) remaining. These will decay to $N_0/4$ after another half-life, then to $N_0/8$ after another – and so on. Figure 15.4 illustrates this with a graph

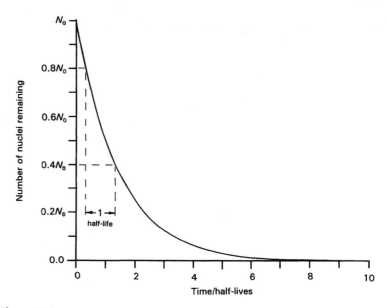

Figure 15.4

Table 15.1

Radionuclide	Approximate half-life
Carbon-14	5.7×10^3 years
Potassium-40	1.3×10^9 years
Iodine-131	8 days
Radium-226	1.6×10^3 years
Uranium-238	4.5×10^9 years

showing the number of nuclei remaining versus time expressed in half-lives. The graph indicates that the argument applies to any starting point; for example, it takes 1 half-life for $0.8N_0$ nuclei to decay to $0.4N_0$, or for $0.6N_0$ nuclei to decay to $0.3N_0$. Table 15.1 gives a few examples of half-lives, which can range from fractions of a second to thousands of millions of years.

Carbon-14, which has a half-life of about 5700 years, is formed in the atmosphere (as we shall see below). From there it makes its way into plants and animals, where it forms a small proportion of their natural carbon content. When an organism dies, the proportion of carbon-14 decreases as a result of β-decay. Archaeologists are able to estimate the age of suitable carbon-based materials by *carbon dating*, which involves measuring their radioactivity to determine the proportion of carbon-14 remaining. Similar methods are used to date certain types of rock. For example, potassium-40 decays to argon-40 with a half-life of about 1.3×10^9 years and their relative proportions provide an estimate of the age of the sample.

The *activity* of a particular sample of a radionuclide is the number of nuclei that disintegrate per unit time. The SI unit of activity is the becquerel (symbol Bq), and 1 Bq is the activity of a radionuclide decaying at an average rate of 1 disintegration per second. Previously, the unit of activity was the curie (Ci) where 1 Ci = 3.7×10^{10} Bq. Assuming that the activity of a given sample of a particular radionuclide is proportional to the number of nuclei remaining undecayed, the half-life of the radionuclide may be determined by measuring the time taken for the activity of the sample to fall by half.

15.5 NUCLEAR REACTIONS

Nuclear reactions are brought about by bombarding the nucleus with energetic particles. For example, oxygen-17 is formed by bombarding nitrogen-14 with α-particles, as follows:

$$^{14}_{7}\text{N} + ^{4}_{2}\alpha \rightarrow ^{17}_{8}\text{O} + ^{1}_{1}\text{p}$$

This can be written in abbreviated form as $[^{14}_{7}\text{N}(\alpha, \text{p})^{17}_{8}\text{O}]$ where

(α, p) indicates that the initial nuclide is bombarded with α-particles and that protons are emitted. As we noted earlier, carbon-14 is formed in the atmosphere; this occurs as the result of bombardment of nitrogen-14 with neutrons (resulting from the action of cosmic radiation) as follows:

$$^{14}_{7}\text{N} + ^{1}_{0}\text{n} \rightarrow ^{14}_{6}\text{C} + ^{1}_{1}\text{p} \qquad \text{or} \qquad [^{14}_{7}\text{N}(\text{n, p})^{14}_{6}\text{C}]$$

Earlier, we saw that the binding energy of a nucleus is the energy that would be required to separate it into its component neutrons and protons. Looking at this the other way round, binding energy is the energy that would be released in forming the nucleus from its component neutrons and protons. Figure 15.2 shows that the highest binding energies per nucleon are associated with mass numbers in the region of 60. If a nucleon below this peak is able to increase its binding energy then there will be a corresponding release of nuclear energy. Figure 15.2 indicates that there are two ways of achieving this, namely nuclear fission and nuclear fusion.

Nuclear fission is a type of nuclear reaction where a heavy nucleus is split into smaller fragments, for example the fission of uranium-235 caused by neutron bombardment. This involves a number of reactions leading to a variety of fission products, including lanthanum-148 and bromine-85 which are formed as follows:

$$^{235}_{92}\text{U} + ^{1}_{0}\text{n} \rightarrow ^{148}_{57}\text{La} + ^{85}_{35}\text{Br} + 3^{1}_{0}\text{n} \qquad \text{or} \qquad [^{235}_{92}\text{U}(\text{n, 3n})^{148}_{57}\text{La}, ^{85}_{35}\text{Br}]$$

Figure 15.2 indicates how a nucleon in a heavy nucleus experiences an increase in binding energy if the nucleus is split and the nucleon becomes part of a smaller nucleus formed as a fission product. As the mass number is reduced and the binding energy per nucleon rises towards the maximum (at around $A = 60$) there will be a corresponding amount of energy released which can be used, for example, to generate electricity. The amount of energy released is enormous – about 8×10^{13} J per kilogram of uranium-235 in the reaction shown above. This is vastly greater than the energy released by chemical reactions; for example, burning a kilogram of coal releases about 3×10^{7} J. The fission of uranium-235 is accompanied by the production of more neutrons, 3 in the case above. This leads to the possibility of a *chain reaction* where the neutrons produced cause the fission of more uranium-235 nuclei, and so on, making possible the sustained release of energy. If the reaction is controlled, as in a nuclear reactor, then energy will be produced at a steady rate. If the reaction escalates, as in an atomic bomb, then it will be explosive.

Nuclear fusion occurs when two light nuclei fuse together to form a heavier nucleus. Figure 15.2 indicates how this process also leads to an increase in binding energy per nucleon, hence the release of energy. For example, two nuclei of hydrogen-2 ($^{2}_{1}\text{H}$, called *deuterium* or *heavy hydrogen*) can fuse to form $^{3}_{2}\text{He}$, as follows:

$$^2_1H + ^2_1H \rightarrow ^3_2He + ^1_0n \quad \text{or} \quad [^2_1H(^2_1H, n)^3_2He]$$

The fusion of a kilogram of deuterium would release a similar amount of energy to the fission of a kilogram of uranium-235 as above. However, fusion requires tremendously high temperatures (of the order of 10^8 °C) to give the positively charged nuclei enough energy to overcome the repulsion between them and allow fusion to take place – hence we use the term *thermonuclear fusion*. (In the case of neutron bombardment, the lack of charge on the neutron allows it to approach the nucleus much more easily.) The practical problems of developing a commercial fusion reactor have yet to be overcome but, in thermonuclear weapons, a fission bomb is used as a trigger to provide the high temperatures required.

Questions

1. Write brief notes explaining the significance of the following with regard to the structure and stability of the nucleus:
   (a) mass defect and binding energy;
   (b) neutron/proton ratio;
   (c) α-, β- and γ-radiation;
   (d) electron capture.

2. (a) Show that a parent nuclide is transformed into a different element as a result of (i) α-decay, (ii) β-decay by both electron and positron emission, and (iii) electron capture.
   (b) 1 α-particle and 2 negative β-particles were emitted by the same nucleus during the course of a succession of disintegrations along a radioactive series. What overall effect did this have on (i) the atomic number, (ii) the neutron number, and (c) the mass number?

3. Complete the following.
   (a) $^{238}_{92}U \rightarrow ^{234}_{90}Th + ?$
   (b) $^{234}_{90}Th \rightarrow ^{234}_{91}Pa + ?$
   (c) $^7_3Li + ? \rightarrow 2^4_2\alpha$
   (d) $? \rightarrow ^{216}_{84}Po + ^4_2\alpha$

4. If the mass of a certain radionuclide falls from 98.56 mg to 12.32 mg in 24.3 days, what is its half-life? What will its mass be after a further 2 half-lives? How long would it take for 75.96 mg of the radionuclide to decay to 18.99 mg?

5. Complete the following.

   (a) $[^{9}_{4}\text{Be}(\alpha, \ ?)^{12}_{6}\text{C}]$

   (b) $[^{27}_{13}\text{Al}(?, \ n)^{30}_{15}\text{P}]$

   (c) $[?(\alpha, \ p)^{17}_{8}\text{O}]$

   (d) $[?(n, \ 2n)^{144}_{56}\text{Ba}, \ ^{90}_{36}\text{Kr}]$

6. Sketch the graph of binding energy per nucleon against mass number and use it to show how nuclear fission and thermonuclear fusion can lead to the production of useful energy.

TOPIC 16 CHEMICAL BONDING

COVERING:

- ionic, covalent and intermediate types of bond;
- metallic bond;
- intermolecular forces;
- relative molecular mass.

We need to have a basic understanding of chemical bonding, because it plays a central role in determining the behaviour of all substances, including engineering materials.

16.1 IONIC BONDING

In Topic 14 we saw that the inert gases have an outer octet of electrons, or a pair in the case of helium, that represents a particularly stable electronic structure. As we shall now see, other elements tend to behave in such a way that they achieve these stable configurations by losing or gaining electrons.

An atom of sodium (Na in group I) will tend to get rid of the single $3s$ electron in its outer shell, thereby achieving the neon configuration and becoming a positively charged sodium *ion* (Na^+) in the process. (Note that it does not become a neon atom, because the nucleus remains the same.) In chemical shorthand

$$Na = Na^+ + e^-$$

where e^- represents the electron. This tendency to form positive ions is characteristic of metallic elements.

By contrast, chlorine, a non-metal from the opposite side of the periodic table (Cl in group VII), is one electron short of the argon configuration. If it can gain an electron from elsewhere, it becomes a negatively charged chloride ion, thus

$$Cl + e^- = Cl^-$$

Both these tendencies are satisfied when sodium and chlorine are combined in sodium chloride, i.e. common salt, as follows:

$$Na + Cl = Na^+ + Cl^-$$

Sodium chloride is a crystalline solid held together by *ionic bonds*.

Ions formed from single atoms can be regarded as charged spheres. A sodium ion and a chloride ion will be drawn together by the attractive force between them due to their opposite charge. The closeness of their approach, however, is limited by their outer orbitals, which cannot interpenetrate, because they would then exceed their quota of two electrons where they occupy the same space. Furthermore, any interpenetration would result in reduced screening of the two nuclei, which would cause repulsion between them. The overall effect is that any attempt to squeeze the ions together results in a repulsive force and the outer shell of each ion behaves rather like a spherical elastic skin.

Figure 16.1 indicates how the attractive and repulsive forces vary with separation. The attractive force predominates at larger separations and the repulsive force only becomes important at smaller separations, where the outer shells approach closely. By summing the two relationships we obtain the net force/separation curve.

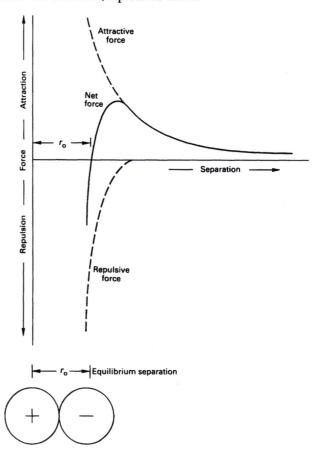

Figure 16.1

There is a point where the two component forces balance so that the net force is zero; this represents the *equilibrium separation* r_o, where the ions would naturally come to rest in the absence of any external forces. If we pull the ions apart from their equilibrium position, then the balance between the component forces is upset and an attractive force arises as we move up the net force/separation curve; the ions will move just far enough to generate an equal but opposite force resisting our effort in pulling them apart. Similarly, if we push the ions together, then the compressive force we apply to them will be opposed by an equal and opposite resistance due to the repulsive force that arises between them as we move down the net force/separation curve. So, as we noted in Topic 2, it is the deformation of chemical bonds that enables materials to resist tension and compression. The portion of the net force/separation curve close to the equilibrium position (corresponding to small deformations) is very nearly straight where it crosses the horizontal axis in Figure 16.1. This provides us with the fundamental basis of Hooke's law, namely that deformation is proportional to load and hence, Young's modulus (which is equal to the stress/strain ratio, as discussed in Topic 2). Furthermore, since the proportionality is maintained across the horizontal axis, Young's modulus applies to both tensile and compressive stresses in materials which follow this model.

Figure 16.2 shows part of a simple ionic crystal structure in 'exploded' form for clarity. In general terms, ionic crystal structures are regular, extended lattice arrangements in which oppositely charged ions are drawn together but like-charged ions stay apart. In this particular case (named the *rocksalt* structure after naturally occurring sodium chloride) each positive ion is surrounded by six negative neighbours, and vice versa, so there are equal numbers of each overall.

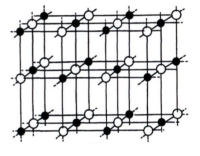

Figure 16.2

Note that group II elements form ions by losing two electrons and group VI elements by gaining two; for example, magnesium oxide ($Mg^{2+}O^{2-}$) is an ionic substance (which, incidentally, also adopts the rocksalt structure).

Many ionic structures are more complex. Electrical neutrality must be preserved; therefore, ions occur in different proportions where their charge magnitude differs. In calcium fluoride, for example, we need twice as many F^- ions as Ca^{2+} ions. The relative size of the ions is also an important factor in determining the way in which they pack together. Furthermore, there are many ions which contain more than one atom; for example, the sulphate ion (SO_4^{2-}) and the nitrate ion (NO_3^-).

16.2 COVALENT BONDING

In the absence of metallic elements, non-metals can achieve stable inert gas configurations by sharing electrons to form *covalent bonds*. For example, two hydrogen atoms can pool their single electrons to form a pair that is shared between them, effectively giving each the helium configuration. The two half-filled atomic orbitals combine to give a single molecular orbital, containing both electrons, which encloses both nuclei as indicated in Figure 16.3.

The electrons spend much of their time in between the nuclei and exert attractive forces on each which serve to tie them together.

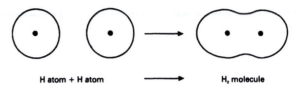

H atom + H atom $\longrightarrow$ H, molecule

Figure 16.3

There is also an opposing repulsive force between the nuclei because of their like charges. The combination of these two forces gives a net force/separation curve of a similar form to that of the ionic bond. This means that the smallest entity of hydrogen that normally maintains an independent existence is the hydrogen *molecule*, H_2, in which two hydrogen atoms are joined by a covalent bond. Hydrogen molecules tend to remain separate under normal conditions and exist as a gas rather than coalesce to form a liquid or a solid.

Chlorine is also one electron short of its neighbouring inert gas configuration (argon), so, like hydrogen, two chlorine atoms share a pair of electrons to form a covalent bond, and the resulting Cl_2 molecules exist as a gas under normal conditions. (Note that electrons involved in forming chemical bonds are called *valence electrons*.)

Carbon is a particularly important element because it forms the basis of a vast number of covalently bonded molecules. As a member of group IV, each atom shares four pairs of electrons with other atoms and achieves the neon configuration. Since carbon forms four covalent bonds, we say it has a *valency* of 4. (The term 'valency' is also applied to the number of charges on an ion; thus, for example, Mg^{2+} has a valency of 2 and Cl^- has a valency of 1.)

Figure 16.4 shows some examples of carbon-based molecules that happen to be *hydrocarbons*, i.e. *compounds* containing only hydrogen and carbon. Compounds are substances which result from the chemical combination of elements. They should not be confused with *mixtures*, which contain more than one individual chemical substance and can be separated by physical methods.

Covalent bonds are often represented by straight lines between the atoms sharing the valence electrons. Figure 16.4(a) shows methane (CH_4), which is the main constituent of natural gas. (Natural gas is a mixture of gases, mostly hydrocarbons.) The four C–H bonds are identical and repel each other because of their identical charge distribution. They spread apart as far as possible, with the result that the methane *molecule* (i.e. the smallest entity of methane) has a tetrahedral form with

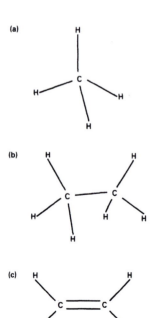

(a)

(b)

(c)

Figure 16.4

equal angles between all the bonds. Figure 16.4(b) shows ethane (C_2H_6), which is also found in natural gas. Figure 16.4(c) shows ethene (C_2H_4), or ethylene, which is an important raw material in the chemical industry. Both ethane and ethene molecules contain two carbon atoms.

In ethane the bonds are distributed round each carbon atom in a tetrahedral configuration, as in methane, and one end of the molecule can rotate relative to the other, like a propeller, about the C–C axis. Ethane (CH_3–CH_3) is the smallest of a family of chain-like molecules which continues with propane (CH_3–CH_2–CH_3) and butane (CH_3–CH_2–CH_2–CH_3) and extends to higher members of great length. Ordinary rubber and many plastics consist of enormously long molecules based on covalently bonded carbon chains. (As we shall see in Topic 21, the elasticity of rubber depends on bond rotation about the axes of the –C–C– bonds along the length of the chains.)

In the ethene molecule the carbon atoms are joined by a *double bond*, where they share two pairs of electrons. This means that, because of its valency of 4, each carbon atom can only bond with two hydrogen atoms. The double bond pulls the carbon atoms closer together than the single bond; furthermore, it is stiffer and stronger, and requires more energy to break it. It also prevents rotation, and all the atoms in the ethene molecule lie in the same plane. (Triple bonds between carbon atoms are also possible but we shall not discuss them here.)

Now we can begin to see important differences between ionic and covalent bonding. The number of covalent bonds an atom can form is dictated by its valency and the bonds are formed specifically between those atoms sharing the valence electrons. This leads to the formation of individual molecules which have definite shapes and sizes and which are capable of existence as separate entities. On the other hand, ionic structures are simply formed from ions packed together, like charged spheres, to form extended crystal structures of no fixed size and in which individual molecules cannot be identified.

Having said this, covalent bonding can also lead to extended crystal structures. For instance, Figure 16.5 shows the structure of diamond, where carbon atoms are joined together so that each has four neighbours arranged around it in a tetrahedral configuration. Like an ionic crystal, this structure has no fixed size and it is not possible to identify an individual diamond molecule.

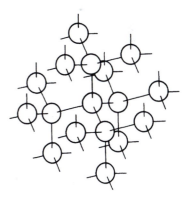

Figure 16.5

16.3 INTERMEDIATE BONDING (IONIC–COVALENT)

Compounds of the types we have been considering are seldom 100% ionic or 100% covalent – most lie somewhere in between. The best way of approaching this idea is to consider ionic bonds with covalent character and covalent bonds with ionic character.

Ions are of different sizes. Table 16.1 shows that there is a decrease in radius along the series from O^{2-} to Al^{3+}. All these ions have the

Table 16.1

Ion	Atomic no.	Radius/nm
O^{2-}	8	0.14
F^-	9	0.13
Na^+	11	0.10
Mg^{2+}	12	0.07
Al^{3+}	13	0.05

neon configuration and the decrease is associated with the increasing positive charge on the nucleus (from 8 to 13) as it holds the skin of outer electrons progressively more tightly. Thus, a highly charged positive ion tends to have a 'hard', compact skin. On the other hand, a highly charged negative ion tends to have a larger, 'soft', more deformable skin with the electrons less under the influence of the nucleus. As Figure 16.6 suggests, if we let two such ions come close together, then the negative ion will be distorted, or *polarised* by the positive ion, so that its centre of negative charge is displaced from its centre of positive charge at the nucleus. The outer electrons will therefore tend to spend more time between the two nuclei. In other words, the ionic bond will take on covalent character. The larger the size and the greater the charge on the negative ion the greater will be its susceptibililty to polarisation. The smaller the size and the greater the charge on the positive ion the greater will be its polarising power.

Polarisation can also occur in covalent bonds and this gives them ionic character. It occurs where the bonded atoms differ in their *electronegativity*, which is their tendency to attract electrons and to form negative ions. (As we might expect from Topic 14, electronegativity tends to increase from left to right across the periodic table and to decrease from top to bottom.)

If two different atoms are joined by a covalent bond, then the valence electrons will tend to be drawn towards the more electronegative atom. That end of the bond will therefore tend to acquire a negative bias, leaving the other end with a positive bias, thereby giving the bond partial ionic character. The greater the difference in electronegativity the greater the polarisation of the bond and the greater its ionic character. We shall meet an example of a polarised covalent bond below, when we discuss the water molecule.

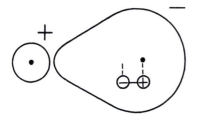

Figure 16.6

16.4 METALLIC BONDING

Metallic elements are described as *electropositive* because they tend to form positive ions by losing electrons. In the absence of non-metallic elements with which to form ionic bonds, the valence electrons join forces to form a negative 'sea' or *electron gas*. This serves to bind the ions together to form a regular extended crystal structure, as suggested in Figure 16.7(a).

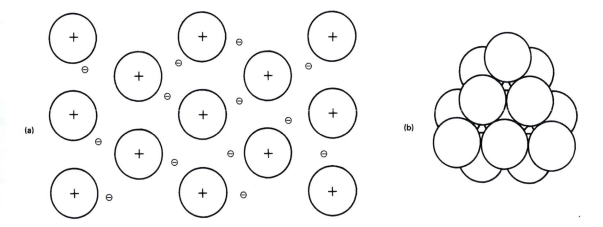

Figure 16.7

The electrons move randomly in between the positive ions, thus providing an attractive force that draws them together. Repulsion effects increase as the ions approach one another and, as before, we can view the bond in terms of a balance between attractive and repulsive forces. The separation between the nuclei of two neighbouring metal ions corresponds to the equilibrium position on the net force/separation curve in Figure 16.1 (page 142). This makes it possible to view metal atoms as elastic spheres and metal crystal structures in terms of three-dimensional geometrical patterns formed by packing spheres together, as in Figure 16.7(b), though without the same charge constraints that apply to ionic structures. (In fact, models of metal crystal structures can be constructed by glueing table tennis balls together.)

As this suggests, metal crystals are of no fixed size and individual molecules cannot be identified within them. Note that the metal ions do not all have to be of the same kind, so *alloys* can be formed in which different metals – copper and nickel, for example – are combined on the same crystal lattice.

As we shall see later, it is the freedom of movement of the valence electrons through the crystal lattice that makes metals good conductors of heat and electricity.

16.5 INTERMOLECULAR FORCES

Summarising very broadly, we have seen that ionic bonds are formed between metallic atoms and non-metallic atoms, covalent bonds are formed between non-metallic atoms, and metallic bonds are formed between metallic atoms. But this does not explain why molecules tend to stick together. For instance, water molecules form a coherent liquid or solid, depending on the temperature. To understand why they do this, we need to examine their structure.

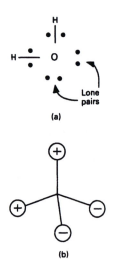

(a)

(b)

Figure 16.8

Oxygen is a group VI element and an oxygen atom shares a pair of electrons with each of two hydrogen atoms. This gives oxygen the neon configuration and hydrogen the helium configuration. The resulting H_2O molecule is a stable entity capable of independent existence (Figure 16.8). The four outer electrons of the oxygen atom that are not involved in bonding form two so-called *lone pairs*. As far as the shape of the molecule is concerned, each lone pair occupies a non-bonding orbital of a similar form to that of the orbitals constituting the covalent bonds. The result is a tetrahedral molecule with an overall shape not unlike that of methane.

Figure 16.8(b) is a simplified representation of the charge distribution in the water molecule. The two lone pairs form negative regions in which there are no nuclei to balance their charge. The two bonding orbitals have a positive bias, because the central oxygen atom is electronegative and draws the valence electrons inwards from the hydrogen nuclei to leave a net positive bias at the outer ends of the bonds. Since there are no screening electrons round the hydrogen nuclei, the effect of this is particularly strong.

A charged plastic comb will attract a thin stream of water running from a tap, because the water molecules orientate themselves so that the orbitals that are oppositely charged to the comb tend to point towards it; the resulting attractive force will bend the stream of water. In a similar way, two water molecules that are close together and free to rotate will tend to orientate themselves so that oppositely charged orbitals point towards one another, with a resulting attractive force that holds the molecules together. This type of intermolecular force, involving hydrogen atoms covalently bonded to strongly electronegative atoms, is called *hydrogen bonding*. It provides a network of attractive forces within a collection of water molecules, and it is this that gives water its coherent nature. In fact, hydrogen bonding is a special case of the attractive forces that generally arise between any *polar* molecules (i.e. those that are permanently polarised).

But inert gases and non-polar molecules (e.g. H_2, Cl_2 and O_2, where the valence electrons are equally shared) will liquefy and solidify if the temperature is low enough. (Oxygen is a group VI element, so O_2 molecules are held together by a double bond $O=O$.) In such cases there can be no permanent polarisation. Figure 16.9(a), which represents a helium atom, suggests the reason.

The electrons are in perpetual motion, so that at any instant, unless they happen to be diametrically opposite to one another, the atom will be temporarily polarised. The magnitude and direction of the polarisation continuously changes as the displacement of the centre of negative charge fluctuates about the nucleus. The atom will tend to induce polarisation in its neighbours, as in Figure 16.9(b); in this case the nucleus of the left-hand atom is attracting the electrons in the right-hand atom or, looking at it the other way, the electrons in the right-hand atom are repelling the electrons in the left-hand atom.

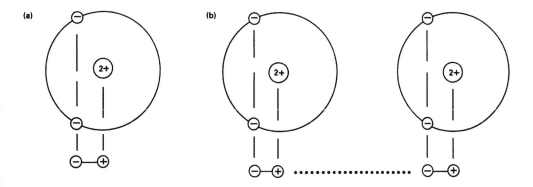

Figure 16.9

Thus, temporary polarisation in the two atoms will tend to fluctuate in sympathy, so that the dipoles are orientated similarly and an attractive force arises between them.

Such forces are called *van der Waals'* forces. They arise wherever atoms and molecules are close together, in addition to any other types of bonding that may be operating. van der Waals' forces are extremely weak in the case of helium but stronger with larger atoms and with molecules where there are more electrons involved. To take an extreme case, polythene consists of very large chain-like hydrocarbon molecules where attraction between the chains is provided by van der Waals' forces.

As a broad generalisation, ionic and covalent bonds are the strongest, followed by metallic bonds, then hydrogen bonds, with van der Waals' forces the weakest.

16.6 RELATIVE MOLECULAR MASS

The mass of a molecule is expressed in terms of its *relative molecular mass* (molecular weight). This is equal to the sum of the relative atomic masses of all the atoms it contains. For example, using the values from Table 14.5 (page 126), the relative molecular mass of water (H_2O) is 18.0 (i.e. (2 × 1.0) + 16.0) and of ethane (C_2H_6) is 30.0 (i.e. (2 × 12.0) + (6 × 1.0)). And following on from the last section of the previous topic, the mass of 1 mol of water is 18.0 g and of ethane 30.0 g.

Although molecules cannot be identified in ionic structures, it is still helpful to use the chemical formulae of such compounds in a corresponding way. Thus, 1 mol of sodium chloride (NaCl) has a mass of 58.5 g (i.e. 23.0 g + 35.5 g).

Relative molecular mass values, or their ionic counterparts, enable us to quantify chemical reactions in terms of the masses of the substances involved. For example, the complete combustion of methane in oxygen yields carbon dioxide (CO_2) and water. In order to form a properly balanced chemical equation, the number of atoms of each element must be the same on each side. In this case the equation is balanced by having two oxygen molecules and two water molecules:

$$\begin{array}{ccccc} CH_4 & + & 2O_2 & = & CO_2 & + & 2H_2O \\ 16.0 & & 2 \times 32.0 & & 44.0 & & 2 \times 18.0 \end{array}$$

The relative molecular masses underneath enable corresponding quantities to be calculated in whatever units are required. Thus, 160 kg of methane needs 640 kg of oxygen for complete combustion. 1.00 g of methane yields (36.0/16.0 =) 2.25 g of water. 22 mg of carbon dioxide is produced by 8 mg of methane, and so on.

Questions

(Use any previously tabulated data as required.)

1. Deduce the general type of chemical structure of the following: caesium chloride (CsCl), carbon tetrachloride (CCl_4), ice, solid argon, carbon dioxide, liquid ammonia (NH_3).

2. Find the relative molecular mass (or ionic counterpart) of each of the following: ethane, calcium carbonate ($CaCO_3$), ethanol (C_2H_5OH), silver nitrate ($AgNO_3$), chloroform ($CHCl_3$).

3. One compound contains 75% carbon and 25% hydrogen by weight and another contains 80% carbon and 20% hydrogen. Suggest what each might be.

4. Find the aluminium content of alumina (Al_2O_3) expressed as a weight percentage.

5. Find how much carbon dioxide results from the complete combustion of 4.5 g of propane (C_3H_8) if the reaction follows the equation

$$C_3H_8 + 5O_2 = 3CO_2 + 4H_2O$$

6. The thermal decomposition of calcium carbonate ($CaCO_3$) yields CaO and CO_2. How much calcium carbonate would be needed to give 1 kg of carbon dioxide?

7. Write the balanced chemical equation for the complete combustion of ethyl alcohol (C_2H_5OH) to carbon dioxide and water, and hence find how much alcohol yields 14.35 g of carbon dioxide.

See also Further Questions on page 252.

TOPIC 17 HEAT AND TEMPERATURE

COVERING:

- heat and internal energy;
- heat capacity and latent heat;
- thermal expansion and contraction;
- thermal stress.

Whether a particular substance exists as a solid, a liquid or a gas at a given temperature and pressure depends upon the strength of the forces of attraction between its constituent atoms, ions or molecules. Thus, at atmospheric pressure and room temperature the van der Waals' forces between the oxygen molecules in air are not strong enough to make them stick together, nor are the hydrogen bonds between water molecules strong enough for them to form ice. On the other hand, nearly all metallic and ionic/covalent materials are solids. Furthermore, simple solids tend to turn to liquids, and liquids to gases, if they are heated. These observations seem to point to the idea that the cohesion due to the forces of attraction between atoms, ions and molecules is opposed by the effect of heat.

Heat is the energy that is transferred between two bodies as a result of a temperature difference between them. Heat will flow from the hotter to the colder body; therefore, *temperature* is the property that determines the direction in which the heat will flow.

First we must recognise that the heat absorbed by a body may increase its internal energy (associated with its temperature and physical state) and enable it to do external work on its surroundings, as, for example, in raising the temperature of a body of gas, enabling it to expand and drive a piston. It follows that the increase in the internal energy of a body equals the heat added to it less any external work that may be done by it. On the other hand, it is possible to increase the internal energy of a body simply by doing work on it without supplying any heat at all. For example, in braking a car, kinetic energy is transformed into an increase in the internal energy of the brakes, thereby raising their temperature. The increased temperature of the air compressed in a bicycle pump is a result of an increase in its internal energy due to work done on it. When a falling body hits the ground, its original potential energy is transformed into increased internal energy of the body and of the ground, and to some extent the air through which it fell, plus some sound energy.

These are all examples of the principle of conservation of energy which we met in Topic 8. They illustrate the need to view any change in the internal energy of a body as an exercise in accountancy: work and heat both represent energy in the process of being transferred either to or from the body. In this topic we shall consider the internal energy changes in a body as it absorbs or emits heat and, for the purposes of this discussion, we shall generally assume that conditions are such that a negligible amount of work is done.

First, we need to recognise that the atoms, ions or molecules that constitute a particular substance are in a state of continual thermal agitation which becomes more energetic if the temperature of the substance is increased. For simplicity we shall base our discussion on the general case of an unspecified model substance consisting of chemically bonded atoms.

In the solid state the atoms vibrate about fixed positions on the crystal lattice, where they are trapped between their neighbours. The thermal energy associated with their vibrations is insufficient to overcome the cohesion due to the forces of attraction between them. If the temperature is raised, the vibrations become more vigorous until, at the melting point, the atoms have sufficient energy to escape from their fixed positions and the substance changes from a solid to a liquid. Although the atoms still remain in contact with their neighbours, they now have the capacity for translational motion relative to one another – hence, liquids can flow. In the gaseous state the atoms have sufficient energy to overcome the forces of attraction and they fill the entire volume of the container that they occupy irrespective of how large that might be. At normal pressures the atoms are widely separated and they move randomly and independently of one another, only interacting when they collide. Although the velocity of individual atoms will vary greatly, their mean velocity remains constant at a given temperature and increases as the temperature is raised.

Let us examine these ideas more closely.

17.1 A SIMPLE MODEL

We shall begin with the assumption that the general form of the net force/separation curve in Figure 16.1 (page 142) can be used to represent the behaviour of any of the types of bond that we met in the previous topic, including that between the atoms in our model substance. Figure 17.1 shows the way in which the corresponding potential energy varies. As before, the horizontal axis gives the position of the centre of one atom relative to the other. The equilibrium position, where the net force between the atoms is zero, corresponds to the potential energy minimum at the bottom of the trough. The shape of the trough reflects the effect of the attractive and repulsive components of the net force/separation curve and is correspondingly asymmetric.

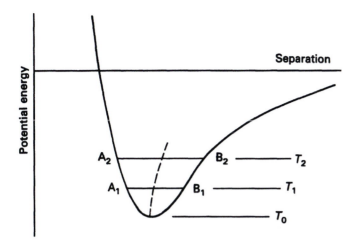

Figure 17.1

As we saw in the previous topic, any deformation of the bond results in an opposing force. The system will therefore tend to return to its equilibrium position in much the same way as the pendulum and the spring in Topic 11. The linear net force/separation relationship close to the equilibrium position suggests that if the atoms are displaced slightly and then released, they will oscillate with simple harmonic motion.

As we noted earlier, the atoms in a solid are normally in a continuous state of oscillation, and this provides us with a picture of the way in which the temperature of a body increases when it absorbs heat. There is a continuous interchange of kinetic and potential energy as the atoms oscillate to and fro between the points A_1 and B_1 in Figure 17.1 at a temperature T_1. The potential energy follows the curve throughout the cycle and is at a maximum at A_1 and B_1, where the kinetic energy is zero and the system is on the point of changing direction. From Topic 11 (see page 88) we would expect the sum of the potential energy and kinetic energy to remain constant throughout the cycle, in which case the horizontal line A_1B_1 represents the total energy. The kinetic energy at any point is therefore equal to the vertical distance between the horizontal line and the potential energy curve.

If heat is added to the system to raise the temperature to T_2, then the total energy increases to a higher level represented by the horizontal line A_2B_2. Conversely, if the system cools from T_2 to T_1, then the difference in internal energy is released to the surroundings as heat.

If the temperature is lowered to T_0, corresponding to the bottom of the trough, the system has no kinetic energy and the atoms come to rest; this represents *absolute zero*, which is the lowest temperature that is theoretically possible. The vertical distance of any one of the series of horizontal lines A_nB_n above the trough can be viewed in terms of the kinetic energy associated with the thermal motion of the system at a particular temperature and the potential energy associated with the interaction between the vibrating atoms. The quantity of heat which a body absorbs in increasing its temperature is known as its *heat capacity*. We shall define this more carefully very shortly. (Readers who know about quantum theory will have noticed that our simple model is valid for an 'average' oscillating atom but not for individuals, and that we have ignored the fact that the system must possess some residual kinetic energy at absolute zero.)

Let us assume that the mid-point of each horizontal line A_nB_n represents the average separation between the centres of the atoms. Because the potential energy curve is not symmetrical, this distance increases as the temperature is raised (following the broken line in the figure). The substance of which the atoms form a representative part therefore undergoes *thermal expansion*. Conversely, it undergoes *thermal contraction* as it cools.

The addition or removal of heat does not always change the temperature of a substance. At the melting point the amplitude of the oscillations is large enough for the atoms to move past their neighbours and escape from their fixed positions on the crystal lattice. In effect, the breaking down of the crystal structure and the freedom of the atoms amounts to an increase in potential energy. The substance pays for this by absorbing heat from its surroundings at constant temperature until it has completely melted. This heat, called the *latent heat of fusion*, is returned to the surroundings if the substance resolidifies. Thus, ice must be supplied with latent heat for it to melt at 0 °C, and this heat is returned if the water refreezes at 0 °C. (The word 'latent' means 'concealed' in this context, because the heat transfer is not revealed as a temperature change.)

In the liquid state the forces of attraction are still able to maintain overall cohesion. The liquid still has a definite volume which is contained within its boundary surface (unlike a gas, which is diffuse and expands to fill its container). The liquid can flow, because its constituent atoms can readily change position relative to one another. If it is heated, the atoms move more vigorously, and if the temperature is raised to the boiling point, they will have enough energy to climb out of the potential energy trough and become independent of one another – thus, the forces of attraction are overcome and the substance changes from a liquid to a gas. During this change of state the temperature of the liquid remains constant as it absorbs its *latent heat of vaporisation*. Because the latent heat of vaporisation of a substance represents the energy required to separate the constituent atoms completely, it is generally considerably greater than the latent heat of fusion absorbed during the melting process.

Note that liquids tend to evaporate below their boiling points and some, such as alcohol and water, can become noticeably cooler in doing so. This is because molecules with sufficiently high kinetic energy are able to escape from the liquid surface. This has the effect of reducing the average kinetic energy of those remaining behind – hence, the liquid becomes cooler.

17.2 HEAT CAPACITY AND LATENT HEAT

Now we need to quantify the amount of heat involved as the temperature or physical state of a substance is changed. For example, how much heat must be supplied to cold water to raise its temperature to boiling point? And how much heat is required to change the boiling water to steam?

Figure 17.2 is an idealised plot of temperature against time which summarises the quantities involved as a substance is heated from the solid right through to the gaseous state.

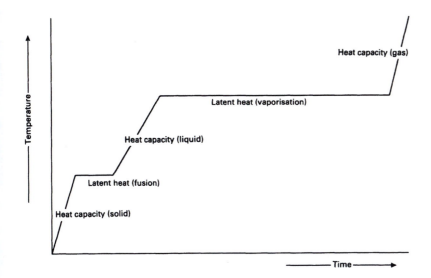

Figure 17.2

For measuring temperature we shall use the Celsius scale (formerly called the centigrade scale), where the values 0 °C and 100 °C are assigned to the freezing and boiling points of water at 1 atmosphere. (1 atmosphere is specified because the freezing and boiling points vary with pressure.) On this basis, absolute zero turns out to be −273.15 °C. The SI unit of temperature is the kelvin (symbol K) but this is much less commonly used than the degree Celsius. 0 K represents absolute zero and a temperature interval of 1 K is the same as 1 °C, so for our purposes °C = K − 273.

Different substances have different heat capacities, in other words, they require different quantities of heat to raise their temperature by the same amount. In general, provided that no change of state occurs,

$$Q = mc\theta \qquad\qquad (17.1)$$

where Q joules of heat are required to raise m kilograms of the substance through a temperature interval of θ K (or θ °C), c is called the *specific heat capacity* and is defined as the heat required to raise the temperature of 1 kg of the substance by 1 K (1 °C); its units are therefore J kg^{-1} K^{-1}. Note that c is often assumed to be constant, although it can vary quite considerably with temperature, depending on the substance and the temperature range involved. Table 17.1 shows some typical approximate mean values. Water has a particularly high value, making it a useful heat transfer medium for heating and cooling purposes.

Table 17.1

Substance	Specific heat capacity/J kg^{-1} K^{-1}
Aluminium	900
Brass	380
Ice	2100
Iron	460
Lead	130
Mild steel	480
Water	4200

Molar heat capacity (symbol C_m) is sometimes used instead of specific heat capacity. This is based on the mole rather than the kilogram and is obtained by multiplying the specific heat capacity by the mass of one mole (kg mol^{-1}) of the substance. The units are therefore

$$\text{J kg}^{-1}\text{ K}^{-1} \times \text{kg mol}^{-1} = \text{J mol}^{-1}\text{ K}^{-1}$$

For many simple solid substances the molar heat capacity is approximately 25 J mol^{-1} K^{-1} (Dulong and Petit's law). For gases the specific and molar heat capacities at constant volume are less than at constant pressure, where extra heat is needed to enable the gas to expand to keep its pressure constant. The specific heat capacity of dry air, for example, is about 0.7 kJ kg^{-1} K^{-1} at constant volume, compared with about 1 kJ kg^{-1} K^{-1} at constant pressure.

Used on its own, the term *heat capacity* (symbol C, with units of J K^{-1}) is applied to particular objects. In terms of Equation (17.1), $C = mc$ for an object of mass m made from a material with a specific heat capacity c.

The latent heat absorbed or emitted by a substance as it changes state at constant temperature is given by

$$Q = ml \tag{17.2}$$

where Q joules of heat are absorbed or emitted by m kilograms of the substance, and l is the *specific latent heat* in J kg^{-1}. (Note that latent heat is sometimes expressed as *molar latent heat*, i.e. J mol^{-1}.) Table

Table 17.2

Latent heat of	Specific latent heat/J kg^{-1}
Fusion (ice–water)	3.3×10^5
Vaporisation (water–steam)	2.3×10^6

17.2 gives approximate values for the specific latent heats of fusion and vaporisation of water, which we shall need below.

Worked Example 17.1

Estimate the difference in water temperature between the top and the bottom of a waterfall 86 m high. (Assume $g = 9.8$ m s^{-2}.)

Assuming that the potential energy of the water at the top of the waterfall (mgh) is all used to raise its temperature, then, from Equation (17.1),

$$Q = mc\theta = mgh$$

Therefore, since $c = 4200$ (Table 17.1),

$$\theta = \frac{gh}{c} = \frac{9.8 \times 86}{4200} = 0.20 \ ^\circ C$$

Worked Example 17.2

Calculate the power required to maintain the temperature of a house that is losing heat at a rate of 28.8 MJ per hour.

A heat loss of 28.8 MJ per hour is offset by a heat input equal to

$$\frac{28.8 \times 10^6}{3600} = 8.00 \times 10^3 \text{ J s}^{-1} = 8.00 \text{ kW}$$

Worked Example 17.3

Starting with 2 kg of ice at -5 °C, find the heat required at each stage to effect the following changes: (a) heat the ice to 0 °C; (b) change the ice to water at 0 °C; (c) heat the water to 100 °C; (d) change the water to steam at 100 °C.

Using data from Tables 17.1 and 17.2:

(a) $Q = mc\theta = 2 \times 2100 \times 5 = 21$ kJ
(b) $Q = ml = 2 \times (3.3 \times 10^5) = 660$ kJ
(c) $Q = mc\theta = 2 \times 4200 \times 100 = 840$ kJ
(d) $Q = ml = 2 \times (2.3 \times 10^6) = 4600$ kJ

Worked Example 17.4

500 g of boiling water is poured into a 1.25 kg aluminium saucepan at 20 °C. Find the final temperature, assuming no heat losses.

Let the final temperature $= T$. The heat gained by the saucepan is given by

$$mc\theta = 1.25 \times 900 \times (T - 20)$$

This is equal to the heat lost by the water, which is given by

$$mc\theta = 0.5 \times 4200 \times (100 - T)$$

Hence,

$$T = 72 \text{ °C}$$

17.3 EXPANSIVITY

Thermal expansion and contraction, if unrestrained, can generate large enough stresses to cause problems in engineering structures such as bridges and railway lines unless proper allowance is made to relieve or accommodate them.

We can calculate the length change of a material from its *linear expansivity*, α. This is the fractional length increase per unit temperature rise as given by

$$\alpha = \frac{L_2 - L_1}{L_1} \times \frac{1}{\theta} \tag{17.3}$$

where the length L_1 increases to L_2 as the result of a temperature rise of θ. This equation can be rearranged to give

$$L_2 = L_1 (1 + \alpha\theta) \tag{17.4}$$

There are corresponding expressions for area and volume expansion:

$$A_2 = A_1 (1 + \beta\theta) \tag{17.5}$$

$$V_2 = V_1 (1 + \gamma\theta) \tag{17.6}$$

where A and V represent area and volume, respectively, and β is the *area* or *superficial expansivity* and γ the *volume* or *cubic expansivity*. (As a reasonable approximation, $\beta = 2\alpha$ and $\gamma = 3\alpha$.) Note that the values of all three expansivities generally vary to some extent with temperature, although they are often taken to be constant unless accurate results are needed. Some typical approximate mean values of α at normal temperature are shown in Table 17.3.

Table 17.3

Substance	Linear expansivity/K^{-1}
Aluminium	24×10^{-6}
Brass	19×10^{-6}
Copper	17×10^{-6}
Glass	9×10^{-6}
Steel	12×10^{-6}

Obviously we cannot measure α or β for liquids, because they flow; instead we measure their volume expansivity in a container. But this, of course, expands too. The volume of the space within a solid container changes just as though it is made from the same material as the container itself (i.e. as though the container is solid throughout). Liquids generally expand more than solids; hence, they generally expand more than the space inside their containers. The apparent expansion of a liquid is therefore less than its absolute (i.e. real) expansion by an amount corresponding to the expansion of the space it occupies inside the container. The volume expansivities are related approximately as follows:

$$\gamma_{absolute} = \gamma_{apparent} + \gamma_{container}$$

The thermal expansion of a material obviously results in a decrease in its density. In the case of liquids and gases, which can flow, this leads to natural convection, which we shall discuss in more detail in the next topic.

Water behaves in a special way. Ice has a very open crystal structure, where each molecule is surrounded tetrahedrally by four others; each of its orbitals attracts an oppositely charged orbital from each of its four neighbours. When ice melts, the open, rigid crystal structure is disrupted and tends to collapse. This allows the molecules to move closer together, effectively increasing the average number of neighbours per molecule. Water therefore occupies less volume in the liquid state and is more dense than ice (hence, icebergs float and frozen water pipes burst). The density continues to increase as the temperature is raised above 0 °C but by 4 °C the normal thermal expansion starts to win and water expands from a maximum density value at 4 °C as its temperature is raised further. As we noted in Topic 4, the maximum density of water is 1000 kg m^{-3} (= 1 g cm^{-3}). To a very close approximation, this is equivalent to 1 kg per litre. Although the litre is not an SI unit, it is very widely used for volume measurement. (The former definition of the litre was the volume occupied by 1 kg of water at 4 °C but now it is 1 cubic decimetre (i.e. 1 × 10^{-3} m^3), which is not quite the same.)

Worked Example 17.5

What is the volume at 89 °C of a glass flask which has a capacity of 1.000 1 (litre) at 15 °C?

For glass $\gamma = 3\alpha = 3 \times (9 \times 10^{-6}) = 27 \times 10^{-6}$ K^{-1}. From Equation (17.6),

$$V_2 = V_1(1 + \gamma\theta)$$

Therefore,

$$V_2 = 1.000(1 + (27 \times 10^{-6} \times 74)) = 1.002\ 1$$

17.4 THERMAL STRESS

Before leaving this topic, let us consider the stress that arises in a solid material, say a metal bar, if it is restrained from expanding when it undergoes a temperature increase θ. From Equation (17.3), the unrestrained expansion of the bar from L_1 to L_2, expressed as a fractional length increase, would have been $(L_2 - L_1)/L_1 = \alpha\theta$. In effect, the restraint acting on the bar compresses it from L_2 to L_1, thereby producing a strain $\varepsilon = (L_2 - L_1)/L_2$. To a close approximation $(L_2 - L_1)/L_1 = (L_2 - L_1)/L_2$, because L_1 and L_2 will generally have very similar values. It follows that the effective compressive strain ε is given by $\alpha\theta$. But we know from Topic 2 that $E = \sigma/\varepsilon$, where E represents Young's modulus and σ the stress. The thermal stress in the bar is therefore given by

$$\sigma (= E \times \varepsilon) = E\alpha\theta \tag{17.7}$$

Thermal stress is the reason why an ordinary drinking-glass is liable to crack when hot water is poured into it; the inside of the glass tries to expand before the outside. (Heat-resistant glass with low expansivity is much less susceptible to thermal stress.)

Worked Example 17.6

A steel bar is restrained from expansion while its temperature is raised from 3 °C to 28 °C. (a) Find the resulting thermal stress, and (b) find the corresponding restraining force if the bar has a diameter of 70 mm. (Assume $E_{steel} = 2 \times 10^{11}$ N m^{-2}.)

(a) From Equation (17.7),

$$\sigma = E\alpha\theta$$

Therefore, since $\alpha = 12 \times 10^{-6}$ (Table 17.3),

$$\sigma = (2 \times 10^{11}) \times (12 \times 10^{-6}) \times (28 - 3) = 60 \text{ MN m}^{-2}$$

(b) force = stress × cross-sectional area, which is equal to

$$(60 \times 10^6) \times \pi(0.035)^2 = 0.23 \text{ MN}$$

Questions

(Use any previously tabulated data as required.)

1. In each case find the temperature rise if the heat produced by a 70 W heater over a period of 1 min is absorbed by 1 kg of the following: (a) water, (b) ice, (c) aluminium, (d) iron and (e) lead.

2. Find the heat evolved as 5 kg of water at each of the following temperatures is changed to ice at 0 °C; (a) 0 °C, (b) 5 °C, (c) 50 °C.

3. Estimate what the theoretical minimum velocity of a snowball at 0 °C would have to be for it to melt completely on impact.

4. A continuous-flow water heater is required to raise the temperature of a water supply from 5 °C to 25 °C at a flow rate of 1.25 litres per minute. Assuming no heat losses, what must the power output of the heater be?

5. A 2 kW electric kettle contains 1.5 kg of water at 10 °C. Assuming the heat capacity of the kettle is 200 J K^{-1}, and ignoring any heat losses, (a) find the time that would be required for the water to come to the boil, and

(b) find the extra time required to change a third of the water into steam.

6. (a) How much water at 0 °C must be added to cool 2.5 kg of water from 40 °C to 20 °C?
   (b) How much ice at 0 °C would have had the same effect?

7. A copper tube is 0.500 m long at 20 °C. Its length increases by 0.68 mm when steam at 100 °C is passing through it. Find the linear expansivity of the copper.

8. A steel measuring tape, correctly calibrated at 10 °C, gives the distance between two points as 30.000 m at a temperature of 30 °C. Find the true distance.

9. At 17 °C a brass sphere has a diameter of 49.95 mm and a steel tube has an internal diameter of 50.00 mm. At what temperature is the sphere an exact fit inside the tube?

10. If a steel strut is 1.6 m long at −10 °C, to what temperature must it be heated to increase its length by 0.5 mm? If this expansion is restrained completely, what would be the magnitude of the compressive stress in the strut? (Assume $E_{steel} = 2 \times 10^{11}$ N m^{-2}.)

11. A steel drum, filled to the brim, holds 200 litres of liquid at 5 °C. If the liquid has a cubic expansivity of 8.7×10^{-4} K^{-1}, how much overflows if the temperature rises to 35 °C?

12. At 20 °C the difference in length between a steel rod and a brass rod is 175 mm. Find the length of each rod if this difference is to remain constant at any normal temperature.

TOPIC 18 HEAT TRANSFER

COVERING:

- conduction, convection and radiation.

There are three principal mechanisms involved in heat transfer. These are *conduction*, *convection* and *radiation*. (Note that other processes such as evaporation and condensation can also be significant.) Heat transfer is central to many areas of engineering, from domestic refrigerators to nuclear power stations, and very important in others.

Conduction involves heat transfer from a hotter to a colder part of a body – for example, through the base of a heated saucepan. *Natural convection* involves the movement of a fluid, such as air or water, which becomes less dense and tends to rise as it is heated. *Forced convection* involves an external agency such as a fan or a pump to move the fluid. Radiated heat, from the sun, for example, is carried by electromagnetic waves (Topic 13) and requires no transfer medium at all.

In practical situations all three processes may operate simultaneously but, to keep things simple, we shall consider them separately.

18.1 CONDUCTION

Heating one end of a solid bar will cause its constituent atoms to vibrate more vigorously about their fixed positions on the crystal lattice. Some of the vibrational energy will be passed on to neighbouring atoms via the chemical bonds and, as the process continues, heat is transferred along the rod from the hot end towards the cold.

Let us assume that a bar of length L has a cross-sectional area A and that its end faces are parallel and maintained at temperatures θ_1 and θ_2, respectively, where $\theta_1 < \theta_2$ as in Figure 18.1. Let us also assume that the bar is lagged along its length to prevent any heat loss from its sides and that the system has been given sufficient time to reach a steady state. ('Steady state' implies that, although the temperature differs at different points within the body, it remains constant at any given point.) The graph at the bottom of the figure shows that the temperature gradient along the bar is $(\theta_2 - \theta_1)/L$. The rate at which heat flows through the bar (from right to left) will be proportional to its cross-sectional area and to the temperature gradient, so that

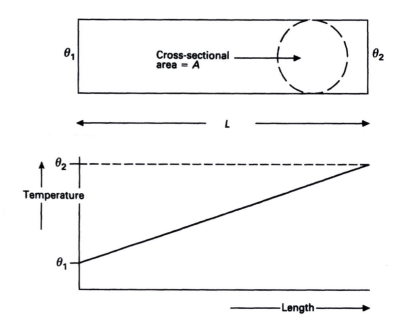

Figure 18.1

$$\frac{Q}{t} = kA\, \frac{(\theta_2 - \theta_1)}{L} \qquad (18.1)$$

where Q/t is the rate of heat flow in J s^{-1} (or W). k, the coefficient of proportionality, is the *thermal conductivity* of the material and has the units W m^{-1} K^{-1} (i.e. J s^{-1} m^{-1} K^{-1}). Table 18.1 shows some typical values. They tend to vary with temperature and other factors such as moisture content. They may vary considerably for different varieties of the same material. For example, the value for concrete ranges from below 0.2 to above 3.5 W m^{-1} K^{-1}, depending on type and moisture content.

Equation (18.1) tells us that the heat flow is proportional to the temperature difference $(\theta_2 - \theta_1)$ and inversely proportional to L/kA. The quantity L/kA is called the *thermal resistance* of the material; as we shall see from Topic 30, it is the thermal analogy of electrical resistance.

Remember that, in practice, the surface temperature of a conductor may not be the same as that of its surroundings. For example, the inside surface temperature of a window pane may be significantly different from room temperature because of the insulating effect of the layer of air immediately adjacent to the glass. Such effects can reduce heat flow very considerably.

Metals generally seem to feel colder than other materials at the same temperature, because they are better at conducting heat away from the body. Their high thermal conductivity stems from the freedom of movement of the valence electrons that maintain the metallic

Table 18.1

Material	Thermal conductivity/W m^{-1} K^{-1}
Metals:	
Aluminium	210
Copper	400
Steel	50
Non-metals:	
Concrete	1.5
Brick	0.6
Glass	1
Ice	2.1
Plaster	0.13
Wood (parallel to grain)	0.38
Wood (perpendicular to grain)	0.15
Water	0.58
Thermal insulators:	
Air	0.02
Glass wool	0.04
Polystyrene (expanded)	0.03

bond throughout the metal crystal structure. The free electrons in the hot part of the metal gain kinetic energy and rapidly transfer it to the colder parts as they migrate there. Some heat is still transferred via the vibrations of the crystal lattice but much less than by the free electrons. There tend to be no free electrons in ionic and covalent substances, so these have to depend on the lattice vibrations; hence, they generally have low thermal conductivity values compared with metals.

Air has very low thermal conductivity, because its constituent molecules are widely separated and can only pass kinetic energy from one to the other during their relatively infrequent collisions. If convection currents can be prevented, then air is a very good thermal insulator. Glass wool contains air trapped between the fibres, which gives it its very low thermal conductivity. Expanded polystyrene relies on the same basic principle and the insulating properties of some of the non-metals in Table 18.1 benefit from small quantities of trapped air. (Note that a vacuum is an ideal insulator in that it allows no possibility of conduction or convection; however, it will not stop the transmission of radiant heat.)

Equation (18.1) tells us that the heat flow through a material is inversely proportional to its thickness. This leads us to an important point about thermal insulation. Let us assume that a 1 m^2 sheet of expanded polystyrene 10 mm thick has one face maintained at -10 °C and the other at 30 °C. The equation tells us that the heat flow through the sheet is 120 W (see Worked Example 18.1 below). If we successively double the thickness of the sheet to 20 mm, 40 mm and 80 mm, the heat flow falls to 60 W, 30 W and 15 W, respectively; in other words, the first increase of 10 mm saves 60 W, a further increase of

20 mm saves an extra 30 W, but the final increase of 40 mm saves only another 15 W. Obviously the law of diminishing returns is operating: there is a critical thickness beyond which the heat saved over a given period of time does not justify the cost of the extra polystyrene used and the space wasted.

In many practical situations it is necessary to consider heat passing through successive thicknesses of different materials – for example, the three layers in a double-gazed window (two of glass and one of air).

Let us consider a simple example where a composite sheet built up from two layers has a thermal gradient between its faces. Since sheets tend to be flat and thin, lateral heat loss to the surroundings through the edge is confined to a relatively small strip round the periphery. Away from the edge, a section through the thickness of the sheet can be regarded in the same way as a lagged bar, because it is surrounded by material with an identical thermal gradient and there is no tendency for lateral heat loss. Figure 18.2 represents such a section through a composite of two materials, A and B. Heat will flow through both materials at the same rate Q/t, so if the temperature at the interface between them is θ_{AB}, and if $\theta_1 < \theta_2$, then, from Equation (18.1),

$$\frac{Q}{t} = k_A A_A \frac{(\theta_{AB} - \theta_1)}{L_A} = k_B A_B \frac{(\theta_2 - \theta_{AB})}{L_B} \qquad (18.2)$$

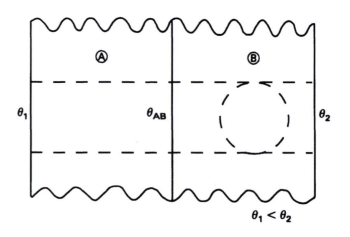

Figure 18.2

Worked Example 18.1

A sheet of expanded polystyrene, 1 m² in area, has one face maintained at $-10\ °C$ and the other at $30\ °C$. Find the heat flow through the sheet as its thickness is successively doubled from 10 mm to 20, 40 and finally 80 mm.

From Table 18.1, $k = 0.03$ W m^{-1} K^{-1}, and, substituting the data into Equation (18.1),

$$\frac{Q}{t} = \frac{0.03 \times 1 \times 40}{L}$$

which gives
  120 W for $L = 0.01$ m,
  60 W for $L = 0.02$ m,
  30 W for $L = 0.04$ m, and
  15 W for $L = 0.08$ m.

Worked Example 18.2

A composite sheet consists of a 30 mm thickness of material A and a 10 mm thickness of material B. (a) Find the heat flow through a 3 × 3 m panel of the composite when the exposed surface of A is maintained at 5 °C and the exposed surface of B at 25 °C. (b) Find the heat flow if the temperatures are reversed. (c) Sketch the temperature gradient in each case. (Assume $k_A = 0.1$ and $k_B = 0.3$ W m^{-1} K^{-1}.).

(a) Letting θ_{AB} represent the temperature at the interface, and substituting the data into Equation (18.2),

$$\frac{Q}{t} = 0.1 \times 3^2 \frac{(\theta_{AB} - 5)}{0.03} = 0.3 \times 3^2 \frac{(25 - \theta_{AB})}{0.01}$$

which gives

$$\theta_{AB} = 23 \text{ °C}$$

and, substituting for θ_{AB} above,

$$\frac{Q}{t} = 0.1 \times 3^2 \frac{(23 - 5)}{0.03} = 540 \text{ W}$$

(b) $\dfrac{Q}{t} = 0.1 \times 3^2 \dfrac{(25 - \theta_{AB})}{0.03} = 0.3 \times 3^2 \dfrac{(\theta_{AB} - 5)}{0.01}$

which gives

$$\theta_{AB} = 7 \text{ °C; hence, } Q/t = 540 \text{ W}$$

(c) See Figure 18.3.

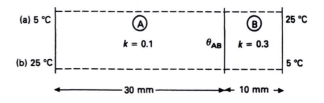

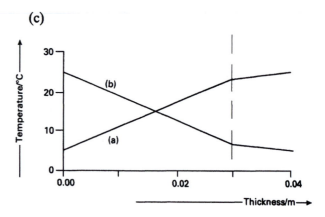

Figure 18.3

Worked Example 18.3

A lake is covered with a thickness of 20 mm of ice which is increasing by 6 mm per hour. Assuming that the water underneath the ice is at 0 °C, estimate the air temperature. (Assume $\rho_{ice} = 920$ kg m^{-3}.)

An increase in thickness of 6 mm over an area of 1 m^2 of ice is a volume increase of $1 \times 0.006 = 0.006$ m^3. The mass of 0.006 m^3 of ice is equal to

$$0.006 \text{ m}^3 \times 920 \text{ kg m}^{-3} = 5.52 \text{ kg}$$

To produce 5.52 kg of ice from water at 0 °C requires the removal of $Q = ml$ joules of heat, where $m = 5.52$ kg and, from Table 17.2 (page 157), $l = 3.3 \times 10^5$ J kg^{-1}. ml joules are removed in 1 h; therefore, the rate of removal is

$$\frac{Q}{t} = \frac{ml}{3600} = \frac{5.52 \times 3.3 \times 10^5}{3600} = 506 \text{ J s}^{-1}$$

From Table 18.1 (page 165) $k_{ice} = 2.1$ W m^{-1} K^{-1}, and, substituting in Equation (18.1), where θ_1 is the air temperature,

$$\frac{Q}{t} = 506 = 2.1 \times 1^2 \frac{(0 - \theta_1)}{0.02}$$

which gives

$$\theta_1 = -4.8 \ °C$$

18.2 CONVECTION

Newton's law of cooling is an empirical law telling us that the rate at which a body loses heat is proportional to its excess temperature over that of its surroundings. The discussion above suggests that the law should be valid for a hot body connected to cooler surroundings via a thermal conductor. However, it is normally viewed in the context of forced convection, although it works fairly well for natural convection if the excess temperature is small.

Let us assume that the temperature of a hot body is θ_2 and that of the surroundings is θ_1. The rate of heat loss Q/t from the body will be roughly proportional to its surface area A and, under conditions where Newton's law of cooling applies,

$$\frac{Q}{t} = k' A(\theta_2 - \theta_1) \tag{18.3}$$

k', the constant of proportionality, depends on the shape, orientation and surface characteristics of the body, and on the nature and flow characteristics of the cooling fluid.

It is useful to remember that if a body loses Q joules of heat, then its temperature will fall by Q/mc °C, where m is its mass and c is the specific heat capacity of the substance from which it is made (see Equation 17.1, page 155). (This implies that the temperature will remain uniform throughout the body as it cools, but in practice there may well be significant temperature variation through a given cross-section at any moment.)

18.3 RADIATION

If a body is completely surrounded by vacuum, then it can only exchange energy with its surroundings by radiation. Thus, the earth receives radiated energy from the sun in the form of heat and light.

In Topic 13 we noted that the essential difference between the various types of electromagnetic radiation is their wavelength. We also noted that black objects appear black because they absorb all the visible wavelengths. Physicists talk about a hypothetical body called a *black body* which has no ability to reflect incident radiation but will absorb it all completely. As a corollary of this property, a black body is the best possible emitter of *thermal radiation*. This is the radiation emitted by all objects by virtue of their temperature and is of particular interest to us in the present context. If the temperature of a black body is increased, it will emit more thermal radiation and, at the same time, the wavelength of the most intense radiation will decrease. Obviously, a black body will cease to appear black when it is so hot that it emits light.

Stefan's law gives the total radiant energy emitted by a black body per unit time (i.e. total radiant power) per unit surface area, as follows:

$$\frac{P}{A} = \sigma T^4 \tag{18.4}$$

where P represents the radiant power emitted by a black body of surface area A, and T is the absolute temperature (K) of its surface. σ is a constant, called Stefan's constant, which has the value 5.67×10^{-8} W m^{-2} K^{-4}. The equation tells us that the body will emit radiation at all temperatures greater than absolute zero.

If the surroundings of the black body are at some lower temperature T_s, then it will receive radiant energy from them at a rate proportional to T_s^4 and its net rate of radiant energy loss P_{net} will be proportional to $(T^4 - T_s^4)$. If $(T - T_s)$ is small, it is a fairly simple mathematical exercise to show that $(T^4 - T_s^4)$ is approximately equal to $4T_s^3(T - T_s)$; thus, the net rate of radiant energy loss is approximately proportional to the temperature difference if this is small. So.

$$\frac{P_{net}}{A} = \sigma(T^4 - T_s^4) \tag{18.5}$$

and

$$\frac{P_{net}}{A} \approx \sigma \times 4T_s^3(T - T_s) \tag{18.6}$$

provided that $(T - T_s)$ is small.

Clearly there will be a net loss of energy from the black body until its temperature reaches that of its surroundings. It will still continue to radiate energy when this happens, but it will then be absorbing it from the surroundings at the same rate.

Real objects are not perfect absorbers and emitters like black bodies – in fact, some are rather poor. The *emissivity*, ε, of a non-black body can be defined as the power emitted per unit area expressed as a fraction of that radiated by a black body at the same temperature. ε ranges from 1 for a perfect emitter down to 0, and Stefan's law can be expressed in the form

$$\frac{P}{A} = \sigma \varepsilon T^4 \tag{18.7}$$

Worked Example 18.4

A 25 mm diameter solid metal sphere is cooling under conditions such that there are negligible heat losses by conduction and convection. If the sphere behaves as a black body, with a net rate of heat loss of 100 W where its surroundings are maintained at 15 °C, then estimate (a) its temperature and (b) the rate at which this is falling. (Assume that the density of the metal $\rho = 9 \times 10^3$ kg m^{-3} and that its specific heat capacity $c = 400$ J kg^{-1} K^{-1}. Assume that Stefan's constant $= 5.67 \times 10^{-8}$ W m^{-2} K^{-4}.)

(a) The surface area of a sphere of radius r is $4\pi r^2$. Substituting the data into Equation (18.5), where T (K) is the temperature of the sphere,

$$\frac{100}{4\pi(0.0125)^2} = 5.67 \times 10^{-8} \ (T^4 - 288^4)$$

which gives

$$T = 975 \text{ K} = 702 \text{ °C}$$

(b) The volume of a sphere of radius r is given by

$$\frac{4}{3} \pi r^3$$

Therefore, its mass m is given by

$$m = \frac{4}{3} \pi r^3 \times \rho$$

and its thermal capacity mc is given by

$$mc = \frac{4}{3} \pi r^3 \times \rho \times c$$

and, substituting the data,

$$mc = \frac{4}{3} \pi (0.0125)^3 \times (9 \times 10^3) \times 400 = 29.5 \text{ J K}^{-1}$$

An object with a thermal capacity of 29.5 J K^{-1} losing 100 J in 1 s will experience an average fall in temperature of

$$\frac{100 \text{ J s}^{-1}}{29.5 \text{ J K}^{-1}} = 3.4 \text{ K s}^{-1}$$

Questions

(Use any previously tabulated data as required. Assume that Stefan's constant $= 5.67 \times 10^{-8}$ W m^{-2} K^{-4}.)

1. For each of the following materials estimate the thickness that would provide equivalent thermal resistance to that of 12 mm of expanded polystyrene:

   (a) glass wool;
   (b) plaster;
   (c) solid glass;
   (d) steel.

2. Two cylinders of identical shape and size, one copper and the other steel, are joined end to end and lagged along their length. The exposed copper face is maintained at 0 °C and the exposed steel face at 100 °C. Under steady-state conditions, (a) what is the temperature at the interface, and (b) what would the relative lengths of the cylinders have to be for an interface temperature of 50 °C?

3. A water tank, made from steel sheet 5 mm thick, has a layer of scale inside that is 0.6 mm thick. When the tank contains water at 90 °C, it loses heat at a rate of 35 kW m^{-2} through its sides. Estimate the temperature of the outside surface of the metal. (Assume that the thermal conductivity of the scale is 1.2 W m^{-1} K^{-1}.)

4. A sheet of aluminium 3 mm thick is sandwiched between two sheets of steel each 1 mm thick. What would be the heat flow per unit area through the thickness of the composite if one face is maintained at 0 °C and the other at 100 °C?

5. An object, hanging from an insulating thread, is cooling in a steady breeze at an air temperature of 15 °C. When it reaches 55 °C, it is losing heat at a rate of 30 W. Assuming negligible radiation losses, estimate the rate at which it will lose heat when it reaches 35 °C.

6. The temperature of each of the following is falling at a rate of 0.75 °C per minute. In each case estimate the rate of heat loss:

(a) a metal object with a heat capacity of 200 J K^{-1};

(b) a metal object of mass 0.5 kg and specific heat capacity 400 J kg^{-1} K^{-1};

(c) a 47.5 mm diameter solid metal sphere with a density of 8900 kg m^{-3} and a specific heat capacity of 400 J kg^{-1} K^{-1}.

7. The temperature of a mild steel can of mass 0.525 kg containing 0.240 kg of water was found to fall at a rate of 1 °C in 2 min 45 s. A sufficient quantity of hot water was added to the can to return it to its original temperature and, under the same cooling conditions, the temperature was found to fall at a rate of 1 °C in 8 min 15 s. Estimate how much water was added to the can.

8. Two ball-bearings, one twice the diameter of the other but otherwise identical, and both at the same temperature, are allowed to cool under identical conditions where Newton's law of cooling is valid. Find their relative initial rates of (a) loss of heat and (b) loss of temperature.

9. A black body, completely isolated in space, is cooling from 650 °C.

   (a) At which temperature does it radiate energy at half its initial rate?

   (b) At what rate does it radiate energy when it has cooled to 0 °C?

   (Assume space is at absolute zero.)

10. A 100 m diameter sphere, completely isolated in space, is radiating energy at a rate of 22.8 MW. If its surface temperature is 127 °C, estimate its emissivity.
    (Assume space is at absolute zero.)

TOPIC 19 GASES

COVERING:

- ideal gas;
- Boyle's law, the pressure law and Charles' law;
- the ideal gas equation;
- real gas and vapour.

In Topic 17 we noted that the constituent particles in a gas are free to move around independently of one another. The constituent particles of most ordinary gases are molecules. For instance, air consists of roughly 80% N_2 and 20% O_2 molecules with small quantities of CO_2 and other molecular substances. (The minor constituents also include the inert gases, which exist as single atoms because of their stable electronic configurations.)

At 0 °C and atmospheric pressure a litre of gas contains approximately 2.7×10^{22} molecules; this means that the average distance between two neighbours is about 3 nm (3×10^{-9} m), which is roughly ten times the size of a small molecule. Gas molecules move around at high speed in a random and disordered fashion, continually colliding with one another and the walls of their container. At ordinary temperatures the molecules in air travel at hundreds of metres per second, with a typical *mean free path* (average distance between collisions) of around 100 nm. The random nature of their motion and their frequent collisions means that at any instant the molecules have a wide range of speeds. However, their average kinetic energy is constant at a given temperarure and proportional to the absolute temperature.

With this simple picture in mind, we can regard the addition of thermal energy to a gas as resulting in an increase in the translational kinetic energy (hence, the speed) of its constituent molecules.

19.1 IDEAL GAS

Scientists use the hypothetical concept of an *ideal gas* in which the individual molecules are assumed to have no volume and experience no intermolecular forces of the kind that we discussed in Topic 16. Their collisions with one another and with the walls of their container are assumed to be perfectly elastic.

Air at ordinary temperatures and pressures behaves more or less like an ideal gas. Anyone who uses a bicycle pump is familiar with the 'springy' nature of air. This is quantified in Boyle's law, which relates the volume V and pressure p of a given mass of ideal gas by

$$V \propto 1/p$$

or

$$pV = \text{constant}$$

Thus, the pressure of the gas is inversely proportional to its volume. That is to say, if its volume is halved, its pressure is doubled, and if its volume is doubled, its pressure is halved, and so on. It is most important to remember that Boyle's law only holds at constant temperature and that it is only strictly valid for an ideal gas.

The pressure that a gas exerts on its container is due to the force arising from the change of momentum as the molecules bounce off the container walls. Because of the enormous numbers of molecules involved, these collisions result in a steady pressure (force per unit area).

Figure 19.1 shows a simple one-dimensional model where the gas in its container is represented by a single molecule moving to and fro perpendicularly between two parallel walls. At a given temperature the molecule moves at a constant speed corresponding to the constant average kinetic energy of the actual gas molecules. If the distance between the walls is halved, the frequency of the collisions is doubled and therefore the pressure is doubled; and if the distance is doubled, the frequency is halved – hence, in one-dimensional terms, $pV = \text{constant}$.

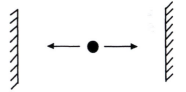

Figure 19.1

If the temperature is raised, there will be a corresponding increase in the kinetic energy of the gas molecules. This is represented by an increase in the speed of the molecule in our model. There will be an increase in the force as it collides with the walls and, assuming the walls are fixed (constant volume), there will be an increase in the frequency of its collisions – hence, there will be a pressure increase. The relationship between the pressure and absolute temperature T of a given mass of ideal gas at constant volume is given by the *pressure law*

$$p \propto T$$

or

$$p/T = \text{constant}$$

In other words, the pressure of a given mass of an ideal gas at constant volume is proportional to its absolute temperature.

If the gas is allowed to remain at constant pressure, by letting it expand as its temperature is raised, then the relationship between its volume and absolute temperature is given by *Charles' law* as follows:

$$V \propto T$$

or

$$V/T = \text{constant}$$

That is to say, the volume of a given mass of ideal gas at constant pressure is proportional to its absolute temperature.

These three laws are combined in the *ideal gas equation*

$$pV = nRT \tag{19.1}$$

where V is in m^3 and p is in Pa. (Remember that the SI unit of pressure is the pascal (Topic 4) and that $1\ Pa = 1\ N\ m^{-2}$.) T is the absolute temperature (K), n is the amount of gas in moles (mol) and R is the *universal molar gas constant* (often abbreviated to *gas constant*), which has the value $8.31\ J\ mol^{-1}\ K^{-1}$.

From Equation (19.1),

$$(nR =) \frac{p_1 V_1}{T_1} = \frac{p_2 V_2}{T_2} \tag{19.2}$$

where the subscripts 1 and 2 denote the pressure, volume and temperature of a given quantity of a gas under two different sets of conditions.

If $T_1 = T_2$, then Equation (19.2) reduces to Boyle's law:

$$p_1 V_1 = p_2 V_2$$

and if $V_1 = V_2$, then it reduces to the pressure law:

$$\frac{p_1}{T_1} = \frac{p_2}{T_2}$$

and if $p_1 = p_2$, it reduces to Charles' law:

$$\frac{V_1}{T_1} = \frac{V_2}{T_2}$$

Furthermore, a plot of p against $1/V$ at constant temperature will be a straight line of slope nRT, and so on.

Worked Example 19.1

An air bubble trebles in volume as it rises from the bottom of a lake to the surface. Estimate the depth of the lake. Assume that $g = 9.8$ m s^{-2}, atmospheric pressure is 101 kPa and the density of water $\rho_{water} = 1000$ kg m^{-3}.)

If p_{atm} and V_{atm} are the pressure and volume of the bubble at atmospheric pressure at the surface of the lake and p_b and V_b are the corresponding values at the bottom then, from Boyle's law, assuming the water temperature is constant,

$$p_{atm} \times V_{atm} = p_b \times V_b$$

but

$$V_{atm} = 3V_b$$

Therefore,

$$p_{atm} \times 3V_b = p_b \times V_b$$

and

$$p_b = 3p_{atm}$$

But p_b results from the pressure due to the depth of water plus atmospheric pressure; therefore, the pressure due to the depth of water alone is equal to

$$p_b - p_{atm} = 3p_{atm} - p_{atm} = 2p_{atm}$$

and, from Equation (4.2) (page 30), this is equal to $\rho_{water}gh$, where h is the depth of the lake.

Rearranging $\rho_{water}gh = 2p_{atm}$ gives

$$h = \frac{2p_{atm}}{\rho_{water}g} = \frac{2 \times 101 \times 10^3}{1000 \times 9.8} = 20.6 \text{ m}$$

Worked Example 19.2

A vertical glass tube, sealed at its bottom end, contains a 144 mm column of air trapped beneath a 126 mm column of mercury. Find the atmospheric pressure if, when the tube is laid horizontally, the air column is 168 mm long.

Vertical position:

The pressure acting on the trapped air column is $(p_{atm} + 126)$ mmHg. Assuming, for simplicity, that the tube is of unit internal cross-sectional area, then

$$pV = (p_{atm} + 126) \times 144$$

Horizontal position:

The pressure acting on the trapped air column is p_{atm} mmHg. Therefore,

$$pV = p_{atm} \times 168$$

From Boyle's law

$$(p_{atm} + 126) \times 144 = p_{atm} \times 168$$

which gives

$$p_{atm} = 756 \text{ mmHg}$$

Worked Example 19.3

The following p–V data were obtained using 1.88×10^{-3} moles of gas at 26 °C:

Pressure/kPa	100	150	200	250	300
Volume/cm^3	46.6	31.1	23.3	18.6	15.5

By non-graphical means (a) confirm that the gas obeys Boyle's law and (b) estimate the value of the universal molar gas constant.

(a) If Boyle's law is obeyed, then pV should be constant. For each pair of values,

$$pV = 4.66, 4.67, 4.66, 4.65 \text{ and } 4.65 \text{ (average } 4.66)$$

(Remember that 1 kPa $= 1 \times 10^3$ Pa and 1 cm$^3 = 1 \times 10^{-6}$ m^3.)

(b) From Equation (19.1) (page 176)

$$R = \frac{pV}{nT} = \frac{4.66}{1.88 \times 10^{-3} \times 299} = 8.3 \text{ J mol}^{-1} \text{ K}^{-1}$$

Note that the units of R are easily derived as follows:

$$R = \frac{pV}{nT} = \frac{\text{N m}^{-2} \times \text{m}^3}{\text{mol} \times \text{K}} = \text{N m mol}^{-1}\text{ K}^{-1} = \text{J mol}^{-1}\text{ K}^{-1}$$

Figure 19.2 is a plot of p against V for a given mass of ideal gas at temperatures T_1 and T_2, where $T_2 > T_1$. Boyle's law is obeyed along each *isotherm* (constant temperature line). The vertical and horizontal arrows, at constant V and constant p, represent changes following the pressure law and Charles' law, respectively.

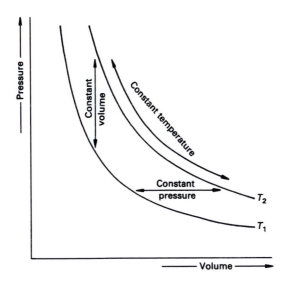

Figure 19.2

Charles' law can be expressed approximately by

$$V = V_0(1 + \theta/273) \tag{19.3}$$

where V_0 is the volume of gas at 0 °C and V is its volume at θ °C. (Compare this equation with Equation 17.6 (page 159), noting that, although the expansivity of different solids and different liquids can vary considerably, the volume expansivity for an ideal gas at 0 °C is approximately $1/273$ K^{-1}.)

Figure 19.3 shows Equation (19.3) plotted as a graph. This again draws our attention to the significance of -273 °C as the lowest possible temperature (absolute zero). Of course, the atoms in a real gas would not have zero volume at absolute zero as the figure seems to suggest; they would form a solid.

The equivalent equation for the pressure law is

$$p = p_0(1 + \theta/273) \tag{19.4}$$

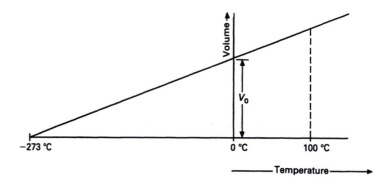

Figure 19.3

and this gives a corresponding graph of the same form as Figure 19.3.

There are two further points that we should note about ideal gases.

Avogadro's hypothesis states that equal volumes of ideal gases at the same temperature and pressure contain the same number of molecules. At *standard temperature and pressure* (S.T.P.) 1 mol of an ideal gas occupies about 22.4×10^{-3} m³, or 22.4 litres. S.T.P., which is fixed at 0 °C and 1 atm pressure (760 mmHg), is a standard condition that makes a useful baseline to which any quantity of ideal gas at any temperature and pressure can be reduced.

Dalton's law of partial pressures states that the total pressure of a mixture of gases equals the sum of the partial pressures of its components. (The *partial pressure* of each component is the pressure it would exert if it was the only occupant of the volume containing the mixture.)

Real gases deviate from ideal behaviour because real atoms and molecules have significant volume and, as we saw in Topic 16, they have forces operating between them.

19.2 REAL GASES

The behaviour of real gases was investigated last century by Andrews. Figure 19.4 gives a broad picture of the relationship between pressure and volume based on his results with carbon dioxide. We shall use this to illustrate our discussion.

At relatively high temperature, represented by the uppermost isotherm in Figure 19.4(a), the cohesive effect of the attractive forces between the gas molecules is small compared with the disruptive effect of thermal energy, and the isotherm is similar to that of an ideal gas. As the temperature is lowered, the effects of intermolecular forces become more significant and the isotherm becomes correspondingly distorted. Before we consider the critical isotherm corresponding to the critical temperature T_c, let us see what happens as we compress the gas along the isotherm XYY'Z corresponding to some temperature below T_c.

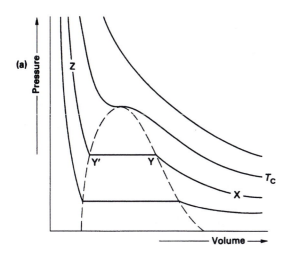

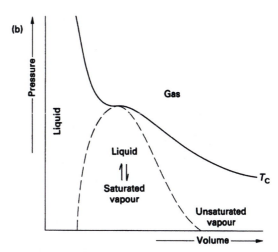

Figure 19.4

From X to Y it behaves as a gas but at Y it starts to condense. From Y to Y′ the pressure remains constant while the volume is reduced, and the gas progressively condenses until at Y′ it has entirely liquefied. Y′ to Z represents the compression of the liquid; liquids are obviously much less compressible than gases, because the molecules are close together, and the graph is correspondingly steep.

The *critical temperature* T_c is the temperature above which a gas cannot be liquefied by increasing its pressure. The *critical isotherm* at T_c therefore represents a boundary above which the substance must exist in gaseous form irrespective of pressure, as shown in Figure 19.4(b). Of course, the gas molecules can still be forced very close together by using high pressure, but above T_c this produces a highly compressed gas, not a liquid.

As Figure 19.4(b) suggests, a *vapour* is a substance in the gaseous state but below its critical temperature; thus, it can be liquefied by pressure alone.

In Topic 17 we noted that liquids evaporate because molecules with sufficiently high kinetic energy escape from the surface. And, as we might expect, if the temperature of the liquid is raised, the number of sufficiently energetic molecules will increase and the evaporation rate will increase. If we enclose a liquid under a vacuum, the molecules which escape will form a vapour above the surface. A number of the vapour molecules will *condense* back into the liquid as they collide with its surface and an equilibrium will be established when the condensation rate is equal to the evaporation rate. Under these conditions the vapour is described as *saturated* and the *vapour pressure* (the pressure exerted by the vapour) is called the *saturated vapour pressure* (s.v.p.). The saturated vapour pressure increases with temperature as the kinetic energy of the molecules is increased; however, it does not depend on the volume of the space above the liquid, because if this is changed and the temperature remains the same (as between Y′ and Y in Figure 19.4), then an imbalance between the evaporation and condensation rates restores the vapour pressure to its original saturated value. If the volume is so large that the liquid evaporates completely before the saturated vapour pressure is reached, then the vapour is described as *unsaturated* (as between Y and X in the figure).

The broken curve in the figure encloses an area within which liquid and saturated vapour coexist; it encloses the horizontal parts of all the isotherms below T_c. The pressure corresponding to each of these represents the saturated vapour pressure at that temperature. If the temperature is raised or lowered, the saturated vapour pressure will increase or decrease accordingly. If the temperature is raised to the point where the saturated vapour pressure is equal to the external pressure, then evaporation will occur throughout the body of the liquid; in other words, the liquid will boil. The boiling point normally quoted for a liquid is that at standard atmospheric pressure; it generally occurs somewhere about two-thirds of the critical temperature on the absolute scale.

If an unsaturated vapour is cooled to the point where it reaches its s.v.p., then it will normally start to condense, like moisture on a cold window, for example. *Relative humidity* is a measure of the extent to which air is saturated with water vapour. Thus, '60% relative humidity' means that the air contains 60% of the moisture that it would contain it if was saturated. The *dew point* is the temperature where the water vapour in the air is just saturated.

Questions

(Use any previously tabulated data as required. Assume ideal gas behaviour. Where necessary, assume atmospheric pressure = 101 kPa and $R = 8.3$ J mol^{-1} K^{-1}. Assume absolute pressure as opposed to gauge pressure (Topic 4) unless otherwise stated or implied. $g = 9.8$ m s^{-2}.)

1. The temperature of 2400 mm^3 of gas is increased from 27 °C to 57 °C. What volume change must be made to keep its pressure constant?

2. The pressure inside a 12 litre gas cylinder fell from 0.82 Mpa to 0.21 MPa, owing to a leaky valve. Estimate the volume of escaped gas at atmospheric pressure.

3. A pressure gauge, used to measure tyre pressures before and after a journey, gave readings of 190 kPa and 210 kPa, respectively. If the initial temperature was 18 °C (and the volume of the tyres remained constant), estimate the air temperature in the tyres after the journey.

4. 7.5 litres of gas at 27 °C and 505×10^3 Pa is allowed to expand to atmospheric pressure. Find its volume at 7 °C.

5. Calculate the volume of 1 mol of N_2 at standard temperature and pressure.

6. If a sample of gas occupies 1.20 m³ at 27 °C and 1950 mmHg pressure, find (a) its volume at STP and (b) its amount in moles.

7. A cylinder with internal dimensions of 100 mm length and 50 mm diameter contains gas at 14 °C and a pressure of 2 atmospheres. Estimate how many gas molecules the cylinder contains.

   (Avogadro constant $= 6.02 \times 10^{23}$ mol^{-1}.)

8. Assuming that air consists of 80% nitrogen (N_2) and 20% oxygen (O_2) by volume, estimate its density at STP.

9. A device like that in Worked Example 19.2 contained a mercury column 120 mm long. When the tube was held vertically, the length of the air column was (a) 79 mm with the open end upwards and (b) 109 mm with the open end downwards. Calculate the atmospheric pressure at the time.

10. Use a graphical method to answer Worked Example 19.3.

11. A device like that in Worked Example 19.2 contained a column of air trapped by water (rather than mercury). With the tube horizontal and atmospheric pressure equal to 759 mmHg, the length of the air column is 235 mm at 10 °C and 340 mm at 60 °C. Assuming that the saturated vapour pressure of water is 9 mmHg at 10 °C (and that Dalton's law applies to the trapped air/water vapour mixture), estimate the saturated vapour pressure of water at 60 °C.

TOPIC 20 LIQUIDS

COVERING:

- non-viscous behaviour (Bernoulli's equation);
- viscous flow (Poiseuille's formula);
- motion in a viscous fluid (Stokes's law);
- surface tension.

We have seen that the forces of attraction operating in a liquid are able to withstand the disruptive effect of thermal energy to the extent that the constituent particles form a coherent mass but remain capable of movement relative to one another. A liquid therefore has a more or less fixed volume contained within its boundary surface and is capable of flowing.

In this book we shall confine ourselves to *laminar flow*. This is flow that can be viewed in terms of the movement of layers (laminae) of liquid where successive particles passing the same point follow the same path (as indicated by the arrows in Figure 20.1). By contrast, in *turbulent flow*, which tends to occur at higher velocities, the flow pattern is broken and irregular, and eddies are formed.

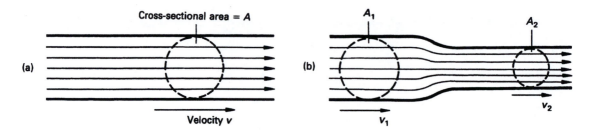

Figure 20.1

Reynolds demonstrated the difference between laminar and turbulent flow in a classic experiment towards the end of the last century. He introduced a 'thread' of liquid dye via a fine nozzle into water flowing through a glass tube. He found that as long as the velocity of the water remained low (where the flow is laminar), the thread of dye remained intact as it was carried along. At higher velocities (where the flow becomes turbulent) the thread broke up and the dye mixed with the water.

Figure 20.1(a) represents laminar flow through a uniform pipe. If the cross-sectional area of the pipe is A and liquid is flowing through it with an average velocity v, then the flow rate, expressed as the volume of liquid passing a given point in one second, will be Av. If the cross-sectional area of the pipe varies, as in Figure 20.1(b), then the velocity varies to maintain a constant flow rate and, assuming the liquid is incompressible,

$$A_1v_1 = A_2v_2 \tag{20.1}$$

This equation, called the *continuity equation*, tells us that, as the pipe narrows, the liquid velocity increases.

20.1 IDEAL (NON-VISCOUS) LIQUIDS

The next step is to consider *Bernoulli's equation*, which describes the behaviour of an ideal liquid. For the purposes of our discussion, an ideal liquid is incompressible and non-viscous and has flow properties that depend only on its density. (As we shall see in the next section, viscosity is a measure of the internal friction in a fluid that gives it resistance to flow.) To identify the parameters involved in Bernoulli's equation, let us consider the case of a liquid flowing through the curiously shaped pipe in Figure 20.2. At any position in the system, p represents the pressure, v represents the velocity of the liquid and h represents the height above some reference level. Let us assume that the liquid is flowing from position 1 to position 2 because of a pressure difference between them. The work done on the fluid by the net force due to the pressure difference results in an increase in its potential energy as it moves uphill and in its kinetic energy because of the increased velocity where the pipe narrows.

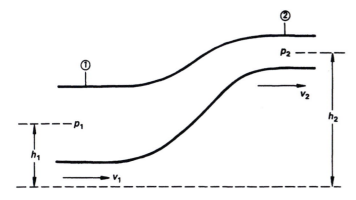

Figure 20.2

Bernoulli's equation tells us that for any small volume anywhere in an ideal liquid the sum of the three parameters involved (pressure, potential energy and kinetic energy) is a constant. In mathematical terms

$$p + \rho gh + \frac{1}{2}\rho v^2 = \text{constant} \qquad (20.2)$$

and applying this to Figure 20.2

$$p_1 + \rho gh_1 + \frac{1}{2}\rho v_1^2 = p_2 + \rho gh_2 + \frac{1}{2}\rho v_2^2 \qquad (20.3)$$

where ρ is the density of the ideal liquid and g is the acceleration due to gravity.

Bernoulli's equation is essentially a statement of the law of conservation of energy, where each term is expressed as energy per unit volume. If this idea is difficult to grasp, just think about what happens if we multiply both sides of Equation (20.3) by unit volume (1 m³). The potential and kinetic energy terms contain density rather than mass, but

$$\text{density} \times \text{m}^3 = \text{kg m}^{-3} \times \text{m}^3 = \text{kg} = \text{mass}$$

which returns them to their more familiar form (mgh and $\frac{1}{2}mv^2$). Furthermore, multiplying pressure by unit volume gives units of energy

$$\text{pressure} \times \text{m}^3 = \text{N m}^{-2} \times \text{m}^3 = \text{N m} = \text{J} = \text{energy}$$

In theory Bernoulli's equation is valid for all fluids (liquids and gases), provided that they are incompressible and have zero viscosity. Real fluids are, of course, compressible, particularly gases. They are also viscous, particularly liquids, and the mechanical work done in overcoming the viscosity of a liquid appears as heat, with the result that there is a decrease in the quantity ($p + \rho gh + \frac{1}{2}\rho v^2$). Nevertheless Bernoulli's equation is extremely important and has wide applications.

In some cases one or other of the parameters may be eliminated. For instance, p_1 and p_2 are both equal to atmospheric pressure in Worked Example 20.1 (below). In this example we find that the theoretical velocity with which liquid escapes from a small hole in the side of a tank is given by $\sqrt{2gh}$, where the hole is a distance h below the liquid surface (Torricelli's theorem). This turns out to be the same velocity as that which a body reaches by falling freely from rest through the same distance. In Question 1 you are asked to use Bernoulli's equation to derive the expression for the pressure at a given depth below the surface of a stationary liquid (Equation 4.2 on page 30). Provided that the liquid is stationary, then $v_1 = v_2 = 0$.

The potential energy term is eliminated for the case shown in Figure 20.3, where a liquid encounters a constriction in a horizontal pipe through which it is flowing. The vertical tubes are simply manometers

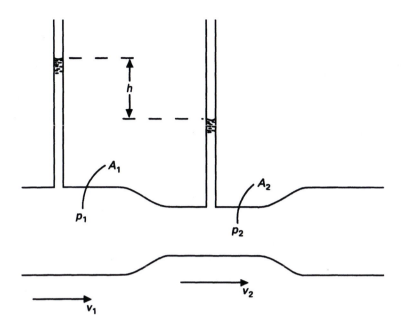

Figure 20.3

(see Topic 4) to indicate the gauge pressures p_1 in the pipe and p_2 in the constriction, where the velocity has increased from v_1 to v_2, owing to the reduction in cross-sectional area from A_1 to A_2. Since the pipe is horizontal, the potential energy terms cancel and Equation (20.3) becomes

$$p_1 + \frac{1}{2}\rho v_1^2 = p_2 + \frac{1}{2}\rho v_2^2 \tag{20.4}$$

and, on rearranging,

$$p_1 - p_2 = \frac{1}{2}\rho(v_2^2 - v_1^2) \tag{20.5}$$

Contrary to what we might expect, the manometer levels in the figure show that the pressure falls where the liquid velocity increases in the constriction, and the equation tells us that the greater the velocity change the greater the pressure drop.

This effect has many applications. For instance, an aircraft wing is designed so that the air flow over the top surface is faster than that over the bottom and the resulting pressure difference provides a lift force. (This example departs from our simple model, because air is compressible.)

In Worked Example 20.2, Equation (20.5) is used as the theoretical basis for the Venturi meter, which is a device used for measuring flow rate through a pipe. Worked Example 20.3 uses Equation (20.4) for the Pitot tube, which is a device used for measuring flow velocity.

Worked Example 20.1

Find an expression for the theoretical velocity v with which liquid escapes from a small hole in the side of a large tank at a distance h below the liquid surface.

Let p_1, h_1 and v_1 be the pressure, height and velocity of the liquid at its upper surface, and p_2, h_2 and v_2 the corresponding values where it escapes from the hole at a distance h below (see Figure 20.4).

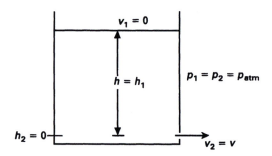

Figure 20.4

The liquid experiences atmospheric pressure p_{atm} at its upper surface and where it emerges from the hole; therefore, $p_1 = p_2 = p_{atm}$.

If we take the level of the hole to be the reference level, then $h_2 = 0$ and $h = h_1$.

If the cross-sectional area of the tank is large compared with that of the hole, then we can assume that the surface level falls at a negligible rate as the liquid escapes; therefore, $v_1 = 0$.

If the liquid density is ρ, then, substituting in Equation (20.3),

$$p_{atm} + \rho g h_1 + 0 = p_{atm} + 0 + \frac{1}{2} \rho v_2^2$$

Therefore,

$$v_2^2 = 2gh_1$$

and, since $v = v_2$ and $h = h_1$,

$$v = \sqrt{2gh}$$

(This is the same as the speed of a body of mass m that has fallen from rest at a height h where its potential energy was mgh; in this case $\frac{1}{2} mv^2 = mgh$.)

Worked Example 20.2

The device illustrated in Figure 20.3 can be used to measure flow rate. Find an expression giving flow rate in terms of the height difference h between the levels in the manometer tubes and the cross-sectional area of the pipe (A_1) and of the constriction (A_2).

From Equation (20.1)

$$A_1 v_1 = A_2 v_2$$

Therefore,

$$v_1^2 = \frac{A_2^2 \, v_2^2}{A_1^2}$$

and, substituting for v_1^2 in Equation (20.5),

$$p_1 - p_2 = \frac{1}{2} \rho \left(v_2^2 - \frac{A_2^2 \, v_2^2}{A_1^2} \right)$$

and

$$p_1 - p_2 = \frac{1}{2} \rho v_2^2 \, \frac{(A_1^2 - A_2^2)}{A_1^2}$$

which rearranges to give

$$v_2 = A_1 \sqrt{\frac{2(p_1 - p_2)}{\rho \, (A_1^2 - A_2^2)}}$$

Since the flow rate is equal to $A_2 v_2$, then, substituting for v_2 from above

$$\text{flow rate} = A_2 \times A_1 \sqrt{\frac{2(p_1 - p_2)}{\rho \, (A_1^2 - A_2^2)}}$$

But if h is the height difference between the manometer levels, then

$$(p_1 - p_2) = \rho g h$$

and

$$\frac{(p_1 - p_2)}{\rho} = gh$$

Therefore,

$$\text{flow rate} = A_1 A_2 \sqrt{\frac{2gh}{(A_1^2 - A_2^2)}}$$

Note that we have assumed an ideal liquid. In practice this device, and the device in Worked Example 20.3 below, would be calibrated (Topic 31) to take into account the effect of friction and other complicating factors associated with real liquids.

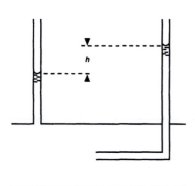

Figure 20.5

Worked Example 20.3

The device illustrated in Figure 20.5 can be used to measure flow velocity through a pipe. Find an expression giving flow velocity v in terms of the height difference h between the levels in the vertical tubes.

The left-hand tube indicates the pressure p_1, where the liquid flows through the pipe with velocity v_1.

The bottom end of the right-hand tube faces into the moving liquid, which is forced up to a height h above the level in the other tube. The liquid at the bottom end is stationary; therefore, $v_2 = 0$ with a corresponding pressure p_2.

Since the liquid flow through the pipe is horizontal, we can ignore the potential energy term in Bernoulli's equation and, from Equation (20.4),

$$p_1 + \frac{1}{2} \rho v_1^2 = p_2 + 0$$

and

$$\frac{1}{2} \rho v_1^2 = p_2 - p_1$$

But

$$p_2 - p_1 = \rho g h$$

Therefore,

$$\frac{1}{2} \rho v_1^2 = \rho g h$$

and

$$v = v_1 = \sqrt{2gh}$$

20.2 REAL (VISCOUS) LIQUIDS

Until now, the only property of a fluid that we have considered is its density. Now we have reached the point where we need to take viscosity into account.

Viscosity is the friction-like resistance to flow that arises from the cohesive forces between the constituent particles in a liquid. If the temperature is raised, the thermal energy of the constituent particles is increased and relative movement between them becomes easier. Thus, the viscosity of most liquids decreases with temperature.

In the absence of viscosity the velocity of a liquid flowing through a pipe would be uniform over the entire cross-section. For a real viscous liquid the velocity profile looks something like that in Figure 20.6, where the length of the arrows represents speed. The liquid molecules in immediate contact with the pipe adhere to its surface and tend to remain stationary. The stationary layer provides resistance to the movement of the layer of molecules adjacent to it and this in turn provides resistance to the movement of the next layer, and so on, giving a velocity profile similar to that in the figure. (Note that the work done in overcoming viscosity ultimately raises the total kinetic energy of the molecules and therefore appears as heat.)

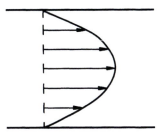

Figure 20.6

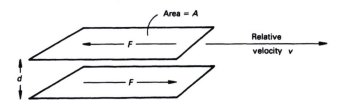

Figure 20.7

Viscosity can be quantified in terms of the friction-like force acting between adjacent layers of liquid moving relative to one another. In Figure 20.7 both layers are moving from left to right but the upper layer moves faster, so that their relative velocity v is equal to (v_{upper} − v_{lower}). The lower layer therefore exerts a retarding force F on the upper layer and experiences an equal and opposite reaction force in the forward direction. F is an example of a shear force and the resulting shear stress is given by F/A, where A is the area of contact between the layers. (We shall look at shear in more detail in the next topic; for the time being, just think of shear forces as acting parallel to a plane, unlike tensile and compressive forces, which act perpendicularly to it.) If d is the distance between the two layers, then for the purposes of our discussion we can say that, for many liquids, F is given by

$$F = \frac{\eta A v}{d} \qquad (20.6)$$

where η is the *coefficient of dynamic viscosity,* often simply called *viscosity.* The SI unit of viscosity is pascal seconds, Pa s (i.e. N s m^{-2}). (Viscosity is sometimes measured in non-SI units called poises, where 10 poises = 1 Pa s.) Some approximate values of η at 20 °C are given in Table 20.1.

Table 20.1

Substance	Viscosity/Pa s at 20 °C
Air	1.8×10^{-5}
Water	1.0×10^{-3}
Machine oil	$1-6 \times 10^{-1}$

From equation (20.6) we see that, for the liquids to which it applies, the shear stress F/A is proportional to the velocity gradient v/d and η is the constant of proportionality; thus, a large shear stress coupled with a small velocity gradient indicates a liquid of high viscosity (for instance, machine oil compared with water).

There are two important laws governing viscous flow. *Poiseuille's formula* gives the flow rate through a cylindrical pipe. If the pipe has an internal radius r and a length l, and the pressure difference between its ends is $(p_2 - p_1)$, then the volume flow rate V of a fluid of viscosity η is given by

$$V = \frac{\pi r^4}{8\eta} \left(\frac{p_2 - p_1}{l} \right) \tag{20.7}$$

Thus, flow rate is proportional to the pressure gradient $(p_2 - p_1)/l$ and inversely proportional to the viscosity of the fluid. Somewhat more surprisingly, it is also proportional to r^4, so, all other things being equal, doubling the radius of the pipe increases the flow rate sixteenfold! Another way of looking at the equation is to say that flow rate is proportional to the pressure difference $(p_2 - p_1)$ and inversely proportional to the resistance to flow as given by $8\eta l/\pi r^4$.

Before moving on to the other important law, we should note that the type of flow under given conditions can be predicted with the aid of the *Reynolds number, Re.* This is a dimensionless quantity given by

$$Re = \frac{v\rho l}{\eta} \tag{20.8}$$

where v, ρ and η are the velocity, density and viscosity of the fluid and l is a linear dimension that is characteristic of the system. If, for a straight uniform pipe, we take v as equal to the volume flow rate divided by the cross-sectional area and l as equal to the internal diameter, then we can normally expect laminar flow if the value of Re is below about 2000 and turbulent flow if it is above about 4000. Either may be possible between 2000 and 4000. Thus, Equation (20.8) tells us that laminar flow is favoured by low values of v, ρ and l and by high viscosity.

The second important law is *Stokes's law*. This gives the viscous resistive force F that acts on a sphere of radius r moving with velocity v through a fluid of viscosity η as follows:

$$F = 6\pi\eta r v \qquad\qquad (20.9)$$

Thus, the resistance experienced by the sphere is proportional to η, r and v. Worked Example 20.4 (below) shows how Stokes's law can be used to determine the viscosity of a liquid by measuring the terminal velocity of a sphere falling through it under gravity (e.g. a small ball-bearing falling through oil); note that for Stokes's law to agree closely with experimental results Re should be less than 0.1, where l is taken to be the diameter of the sphere and v its terminal velocity. Furthermore, the liquid container needs to be large, so that the walls and the bottom have an insignificant effect on the velocity of the sphere.

Worked Example 20.4

Assuming Stokes's law, find an expression whereby the viscosity η of a fluid of density ρ_f may be obtained from the terminal velocity v_t of a sphere of radius r and density ρ_s falling through it under gravity.

The downward force acting on the sphere (i.e. its weight) is opposed by the upthrust due to the fluid it displaces and by the viscous resistive force it experiences. When the sphere has reached its terminal velocity v_t, where its acceleration is zero (Topic 5), then the net downward force acting on it is zero (because $F = ma = m \times 0$). Now

$$\text{mass of sphere} = \text{volume} \times \text{density} = \frac{4}{3}\pi r^3 \times \rho_s$$

and

$$\text{mass of displaced fluid} = \frac{4}{3}\pi r^3 \times \rho_f$$

and, from Equation (20.9),

$$\text{viscous resistive force} = 6\pi\eta r v$$

At terminal velocity v_t the net downward force is equal to

$$\text{weight of sphere} - \text{upthrust} - \text{resistive force} = 0$$

Therefore,

$$\frac{4}{3} \pi r^3 \rho_s g - \frac{4}{3} \pi r^3 \rho_f g - 6\pi\eta r v_t = 0$$

which rearranges to

$$\eta = \frac{2r^2 (\rho_s - \rho_f)g}{9v_t}$$

20.3 SURFACE TENSION

Surface tension is the property that makes a liquid behave as though it has an elastic skin. It is the reason why water forms drops and why its surface can support small objects such as sewing needles whose density would otherwise cause them to sink.

Surface tension has a molecular basis. A molecule in the body of a liquid is completely surrounded by neighbours and therefore experiences attractive forces more or less uniformly in all directions. But a surface molecule experiences a net attractive force into the body of the liquid, since it only has neighbours on that side (apart from a few in the surrounding vapour). Because the surface molecules tend to be pulled inwards, the liquid will tend to adjust its shape to minimise its surface area. Since the sphere is the geometric shape with the smallest surface area for a given volume, liquids tend to form spherical drops, although factors such as gravity and the effect of other surfaces normally prevent this.

Surface tension is given the symbol γ and can be defined as the force in the liquid surface acting perpendicularly to a line of unit length lying in the same plane. The units are therefore N m^{-1} (and for water at ordinary temperatures γ is about 0.073 N m^{-1}). To make this idea clearer, Figure 20.8 shows a film of liquid, such as a soap solution, stretched across a wire frame which has one side, of length l, that can be moved without any frictional resistance. Because the film tries to retract in order to reduce its surface area, a force F must be applied to the movable side to hold it stationary. By our definition above $F = 2l \times \gamma$. (The factor 2 is necessary because the film has two surfaces, an upper and a lower, both of which are pulling on the frame.)

If we stretch the film by pulling the movable side against the surface tension through a distance d, then the work done, equal to $2l\gamma \times d$, is stored as *surface energy*. Surface energy is given the symbol σ and has units of J m^{-2}. In terms of Figure 20.8, the area of the newly created surface is $2ld$; therefore,

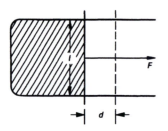

Figure 20.8

$$\sigma = \frac{2l\gamma d}{2ld} = \gamma$$

so γ and σ have the same numerical value. (Note that $\sigma =$ J m$^{-2} =$ N m m$^{-2} =$ N m$^{-1} = \gamma$.)

When a liquid comes into contact with a solid, forces of attraction between the liquid molecules and the solid surface cause adhesion between the two. A detailed discussion is beyond the scope of this book, but we need a broad picture of some of the ideas involved.

The adhesion between water and clean glass is much greater than the cohesion between water molecules themselves. Water therefore tends to form as large an interface with glass as possible, and spreads out over its surface and wets it. If the glass is vertical, then the water will tend to climb up it and form a concave meniscus similar to that in Figure 20.9(a). By contrast, the adhesion between mercury and clean glass is smaller than the cohesion in liquid mercury. Mercury therefore forms discrete drops on a horizontal glass surface, and a convex meniscus on a vertical one, as in Figure 20.9(b), in order to minimise the area of its interface with the glass.

The *angle of contact*, θ in the figure, is the angle measured through the liquid between the liquid and solid surfaces where they meet. The smaller the value of θ the greater the tendency of the liquid to wet the solid surface. In fact, θ is zero for water on clean glass and about 140° for mercury on clean glass.

Wetting agents are used to reduce θ for many purposes: for example, fluxes are used to promote the wetting of metal surfaces with molten solder. Surfaces are sometimes waterproofed by treating them with substances to increase their contact angle with water. Capillarity, which causes liquids to rise in narrow tubes where $\theta < 90°$, is due to surface tension.

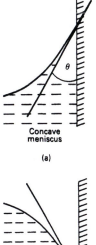

Concave meniscus

(a)

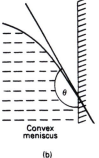

Convex meniscus

(b)

Figure 20.9

Questions

(Use any previously tabulated data as required. $\rho_{water} = 1000$ kg m^{-3}. $g = 9.8$ m s^{-2}.)

1. The gauge pressure p at a depth h below the surface of a stationary liquid of density ρ is given by the expression $p = \rho g h$, where g is the acceleration due to gravity. Derive this expression from Bernoulli's equation.

2. A large steel tank floating in water has a 24 mm diameter hole sealed with a plug 250 mm below the water line. Estimate the initial flow rate into the tank if the plug is removed.

3. Express the flow rate from the previous question in terms of molecules per minute.
    (Avogadro constant $= 6.02 \times 10^{23}$ mol^{-1}.)

4. By consideration of the units on their respective right-hand sides confirm that (a) Equation (20.7) gives a volume flow rate, (b) Equation (20.8) gives a dimensionless number and (c) Equation (20.9) gives a force.

5. An oil drop of density 900 kg m^{-3} and of radius 2.5 × 10^{-6} m fell steadily through a distance of 10.0 mm, in air, in 14.7 s. Estimate the viscosity of the air, assuming that its density may be ignored.

6. An open-ended horizontal tube 500 mm long with an internal diameter of 2 mm was sealed into the bottom of a water tank. 139 cm^3 of water flowed from the tube over the course of 1 min. (a) Check whether the water would have experienced laminar flow through the tube and (b) estimate the depth of water in the tank.

7. A workman accidentally drilled a horizontal hole in a water pipe 3.29 m above ground level. The escaping water travelled a horizontal distance 4.92 m before hitting the ground. Find the gauge pressure in the pipe.

8. A clean rectangular glass plate, measuring 75 × 16 mm and 1.9 mm thick, is suspended so that its long edges are horizontal, its faces are vertical and it just makes contact with a horizontal water surface. Assuming that $\gamma_{water} = 0.073$ N m^{-1}, estimate the force due to surface tension that must be overcome to separate the plate from the water.

9. Estimate the energy required to divide a 2 mm diameter raindrop into ten million identical droplets. (For a sphere of radius r, surface area = $4\pi r^2$ and volume = $4\pi r^3/3$. Assume $\gamma_{water} = 0.073$ N m^{-1}.)

TOPIC 21 SOLIDS

COVERING:

- elastic deformation and modulus of elasticity;
- stress/strain relationships;
- plastic deformation;
- brittle behaviour.

We have already seen that the constituent atoms, ions or molecules in a solid material are trapped betwen their neighbours because they have insufficient thermal energy to escape. They are confined to fixed positions on a crystal lattice, held by a network of cohesive forces that tends to oppose any attempt to deform it, and it is this that gives crystalline solids their characteristic rigidity. We shall begin by considering elasticity, which is the property of a solid that tends to return it to its original dimensions when it has been deformed.

21.1 ELASTIC DEFORMATION

We shall start with a model solid in which the constituent atoms are held together by chemical bonds represented by the net force/separation curve in Figure 21.1. In the absence of external forces the atoms will adopt the equilibrium separation r_0, where the attractive and repulsive components in the bond are balanced (see Topic 16). If we apply an external force to pull the atoms apart or push them together, then an opposing force of equal magnitude will be generated within the bond by the imbalance between the attractive and repulsive components as we move up or down the net force/separation curve. At small displacements the net force/separation curve is virtually linear; hence, it provides the basis of Hooke's law, as we saw in Topic 16 (see page 143). If the external force is removed, then the atoms will return to their equilibrium separation r_0.

Figure 21.2 represents the mechanical deformation of a model material. In tension and compression equal and opposite forces act along the same line and the length of the specimen changes accordingly. In shear (to the right of Figure 21.2) the forces are out of alignment and this results in a twisted deformation of the specimen. (Note that in this case, as the figure stands, additional forces that are needed to prevent rotation have been omitted for simplicity.)

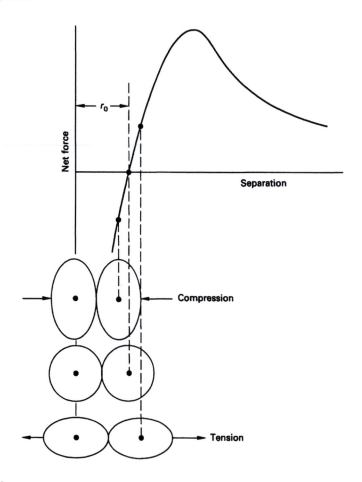

Figure 21.1

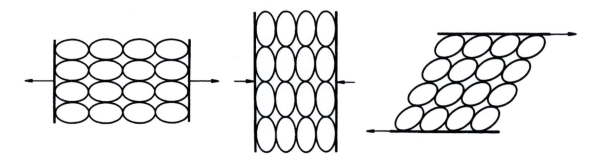

Figure 21.2

In Topic 2 we found the *Young's modulus* of a material as follows:

$$E = \frac{\text{stress}}{\text{strain}} = \frac{F/A}{\Delta l/l_o} = \frac{\sigma}{\varepsilon} \tag{21.1}$$

where Δl represents the extension and l_0 the undeformed length, as indicated in Figure 21.3(a). Some approximate Young's modulus values are given in Table 21.1. (Remember that 1 GPa $= 1 \times 10^9$ N m^{-2}.)

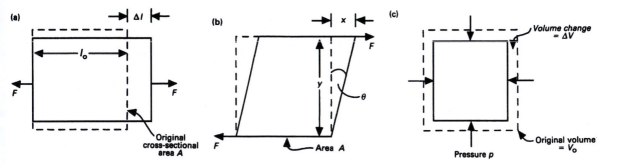

Figure 21.3

Table 21.1

Material	Young's modulus/GPa
Aluminium	70
Brass	110
Copper	110
Glass	70
Nylon	3
Steel	200

As Figure 21.3(a) suggests, materials stretched under tension tend to become thinner (fatter under compression) and in most cases there is a change in volume. The transverse strain ε_t accompanying the longitudinal strain ε_l is given by the Poisson's ratio of the material, μ, where

$$\mu = -\frac{\varepsilon_t}{\varepsilon_l} \qquad (21.2)$$

The negative sign is needed because a positive longitudinal strain (extension) gives a negative transverse strain (contraction), and vice versa.

Now let us consider the equivalent of Young's modulus in shear. In Figure 21.3(b) the shear stress (symbol τ) is given by F/A, where F is the tangential force along the plane of area A over which it is applied. The angle of shear θ is a measure of the shear strain and $x/y = \tan \theta = \theta$ radians if θ is small. Thus, at small strains the *shear modulus G* is given by

$$G = \frac{\tau}{\theta} \qquad (21.3)$$

(If we think of G as F/A divided by x/y, then we have a parallel with viscosity – see Equation 20.6 on page 191.)

The third type of deformation, shown in Figure 21.3(c), is the change in volume ΔV which occurs when an object is subjected to hydrostatic pressure – for example, at the bottom of the sea. In this case the stress is the pressure p. (Remember that pressure is the normal force per unit area acting on a surface (Topic 4).) The volume strain is the fractional volume change $\Delta V/V_0$, where V_0 is the original volume. In this case the modulus is the *bulk modulus K*, given by

$$K = -\frac{p}{\Delta V/V_0} \tag{21.4}$$

The negative sign is needed because a positive pressure change gives a negative volume change, and vice versa.

Note that strain is a dimensionless quantity, so all three moduli have the units of stress, N m^{-2} (or Pa).

For isotropic materials (i.e. those with uniform properties in all directions) the elastic moduli and Poisson's ratio are related by

$$G = \frac{E}{2(1 + \mu)} \tag{21.5}$$

and

$$K = \frac{E}{3(1 - 2\mu)} \tag{21.6}$$

Many solids have a value for Poisson's ratio of between about $\frac{1}{3}$ and $\frac{1}{4}$, which, when substituted in the equations above, gives values for G of about $0.4E$ and for K between about $0.7E$ and $1.0E$.

Note that most solids behave elastically at only very low strains, generally less than 1%, above which they either break in a brittle fashion or deform plastically before fracture. Rubber is exceptional in that it remains elastic up to very large strains, sometimes several hundred per cent, although the stress/strain relationship is not linear, as we shall see later.

Figure 2.1 (page 12) shows the load/extension plot for a particular wire that happened to be 1.72 m long and 0.40 mm in diameter. A more fundamental approach would be to present the stress/strain plot for the material from which the wire was made. In the worked example below, the original load/extension data for the wire is reprocessed and stress is plotted against strain. (Note that strain can be represented as a straightforward ratio, but it is often expressed as a percentage by multiplying it by 100. Since strains in real structures are often very small, engineers sometimes prefer to work in *micro-strain*, which is strain multiplied by a million. Thus, an extension of 1 mm in 1 m can be expressed as 0.001 strain, 0.1% strain or 1000 microstrain.)

Worked Example 21.1

The following load/extension data were obtained for a wire 1.72 m long and 0.40 mm in diameter. Convert the data into stress/strain form and plot them to obtain the Young's modulus for the material from which the wire was made.

Load/N	10	20	30	40	50	60
Extension/mm	0.7	1.5	2.1	2.9	3.6	4.3

Table 21.2

Load/N	Extension/m $\times$ 10^{-3}	Stress/MPa	Strain/$\times$ 10^{-3}
10	0.7	80	0.41
20	1.5	159	0.87
30	2.1	239	1.22
40	2.9	318	1.69
50	3.6	398	2.09
60	4.3	477	2.50

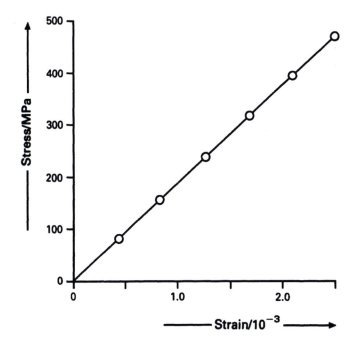

Figure 21.4

Table 21.2 shows load divided by the original cross-sectional area ($\pi \times 0.0002^2$) to give tensile stress, and extension divided by the original length to give strain. Figure 21.4 shows stress plotted against strain.

Since the graph is a straight line and passes through the origin, then, from the figure, estimating the stress to be 480 MPa at a strain of 2.5 × 10^{-3},

$$E = \frac{480 \times 10^6}{2.5 \times 10^{-3}} = 190 \text{ GPa}$$

Figure 21.5 shows a plot that was obtained in basically the same way as Figure 21.4 but, in this case, an ordinary rubber band was used instead of a metal wire. It is clear that it does not obey Hooke's law. Furthermore, we can see from the scales on the two figures (and, of course, we know from experience) that a rubber band stretches very much more easily than a metal wire. Clearly the mechanism of rubber-like elasticity is fundamentally different.

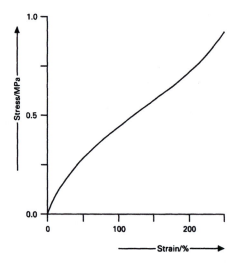

Figure 21.5

21.2 RUBBER-LIKE ELASTICITY

Rubber consists of enormously long chain-like molecules which are randomly tangled together in a disordered mass rather like a plate of spaghetti. At normal temperatures there is sufficient thermal energy for rotation to occur about the covalent bonds along the length of the chains (see Topic 16). The result of this is that the chains are in a constant state of random wriggling motion, continuously changing their shape. On application of an external force to stretch the rubber, the chains tend to straighten by untwisting through bond rotation so that they are brought into partial alignment. On releasing the external force, the chains wriggle back into their disorganised and more compact configurations so that the rubber as a whole retracts. Deformation by bond rotation processes of this kind is much greater, and much smaller forces are involved, than in the bond distortion mechanism we discussed earlier (Figure 21.1).

21.3 PLASTIC DEFORMATION

All solid materials subject to a continuously increasing tensile stress break sooner or later. But before they do, many of them undergo a significant amount of *plastic deformation*, which is a permanent deformation that results from internal structural changes. For instance, many metals stretch elastically, so that, on unloading, they return to their original lengths, but if they have been stretched too far, they are left with a permanent length increase. In other words, they have an elastic limit corresponding to a stress level above which some of the strain is permanent.

As we shall see in Topics 24 and 25, plastic deformation is a complicated process. For the time being we shall look at it briefly in terms of a very simple atomic model of a metal. This is based on the fact that adjacent planes of atoms in a metallic crystal structure can slip over one another under the influence of stress. This is illustrated in idealised form in Figure 21.6(a). Provided that the stress is large enough to move each atom over its neighbour in the adjacent plane (Figure 21.6b), then there is nothing in the nature of the metallic bond to prevent this process from continuing step by step. (The existence of such planes is implied in Figure 16.7 on page 147.) Bonding between the planes is not interrupted, so the force required is less than that needed to pull them apart completely. Our simple model therefore suggests that metals tend to deform plastically in this way rather then snap in a brittle fashion like glass. We cannot, of course, continue stretching the specimen indefinitely. Sooner or later, depending on the material, it will break, although the fracture processes involved are complex and beyond the scope of our discussion.

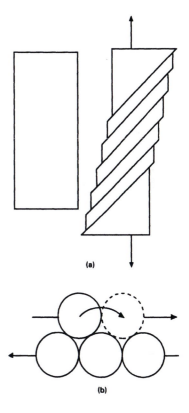

Figure 21.6

21.4 BRITTLE FRACTURE

Brittle materials, like glass, are characterised by their inability to undergo plastic deformation. If a brittle material is fractured, the broken pieces can often be glued back together, so the repaired object is more or less restored to its original dimensions.

This is consistent with our simple two-atom model of elasticity. If we apply stress to a material, it responds by deforming elastically. If the stress is increased, the chemical bonds are progressively deformed until those that are the most highly stressed cannot support the load and the material breaks. The bonds within the broken pieces are no longer subjected to external stress and they return to their equilibrium positions. We can therefore reconstruct the original object by glueing the pieces back together.

Ceramics are non-metallic inorganic materials that include, for example, glass and fired clay products. They often contain compounds of metals and non-metals and they are inclined to be brittle because the ionic–covalent bonding on which they generally depend tends to resist the slip processes that readily occur in metals. At one end of the scale the covalent bond is rigid, directional and specific between the bonded atoms. And at the other end, as Figure 21.7 indicates, relative movement between the planes in an ionic structure as shown would bring positive ions into contact with positive ions and negative with negative. This would lead to strong repulsive forces causing the planes to separate rather than slip over one another.

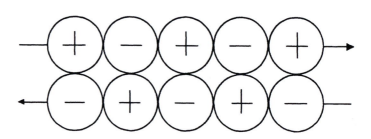

Figure 21.7

Glasses are *amorphous solids* – that is to say, they are non-crystalline, as suggested in Figure 21.8. Like a liquid, their structure has no long-range regularity and such materials are sometimes called *supercooled liquids*.

Rubber only exhibits its special elastic behaviour above its *glass transition temperature*. Below this, there is insufficient thermal energy for bond rotation and the chains are frozen into fixed configurations. The material loses its rubbery characteristics and behaves in a similar way to glass. A number of hard, brittle plastics are based on chain-like molecular structures which have their glass transition temperatures above room temperature.

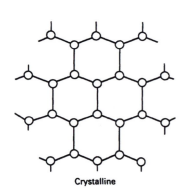

Crystalline

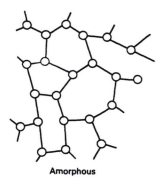

Amorphous

Figure 21.8

Questions

(Use any previously tabulated data as required and assume that Hooke's law is obeyed. Assume that $g = 9.8$ m s^{-2} and that Poisson's ratio for steel is 0.3.)

1. A 1.5 mm diameter wire supports a 9 kg load. A column of 1.4 m square cross-section supports a 10×10^6 kg load. Find the longitudinal stress in each case.

2. An 8 kg load is suspended by a steel wire originally 1 mm in diameter and 5 m in length before loading.

Find (a) the tensile stress in the wire and (b) by how much the load has stretched it.

3. Find the tensile stress in a steel rod originally 2 m long that has been stretched by 1 mm.

4. A 5.0 kg mass is to be suspended from a 2.0 mm diameter wire or thread, 3.0 m long, made from either (a) brass, (b) steel or (c) nylon. Predict the extension in mm in each case.

5. A copper wire has a diameter of 2 mm and a length of 6 m. An aluminium wire has a rectangular cross-section 1.5 × 2.5 mm and a length of 7.5 m. Find the total extension if the wires are joined end to end and used to suspend a 10 kg load.

6. A copper wire 2.2 m long and an aluminium wire 1.4 m long, both 2 mm in diameter, are joined end to end. What tensile load would produce a total extension of 1 mm?

7. An unstressed steel strip is 100.00 mm in width. Find the width of the strip when it is subjected to a longitudinal tensile stress of 400 MPa.

8. Find the percentage volume change of a steel cylinder after it has been subjected to a longitudinal tensile stress of 400 MPa.

9. A solid steel object sinks to a depth of 5 km below the surface of the sea. Estimate the percentage volume change it experiences, assuming the density of sea-water is 1030 kg m^{-3}.

10. A bar made from an unidentified material is initially 100.000 mm in diameter and 1000.00 mm long. When subjected to a longitudinal tensile force of 3.927 MN, its diameter becomes 99.925 mm and its length 1002.50 mm. Estimate (a) the shear modulus and (b) the bulk modulus of the material.

See also Further Questions on page 252.

TOPIC 22 STRUCTURE OF SOLIDS

COVERING:

- metallic, ionic and covalent crystal structures;
- some crystal structures with more than one type of bonding;
- some crystal structures that depend on intermolecular forces;
- amorphous structures (glasses).

In Topics 23–25 we shall discuss the nature of ceramics, metals and polymers but, before we do, we need to consider how the internal structure of materials is influenced by the type (or types) of chemical bonding involved in holding their constituent atoms together. We shall begin with metals because they are basically very simple.

22.1 METALLIC CRYSTAL STRUCTURES

In Topic 16 we saw that we can view metallic crystal structures in terms of packing spheres together. We shall now develop this model further, starting in two dimensions by considering some simple arrangements of coins on a table top.

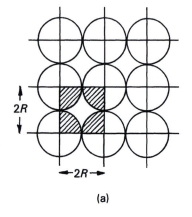

(a)

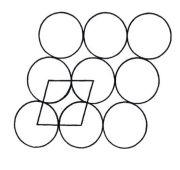

(b)

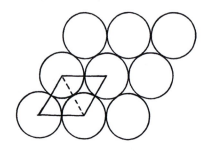

(c)

Figure 22.1

Figure 22.1(a) shows a square arrangement. Vertical and horizontal lines have been drawn through the centres of the coins to form a lattice dividing the structure into identical square cells. If the radius of each coin is R then each cell has an area of $(2R)^2$ and contains four quarters of a coin, making one complete coin, with a total area of πR^2. The fraction of the area of each cell covered by coins is therefore given by

$$\frac{\pi R^2}{(2R)^2} = \frac{\pi}{4} = 0.79$$

We can reproduce this structure by copying one cell as many times as we need and fitting the copies together. Each cell is therefore representative of the structure as a whole and identifies it completely; it follows that a large number of coins arranged in this way will cover 79% of a flat surface.

In Figure 22.1(b) the structure is tilted, with the result that the four-sided spaces between the coins are elongated and reduced in area. Each cell still has sides of length $2R$, and still encloses a total of one coin, but its overall area decreases as the tilt increases and the space between the coins is reduced. The area reaches its minimum value in Figure 22.1(c) where each four-sided space splits into two three-sided spaces and the structure can tilt no further. The central coin now has six immediate neighbours in contact, rather than four, and it is impossible to fit any more around its circumference. This corresponds to the closest packed structure possible, where 91% of the surface is covered. (91% is obtained in the same way as the 79% for the square-packed arrangement, except that the overall area of the cell is smaller; as the broken line in Figure 22.1(c) shows, the overall area is equal to the area of two equilateral triangles with sides of length $2R$.)

In Topic 16 we saw that the valence electrons in metals act collectively and are not confined to specific atoms. The metallic bond therefore tends to be non-specific and non-directional so, as far as any particular atom is concerned, one neighbour is as good as any other, irrespective of direction. We would therefore expect metal atoms to pack together as closely as possible, each surrounded by the maximum possible number of neighbours, as in Figure 22.1(c).

Now let us extend this two-dimensional model by treating Figure 22.1(c) as part of a single layer of close-packed spheres above and below which we can stack identical layers to form a three-dimensional structure. In Figure 22.2(a) each of the spheres is labelled B and the six triangular spaces between them are alternately labelled A and C. Figures 22.2(b) and (c) show that if we start to build a second layer on top of the B layer then the upper spheres can sit in either the A spaces or the C spaces. In both instances, the centres of the filled spaces are exactly one diameter apart. Note that we cannot fill adjacent A and C spaces because their centres are less than 1 diameter apart.

Figure 22.2(d) shows that, by adding more spheres, we obtain a layer of close-packed spheres on top of the original layer. This happens whether we fill the A spaces or the C spaces.

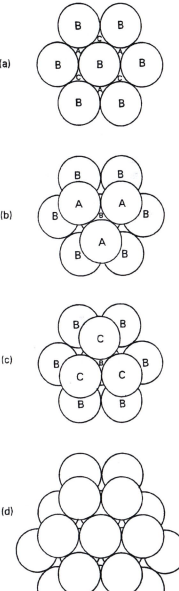

(a)

(b)

(c)

(d)

Figure 22.2

If we add a third close-packed layer underneath the original layer then this occupies the undersides of either the A spaces or the C spaces. We can therefore obtain two types of regular crystal structure, one with close-packed layers stacked in alternating positions (ABABAB ... and so on) with the C positions unoccupied, and the other with all three positions repeated in sequence (ABCABC ... and so on).

The ABABAB arrangement is called the *hexagonal close-packed* (HCP) structure. It can be regarded as being built up from many identical *unit cells* which have a hexagonal form that gives the structure its name. Figure 22.3 shows how we can picture these unit cells and the way they fit together, like bricks, to form the HCP structure. (The cells that we identified in Figure 22.1 are the two-dimensional equivalents of unit cells.)

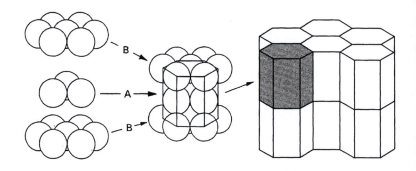

Figure 22.3

The ABCABC arrangement is called the *face-centred cubic* (FCC) structure because the unit cell is a cube with a sphere at each corner and one at the centre of each face, as in Figure 22.4. Note that the FCC unit cell is formed by stacking the close-packed planes perpendicularly to the diagonal from one corner to the other through its centre.

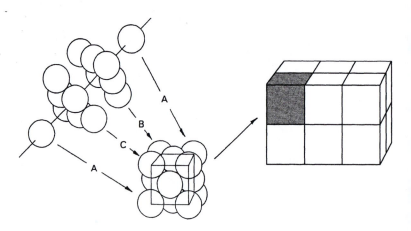

Figure 22.4

(a)

(b)

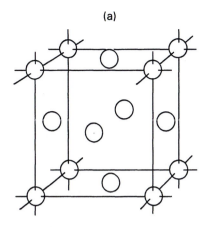

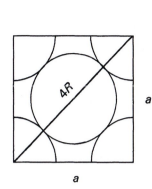

Figure 22.5

It is important to recognise that a unit cell contains incomplete spheres where these are shared with adjacent unit cells. For example, Figure 22.5(a) shows an 'exploded' view of part of the FCC structure, and Figure 22.5(b) shows one of the 6 identical faces of the unit cell. There is one-eighth of a sphere in each corner of the cube, where 8 unit cells join, and one-half a sphere at the centre of each face, where 2 unit cells join. Since there are 8 corners and 6 faces per cube there is a total of 4 spheres per unit cell as follows

$$(8 \times 1/8) + (6 \times 1/2) = 1 + 3 = 4$$

Note that in both HCP and FCC structures each sphere has a *co-ordination number* of 12, that is to say it is surrounded by 12 immediate neighbours. There are 6 in the same close-packed plane, 3 in the plane above, and 3 in the plane below regardless of whether the stacking is ABABAB or ABCABC. The 12 neighbours form a tightly packed shell around the central sphere, with no room for more. These two arrangements are therefore the most densely packed possible. They have an *atomic packing factor* of 0.74, which means that the spheres occupy 74% of the total volume, leaving the remaining 26% as space between them.

Worked Example 22.1

(a) Estimate the density of copper, assuming that it has the FCC structure and that the copper atom has a radius of 0.128×10^{-9} m and a relative atomic mass of 63.5. (1 u = 1.66×10^{-27} kg.)

(b) Calculate the atomic packing factor for the FCC structure given that the volume V of a sphere is related to its radius R by the formula

$$V = \frac{4}{3}\,\pi R^3$$

(a) Figure 22.5(b) shows one face of the FCC unit cell where a represents the length of each edge of the cube and R represents the radius of each atom. From Pythagoras' theorem

$$a^2 + a^2 = 16R^2$$

hence

$$a = \sqrt{8}\,R$$

Since $R = 0.128 \times 10^{-9}$ m for copper, the volume of the unit cell is

$$a^3 = (\sqrt{8} \times 0.128 \times 10^{-9})^3 = 4.75 \times 10^{-29} \text{ m}^3$$

There is a total of four atoms in the FCC unit cell (see above), and the total mass of four copper atoms is

$$4 \times 63.5 \times 1.66 \times 10^{-27} = 4.22 \times 10^{-25} \text{ kg}$$

The estimated density of copper is therefore

$$\frac{4.22 \times 10^{-25}}{4.75 \times 10^{-29}} = 8900 \text{ kg m}^{-3}$$

(b) The total volume of the four atoms in the FCC unit cell is

$$4 \times \frac{4}{3}\,\pi R^3$$

and, from part (a) above, the volume of the FCC unit cell is

$$a^3 = 8\sqrt{8}\,R^3$$

The proportion of the total volume occupied by the atoms is therefore given by

$$4 \times \frac{4}{3}\,\pi R^3 \times \frac{1}{8\sqrt{8}\,R^3} = \frac{2\pi}{3\sqrt{8}} = 0.74$$

hence the atomic packing factor is 0.74.

Some metals adopt the more openly packed *body-centred cubic* (BCC) structure shown in Figure 22.6. The BCC unit cell is a cube with a sphere at each corner and another at the body-centre, hence the name of the structure. Each sphere has only eight immediate neighbours, which do not touch each other. The structure is therefore less densely packed than the HCP and FCC structures, the atomic packing factor being 0.68 for body-centred cubic packing. Figure 22.7(a) shows an 'exploded' view of the BCC structure and Figure 22.7(b) shows one of the faces of the unit cell. There is one-eighth of a sphere in each corner, like the FCC unit cell, but the body-centred sphere is completely enclosed within the unit cell and not shared with any others. The BCC unit cell therefore contains a total of only two spheres. Figures 22.7(c) and (d) show the plane through the body-centre between

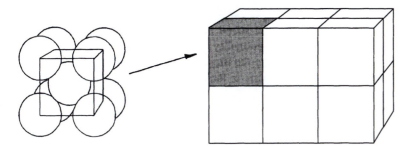

Figure 22.6

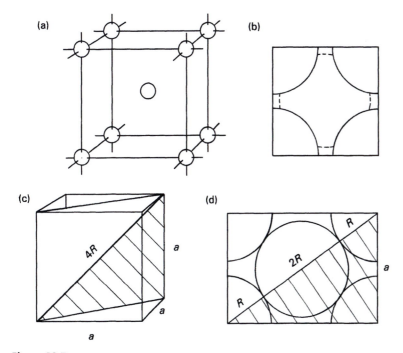

Figure 22.7

diagonally opposite edges of the cube. The unit cell size is determined by the sphere-to-sphere contact from corner to corner through the body-centre, rather than through the face-centre as in the FCC structure (see Figure 22.5). The atomic packing factor for the BCC structure can be found by applying Pythagoras' theorem to the shaded triangle in Figures 22.7(c) and (d), though to do this it is first necessary to find the length of its bottom side. (This can be done by applying Pythagoras' theorem to the cube face diagonal that forms the hypotenuse of the horizontal triangle at the base of the cube in Figure 22.7(c). You will need to do this when you calculate the atomic packing factor for the BCC structure in Question 4 at the end of the topic.)

There are complicated reasons why some metals adopt the less closely packed BCC structure. In some cases the strength of the metallic bond is not great enough to restrain the thermal vibration of the atoms sufficiently for them to adopt close-packed structures at ordinary temperatures. In others there appears to be scope for partial covalent character in the metallic bond, hence directionality, which would inhibit close-packing.

HCP, FCC and BCC are the most common types of crystal structure in metals and, as we shall see in the next topic, their different geometries lead to fundamental differences in their capacity for plastic deformation. Table 22.1 lists the structures of some well-known metals at room temperature.

Table 22.1

Face-centred cubic	Hexagonal close-packed	Body-centred cubic
Aluminium	Beryllium	Sodium
Nickel	Magnesium	Potassium
Copper	Titanium	Vanadium
Silver	Cobalt	Chromium
Platinum	Zinc	Iron
Gold	Zirconium	Molybdenum
Lead	Cadmium	Tungsten

22.2 IONIC CRYSTAL STRUCTURES

The ionic bond, like the metallic bond (but unlike the covalent bond), is non-specific and non-directional; we can simply think of ionic crystal structures in terms of packing spheres together, remembering that they are electrically charged and that opposite charges attract one another and like charges repel. Each ion tends to surround itself with as many oppositely charged neighbours as possible but the extent to which it can do so is limited by the repulsive forces between the neighbours. On this basis, for example, we would expect the two-dimensional structure in Figure 22.8(a) to be unstable because adjacent *anions* (negative ions) are in contact with one another. On the other hand, we would expect the structure in Figure 22.8(b) to be stable because the number of anions

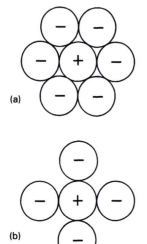

(a)

(b)

Figure 22.8

is less and they can spread out around the central *cation* (positive ion) to avoid touching one other.

However, this picture is complicated by the fact that ions vary in size, anions generally being larger than cations (see Topic 16). In Figure 22.9(a) the cation is large enough to hold the anions apart. In Figure 22.9(b) it is just small enough to allow the anions to touch one another, making it necessary to reduce their number, as in Figure 22.9(c), in order to form a stable arrangement.

As this suggests, the crystal structure adopted by simple ionic compounds is influenced by the *radius ratio*, r/R (where r and R are the radii of the smaller and the larger ions respectively). Assuming that the number of cations and anions is equal then, in terms of the two-dimensional model, Figure 22.10(a) represents part of an extended square-packed structure, corresponding to the arrangement in Figure 22.9(a), in which each ion is surrounded by four neighbours of opposite charge. Similarly, the structure in Figure 22.10(b) corresponds to the triangular arrangement in Figure 22.9(c) where the radius ratio limits the number of neighbours to three. Further complications arise if the magnitude of the charge on the cations and anions is different. For example, in calcium fluoride we need twice as many F^- ions as Ca^{2+} ions to balance the positive and negative charges.

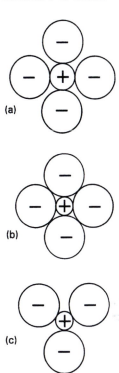

Figure 22.9

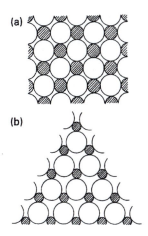

Figure 22.10

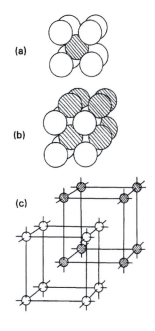

Figure 22.11

Now let us extend this two-dimensional model to some simple three-dimensional ionic structures, beginning with cations and anions of equal size ($r/R = 1$). On size considerations alone, we could surround each ion with 12 neighbours, as in the close-packed metal structures, but this would be unstable because it would bring like-charged ions into contact with one another. However, we can surround each ion with 8 neighbours. Figure 22.11(a) shows that if one ion is placed at the centre of a cube, then 8 oppositely charged neighbours placed at the corners will be separate from one another. Figure 22.11(b) shows that

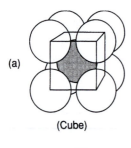

(a)

(Cube)

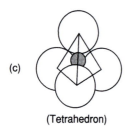

(b)

(Octahedron)

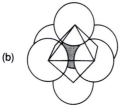

(c)

(Tetrahedron)

Figure 22.12

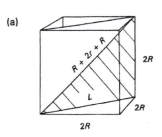

(a)

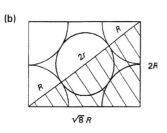

(b)

Figure 22.13

each of these ions can form the centre of an interpenetrating cube. The 'exploded' view in Figure 22.11(c) indicates how this arrangement forms the basis of an extended structure consisting of interpenetrating simple cubic lattices of oppositely charged ions. In this structure, which is called the caesium chloride structure after the ionic compound of that name, each ion has a co-ordination number of 8. (The term 'body-centred cubic' is not used in this case since it tends to be reserved for structures consisting of identical constituents, for example metal atoms.)

What happens if we reduce the radius ratio? If the central ion in Figure 22.12(a) is progressively reduced in size, then the outer ions move in closer together. In Figure 22.13(a) we have reached the limiting value of the radius ratio, where the outer ions touch each other, making the length of the cube edge $2R$. Application of Pythagoras' theorem to the shaded triangle shows this limiting value to be 0.73, as we shall now see.

The length of the hypotenuse is $(R + 2r + R)$ because this is the line of contact between the large and the small ions through the body-centre of the cube. First we obtain the length L of the bottom side of the shaded triangle by applying Pythagoras' theorem to the horizontal triangle at the base of the cube, as follows:

$$L = \sqrt{(2R)^2 + (2R)^2} = \sqrt{8R^2} = \sqrt{8}\,R$$

Applying Pythagoras' theorem to Figure 22.13(b) we can write

$$(R + 2r + R)^2 = 4R^2 + 8R^2$$

which simplifies to

$$R + r = \sqrt{3}\,R$$

hence

$$\frac{r}{R} = \sqrt{3} - 1 = 0.73$$

If we now reduce the number of outer ions to 6 then they can separate from one another and form the stable arrangement in Figure 22.12(b). This corresponds to the *rocksalt* structure, which is shown in Figure 16.2 on page 143. Careful examination of Figure 16.2 shows that this structure can be regarded either as interpenetrating face-centred cubic lattices of oppositely charged ions, or as a simple cubic lattice with alternate positive and negative ions. The co-ordination number is 6 and each ion occupies the octahedral space between its six neighbours.

If we progressively reduce the size of the central ion in the octahedral space, the outer ions move together until they touch when

$r/R = 0.41$. This is shown in Figure 22.14 where the broken circle indicates the positions of the fifth and sixth outer ions, above and below the central ion in the octahedral space. Again, the limiting radius ratio is obtained using Pythagoras' theorem. (You are asked to do this in Question 5 at the end of the topic.)

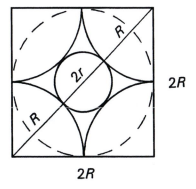

Figure 22.14

If the number of outer ions is now reduced to four they can separate to form the tetrahedral arrangement shown in Figure 22.12(c). Later we shall refer to an important structure of this kind, called the *zinc blende* or *sphalerite* structure, in which zinc ions (Zn^{2+}) are surrounded tetrahedrally by sulphide ions (S^{2-}) and vice versa.

This discussion suggests that we should be able to predict the structure of simple ionic compounds from the relative size of the ions involved. For example, the radii of caesium (Cs^+) and chloride (Cl^-) ions are about 0.167 and 0.181 nm respectively, giving a radius ratio of 0.92. Since the arrangement in Figure 22.12(a) is stable for radius ratios between 0.73 and 1.00 then this is the type of structure that we would expect caesium chloride to adopt, which indeed it does. Similarly, the radius of the sodium ion (Na^+) is about 0.097 nm, which is 54% of the radius of the chloride ion; as we might expect, sodium chloride adopts the arrangement in Figure 22.12(b) which is stable for radius ratios down to 0.41.

These arguments suggest that we can view ionic structures in terms of packing ions together as tightly as possible within the limitations imposed by the radius ratio. However, in practice, the radius ratio only gives a guide to the type of structure adopted. There are many exceptions, possible causes for which include, for example, the directional restrictions imposed by partial covalent character in the ionic bond. Nevertheless the concept is a useful one and helps us to understand the nature of ionic materials.

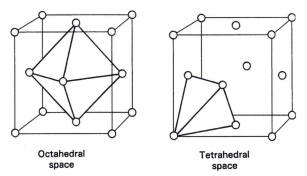

Octahedral space   Tetrahedral space

Figure 22.15

Another helpful approach is to think of ionic structures in terms of a lattice of anions held together by the mutually attractive forces with cations occupying the *interstices* or spaces between them (and vice versa). For example, a face-centred cubic lattice contains both octahedral and tetrahedral spaces such as those shown in Figure 22.15. On this basis, Figure 16.2 shows that we can regard sodium chloride as a face-centred cubic lattice of chloride ions held together by sodium ions

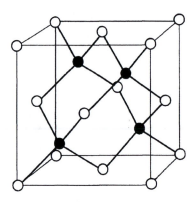

Figure 22.16

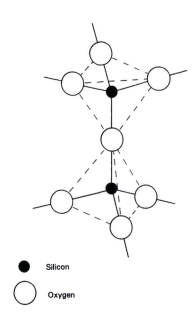

● Silicon

○ Oxygen

Figure 22.17

in the octahedral spaces between them. Similarly, the *zinc blende* or *sphalerite* structure, which is shown in Figure 22.16, can be viewed as a face-centred cubic lattice of sulphide ions with four of the tetrahedral spaces in each unit cell occupied by zinc ions. (Note that the bonding in zinc blende has a high degree of covalent character.)

22.3 COVALENT CRYSTAL STRUCTURES

In Topic 16 we met the idea that the covalent bond has a specific and directional nature; electrons are shared between particular atoms and the resulting bond is orientated in a particular direction. For instance, diamond is a form of carbon in which each atom is covalently bonded to four neighbours to give the very open tetrahedral structure shown in Figure 16.5 on page 145. It is important to recognise that this tetrahedral arrangement arises because carbon, as a group IV element, forms four covalent bonds; it has nothing to do with radius ratios in this case.

In fact the diamond structure can be regarded as cubic, like zinc blende except that all the Zn^{2+} and S^{2-} sites are occupied by covalently bonded C atoms. Silicon and germanium, which are also group IV elements, adopt the same type of structure.

Silicon and oxygen, between them, account for about three-quarters of the mass of the earth's crust. They form the basis of many minerals of commercial importance and a variety of manufactured materials. In *silica* (silicon dioxide, SiO_2) silicon atoms are joined together via oxygen atoms to form extended crystal structures. The Si–O bonds are strong, and for our purposes we can treat them as essentially covalent though they do possess some ionic character. Figure 22.17, which represents a small portion of an extended silica structure, shows how the twofold valency of oxygen is satisfied by two neighbouring silicon atoms and the fourfold valency of silicon by four neighbouring oxygen atoms (hence we have the formula SiO_2). As we shall see in the next section, it is important to recognise that the structure can be regarded as being built up from SiO_4 tetrahedra (shown as broken outlines in the figure). The tetrahedra are joined to their neighbours at each apex via the oxygen atom shared between them. Silica has a number of polymorphs, that is to say it exists in a number of physical forms. In *cristobalite*, for example, the silicon atoms adopt the same relative positions as in elemental silicon, but with oxygen atoms in between.

For our purposes, these examples tell us as much as we need to know about covalent crystal structures.

22.4 SOME CRYSTAL STRUCTURES WITH MORE THAN ONE TYPE OF BONDING

So far we have considered relatively simple structures, each involving only one type of chemical bond. In this section we shall consider examples of some rather more complicated structures containing bonds of more than one type. We shall begin with graphite.

Graphite is an *allotrope* of carbon, that is to say it is one of the several different forms in which the element exists. (The word 'allotrope' is used in a similar way to 'polymorph' except that it refers specifically to elements.) Diamond, which is of course another allotrope of carbon, is an extremely hard material which is normally a poor conductor of electricity. This is because each carbon atom is held firmly in place between its neighbours by strong covalent bonds from which the valence electrons can only escape with difficulty. By contrast, graphite is so soft that it is used for making pencil 'leads', and it conducts electricity. The contrast arises because the structures of the two materials are different.

Figure 22.18(a) shows how the carbon atoms in graphite are arranged in hexagonal patterns to form planes in which each atom is covalently bonded to three neighbours. This involves three of the four outer electrons of each carbon atom. The fourth electron becomes *delocalised*, which means that it is no longer tied to a particular atom; instead, it occupies a special kind of orbital which extends above and below the entire plane, rather like a sandwich. The delocalised electrons are free to move within the plane and are therefore able to conduct electricity along it like the free electrons in a metal. Figure 22.18(b) shows how the planes themselves are stacked on top of one another, held together by van der Waals' forces. Since these forces are weak (see Topic 16) the planes are easily persuaded to slide over one another, making graphite soft enough to be used in pencil leads and as a lubricant.

Van der Waals' forces, like metallic and ionic bonds, are non-specific and non-directional. (This is illustrated by the fact that inert gases adopt close-packed structures when they solidify.) However, intermolecular forces that arise from the permanent polarisation of molecules can lead to directional constraints in the formation of crystals. For example, in Topic 16 we saw that the water molecule has a tetrahedral form, with two orbitals positively biased and two negatively biased. Neighbouring water molecules tend to orientate themselves so that oppositely charged orbitals from each are directed towards one another, giving rise to hydrogen bonds between them. Each molecule in ice is rigidly bonded to four neighbours, reflecting its tetrahedral form and giving ice a very open crystal structure containing a large proportion of empty space (see Figure 22.19). If ice is melted the rigid structure is disrupted and tends to collapse, allowing an increase in the average number of neighbours per molecule and a reduction in the total volume. For this reason the density of water is greater than that of ice.

Finally we shall consider *silicates*. These compounds, some of which are very important in the engineering context, are based on SiO_4 tetrahedra (Figure 22.17) and can be classified according to the way in which the tetrahedra are joined together. In the case of silica itself, the tetrahedra are joined to their neighbours via the shared oxygen atom

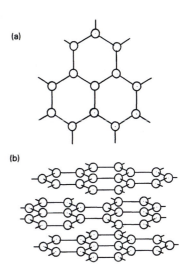

(a)

(b)

Figure 22.18

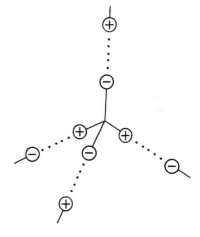

Figure 22.19

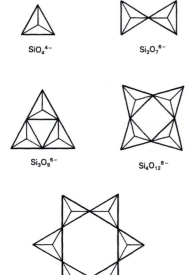

SiO_4^{4-}

$Si_2O_7^{6-}$

$Si_3O_9^{6-}$

$Si_4O_{12}^{8-}$

$Si_6O_{18}^{12-}$

Figure 22.20

at each apex and, as we saw in the previous section, the valency requirements of both the silicon and oxygen atoms are satisfied. However, in the case of an isolated SiO_4 tetrahedron each oxygen atom has only seven outer electrons rather than a complete octet. This deficiency can be overcome by forming ionic bonds with metal atoms so that the tetrahedron becomes the negative silicate ion SiO_4^{4-}. For example, *forsterite* (Mg_2SiO_4) is a silicate in which SiO_4^{4-} tetrahedra are linked together by Mg^{2+} ions.

Figure 22.20 shows, in schematic form, various other silicate ions that are formed from tetrahedra joined together via the oxygen atom at their shared apex. These include 3-, 4- and 6-membered rings with the formulae $Si_3O_9^{6-}$, $Si_4O_{12}^{8-}$ and $Si_6O_{18}^{12-}$ respectively. Note that the valency of each of these rather complicated ions corresponds to the total number of unjoined oxygen apexes, each of which has gained an electron in order to complete the octet of outer electrons. The electrons may be acquired from a variety of metal atoms which, in effect, then become metal ions that hold the silicate ions together. For instance, in the precious stone emerald, $Al_2Be_3(Si_6O_{18})$, the electrons are provided by aluminium and beryllium atoms.

Figure 22.21 shows how large numbers of tetrahedra can join together to form long single and double silicate chains. Various types of asbestos are based on silicate chains, and their structure is reflected in their fibrous nature.

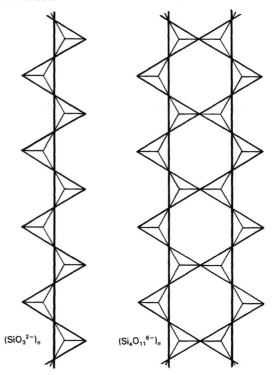

$(SiO_3^{2-})_n$

$(Si_4O_{11}^{6-})_n$

Figure 22.21

Figure 22.22 shows how each tetrahedron can join to three others to produce sheet-like structures which occur in clay minerals, talc and mica, among other substances. The plastic behaviour of clay/water mixtures, the lubricating properties of talc and the cleavage properties of mica reflect their sheet-like structures.

If the tetrahedra are linked to their neighbours by all four apexes, then we have three-dimensional *framework* structures, such as cristobalite which we met in the previous section.

22.5 AMORPHOUS STRUCTURES (GLASSES)

In Topic 21 we noted that the term *glass* refers to amorphous (i.e. non-crystalline) solids and that such materials are sometimes called supercooled liquids.

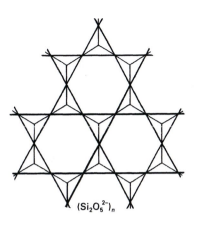

Figure 22.22

When a liquid is cooled, it generally reaches a point where the thermal energy of its constituent atoms, ions or molecules is low enough to allow them to pack together to form an ordered crystalline solid under the influence of the chemical bonding forces that operate between them. However, if they are somehow prevented from settling into their respective positions on the crystal lattice, then they may form an amorphous structure. Figure 21.8 shows, in schematic form, the difference between a crystalline and an amorphous structure.

Many types of commercial glass are made by cooling molten silica (usually containing various additives). The Si-O-Si bonding is strong, and relative movement between the tetrahedra in molten silica is difficult; the viscosity is therefore high and, on cooling, reordering of the molten silica into a crystalline structure is restricted. Under manufacturing conditions the melt is cooled too quickly to allow recrystallisation to occur, and an amorphous stucture is formed. Even though the cold glass is solid, in the sense that relative movement between its constituent atoms can no longer occur, it retains the liquid-like amorphous structure – hence its description as a supercooled liquid. We shall discuss silica glass in more detail in Topic 23, and in Topic 25 we shall see that polymers can show glassy characteristics.

Questions

(Assume 1 u = 1.66×10^{-27} kg.)

1. Show that coins arranged as in Figure 22.1(c) will cover 91% of the area of a flat surface.

2. Estimate the radius of the zinc atom, given that zinc has a density of 7100 kg m^{-3} and a relative atomic mass of 65.4. Assume that zinc atoms fill 74% of the total space that they occupy.

3. Estimate the density of iron given that it has a BCC structure and that the iron atom has a radius of 0.124 nm and a relative atomic mass of 55.8.

4. Calculate the atomic packing factor for the BCC structure.

5. Calculate the limiting value of the radius ratio for the rocksalt structure.

6. With reference to Figure 16.2, and given the following data, estimate the density of sodium chloride.

	Approximate ionic radius	Relative atomic mass
Sodium	0.097 nm (Na^+)	23.0
Chlorine	0.181 nm (Cl^-)	35.5

See also Further Questions on page 252.

TOPIC 23 THE NATURE OF CERAMICS

COVERING:

- brittleness;
- stress concentration;
- crack propagation;
- Griffith's equation;
- glass;
- fired clay ceramics;
- concrete.

At one time the term *ceramics* was normally confined to pottery and similar fired clay products. Modern usage of the word often includes artificial non-metallic inorganic materials in general (mostly compounds of metallic and non-metallic elements) including glass, brick, cement and concrete, and many rocks and minerals. Ceramics generally rely on ionic–covalent bonding, which means that the valence electrons are localised. Such materials therefore tend to be poor conductors of heat and electricity. The bonding is relatively strong, in many cases stronger than metallic bonding, with the result that ceramics tend to be resistant to heat and chemicals. A major drawback is their brittleness.

23.1 BRITTLE BEHAVIOUR AND STRENGTH OF CERAMICS

Figure 21.6 (page 203) shows how slip processes enable metals to deform plastically. Ceramics are inclined to be brittle because slip is inhibited by their structure. Figure 21.7 (page 204) shows how relative movement between the planes in an ionic structure brings like-charged ions closer together; ultimately, this leads to strong repulsive forces which cause the planes to separate rather than slip over one another. Covalent structures resist slip because the covalent bond is rigid, directional and specific between the bonded atoms. Similar arguments apply to the intermediate types of bond discussed in Topic 16. Ceramic materials therefore tend to fail in a brittle fashion, with little or virtually no plastic deformation; this means, for example, that the pieces of a broken teacup can be fitted back together like a jigsaw puzzle.

Since slip is inhibited in ceramics, we might expect to be able to estimate their tensile strength from the net force/separation curve

(a)   (b)

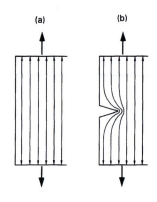

(c)

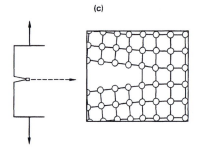

Figure 23.1

(Figure 21.1 on page 198). The peak force represents the force needed to break the chemical bond between two atoms so, allowing for the number of chemical bonds that need to be broken within a given cross-sectional area, we should be able to estimate the tensile strength of the material as a whole.

In practice, the actual measured strengths tend to be very much lower than the theoretical strengths, typically by factors of around a hundredth to a thousandth or even less. This is because real brittle materials contain defects which weaken them. For example, the tensile strength of a freshly drawn glass fibre can approach the theoretical value because the surface is extremely smooth; but handling glass introduces minute surface scratches which, although invisible to the naked eye, reduce its strength dramatically.

Figure 23.1 represents the surface layers of (a) a perfect material and (b) the same material with a small surface crack. (It is important to recognise that the size of the crack is negligible in comparison with the cross-sectional area of the material as a whole.) In both cases the material is subjected to tensile stress. The stress lines (the double-headed arrows drawn inside the material) represent the way in which the load is carried from one end to the other via the chemical bonds. Figure 23.1(b) indicates how the stress lines crowd together at the tip of the crack in order to carry the load around it; *stress concentration* therefore occurs in the material at the crack tip where it has to carry more than its fair share of the load. If the crack is long, then there are more stress lines to be diverted around its tip and the stress concentration is greater. In a very narrow crack the tip might be closed by only a small number of chemical bonds or, conceivably, even just one as shown in Figure 23.1(c). Taking this extreme case as an example, the bond at the crack tip carries a very much greater load than the average throughout the material. If the stress concentration is great enough, the bond will reach its load-bearing capacity and break. Once it has broken the burden will fall on the next bond. But the crack is now longer and the stress concentration greater. If the original overall load is maintained then this bond will break, then the next – and so on. This means that a crack in a strong material may *propagate* or grow at a relatively low general stress level in the material as a whole, owing to the local stress concentration that it produces.

It can be shown that the stress concentration factor at the crack tip (the factor by which the stress is increased there) is approximately $2(c/r)^{1/2}$ where r is the radius of curvature of the crack tip and c is the length of a surface crack or half the length of an internal crack (as in Figure 23.3 on page 224). The stress concentration is therefore greatest for relatively long, sharp cracks with small tip radii. For example, if $c = 1$ μm and $r = 1$ nm the estimated stress concentration would be

$$2\left[\frac{c}{r}\right]^{1/2} = 2\left[\frac{1 \times 10^{-6}}{1 \times 10^{-9}}\right]^{1/2} = 63$$

Such cracks are invisible to the naked eye, but this simple calculation shows that even defects of this size are capable of causing local stresses equal to the theoretical strength at relatively low overall stress levels in the material as a whole.

Next we need to consider the conditions under which a crack will propagate through a brittle material. In about 1920, A.A. Griffith developed his theory of crack propagation, which showed that the surface energy needed for a crack to propagate can be provided by the strain energy stored in the material.

Topic 20 shows how energy must be provided to increase the surface area of a liquid. Similarly, energy must be provided to create the new surfaces formed when a crack propagates through a solid material; in effect, this is the energy required to break the bonds originally holding the two surfaces of the crack together.

Topic 8 shows how strain energy is stored in a material under mechanical stress. Strain energy can be converted into other forms of energy, for example the strain energy stored in the stretched rubber of a catapult can be converted into the kinetic energy of a stone. As we shall now see, strain energy can be converted into surface energy; and this provides a mechanism whereby cracks propagate through materials under stress.

Figure 23.2(a) shows the edge of a thin, flat, sheet of material. Horizontal lines have been drawn on the sheet to enable us to visualise the strain in the material when a load is applied to it. Figure 23.2(b) shows how a tensile load causes the material to stretch; the horizontal lines move apart as the strain increases and a corresponding amount of strain energy is stored in the stressed material. If we keep the load constant and cut a notch at the edge of the sheet, then the notch will gape open; the horizontal lines above and below it will close up together as the material relaxes where the chemical bonds that have been cut are no longer able to transmit the load. (The strain is greatly exaggerated in the figure and, as before, the size of the notch is negligible in comparison with the cross-sectional area of the material as a whole.) Strain energy stored in the material will be released in the region behind the

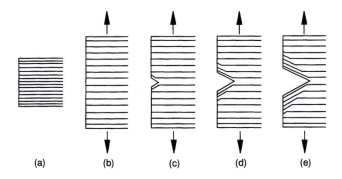

(a)   (b)   (c)   (d)   (e)

Figure 23.2

notch tip where the material has relaxed. Figures 23.2(c) to (e) show that, as the depth of the crack is increased, the area over which strain relaxation occurs will also increase. The area of strain relaxation (hence the amount of strain energy released) increases at a progressively greater rate as the crack deepens, rather like the ripples spreading out behind a boat crossing a smooth pond. In fact, the amount of strain energy released is proportional to c^2, where c is the length of the crack. This means that as the crack length doubles, then trebles, the strain energy released increases by a factor of 4, then 9 and so on.

We are now in a position to be able to examine the crack propagation process in terms of an energy balance. Firstly, the surface energy needed to create a crack is proportional to its length c. But, as we have just seen, the strain energy released by the crack is proportional to c^2. This means that, for a crack to grow, it must be provided with surface energy at a constant rate with respect to its length increase while, at the same time, it releases strain energy at an ever-increasing rate. There comes a point, at some critical length, where the strain energy released by a further increase in length is sufficient to pay for the surface energy needed for this to occur, and the crack will be able to propagate on its own.

So, for any given overall stress level in an ideal brittle material there will be a critical value of c at which the crack will propagate. The critical crack length will vary depending on the stress level in the material; a long crack will propagate at a relatively low stress whereas a short one will need a relatively high stress. These ideas are embodied in Griffith's equation, which gives the tensile stress σ, perpendicular to the crack, that is necessary to make it propagate, as follows:

$$\sigma = \left[\frac{2E\gamma}{\pi c}\right]^{1/2} \tag{23.1}$$

E is Young's modulus and γ is the energy required to create unit area of new surface (i.e. J m^{-2}) which, in a truly brittle solid, is the energy needed to break the chemical bonds formerly bridging the fractured surfaces. In practice, the value of γ is often much larger than this because it includes energy consumed in processes such as plastic deformation.

In effect, σ gives us the tensile strength of a material containing a surface crack of length c or an internal crack of length $2c$, as in Figure 23.3. The strength of a particular material (where E and γ have fixed values) is proportional to $1/c^{1/2}$, that is to say its strength depends on the size of the largest crack it contains. Calculations based on these ideas suggest that the size of cracks and scratches that are responsible for the normal strength of glass are of the order of a thousandth of a millimetre deep. (The general principle can be put to good use in cutting a sheet of glass to size; fracture of the glass can be controlled by scratching its surface with a glass cutter to make a large crack which then provides a line of weakness.)

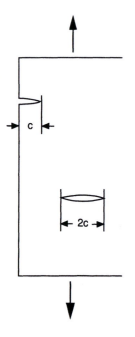

Figure 23.3

The strength-reducing effect of small cracks and other defects tends to make the tensile behaviour of brittle materials somewhat unpredictable. To avoid propagation, tensile stresses must be low enough to keep the critical length longer than any cracks actually present. Brittle materials are inclined to be stronger and more reliable in compression. For instance, the cracks in Figure 23.3 would close under compression, allowing the load to be carried across their faces. Of course, real materials contain defects orientated in all directions, nevertheless their compressive strengths still tend to be greater than their tensile strengths.

In practice, it can be difficult to measure the tensile strength of brittle materials directly because they tend to break in the grips holding the ends of the test piece. Bending tests, using beam-shaped test pieces, are often used to solve this problem. Figure 23.4 shows that, in bending, the bottom part of the beam is under tension where the loading forces make it stretch. If the load is progressively increased then fracture will eventually begin at the bottom when the tensile stress becomes too high. The upper part will be in compression but, since the compressive strength of this type of material is generally higher than its tensile strength, the bottom part will tend to fail first. True tensile strength values obtained in direct tension are generally lower than estimated values obtained indirectly from bending tests, partly because the whole test piece is subjected to the full tensile load in direct tension and the likelihood of a critical flaw being present is correspondingly higher.

It is important to recognise that brittle materials under stress are able to store considerable amounts of strain energy, some of which may be converted into kinetic energy of flying fragments after fracture occurs; appropriate safety precautions must always be taken, for example the use of safety guards during mechanical tests.

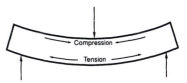

Figure 23.4

23.2 GLASS

Ordinary glass is an important ceramic material made from silica (SiO_2). As we saw in the previous topic, it has a non-crystalline, liquid-like structure which results from cooling silica too rapidly from the liquid state to allow it to recrystallise. The amorphous structure, and the essentially covalent nature of the Si–O bond, makes glass resistant to plastic deformation and therefore brittle. Figures 23.5(a) and (b) are two-dimensional models of crystalline silica and silica glass respectively. (For simplicity, each silicon atom is represented as being surrounded by three oxygen atoms, rather than by four as in the real three-dimensional structures.) To help commercial processing, the melt viscosity of pure silica may be reduced by incorporating modifiers such as sodium oxide (Na_2O) and calcium oxide (CaO) to partly break down the three-dimensional network. Figure 23.5(c) illustrates the result; some of the Si–O–Si links have been broken and the loose ends sealed by

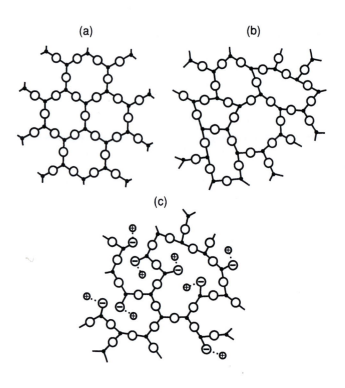

(a)

(b)

(c)

- Silicon
- O Oxygen
- ⊕ Metal ion

Figure 23.5

means of an ionic bond between the terminal oxygen atom and a metal ion. If sufficient breaks are made the liquid glass becomes more mobile and workable.

Glass is not normally used where it would be subjected to high stresses, and accidental surface scratches are therefore not usually a serious problem. But, in certain applications, sheet glass needs to have resistance to surface crack propagation. For some purposes this may be achieved by means of an industrial process called *thermal tempering*, which involves heating the glass sheet to a high enough temperature to allow some relative movement between the atoms in its structure. The surface is then rapidly cooled with air jets or in oil; as it cools it becomes rigid and contracts while the interior remains relatively mobile so that its structure is able to adjust to the contraction in the surface. As the interior begins to cool and contract, it pulls the surface layers into compression which now cannot be relieved because the surface material is no longer hot enough for its structure to readjust. Consequently, when the glass is cold the surface layers are permanently locked into compression. If a tensile load is applied then the built-in compressive stress must be overcome before a tensile stress can be developed in the surface. The applied load needed to cause the propagation of any surface cracks is therefore increased and, in effect, the glass is strengthened. At one time tempered glass was widely used in motor car windscreens but these had the disadvantage that, once the compressive skin had been penetrated, by a stone for example, they tended to fail catastrophically with the rapid propagation of cracks owing to the tensile stress inside.

Another way of improving the strength of glass is to draw it from the melt into fine fibres, which have much smoother surfaces than ordinary glass in bulk form. Commercial glass fibres can be very strong, with tensile strengths as high as 2000–4000 MPa compared with indirect values of the order of about 100 MPa for bulk glass (see Table 25.1 on page 250).

23.3 COMPLEX CERAMICS

Many of the ceramics used in engineering applications have complex structures containing several constituent phases, sometimes with a significant degree of porosity. The role of porosity in determining their properties is complicated, however it generally brings about a reduction in mechanical performance because, in effect, the pores reduce the effective load-bearing cross-sectional area of the material. In some cases they may also act as stress concentrators.

In this section we shall briefly consider two particularly important materials of this type, namely fired clay products and concrete.

Fired clay products include bricks, drainage pipes, tiles of various types and porcelain. The clay is often mixed with other minerals, and with the correct amount of water to give it the necessary plasticity, then moulded to the required shape and dried to remove some of the water. Finally it is fired at high temperature, normally somewhere in the range between about 800°C and 1450°C. The raw materials used, their relative proportions, and the precise firing conditions vary, depending on the finished product required.

The reactions that occur during firing are beyond the scope of this book, but the final products consist essentially of crystalline phases which are bound together in a matrix of glassy material. The glass is formed during the firing process and spreads around the surfaces of the crystalline particles, generally leaving a significant proportion of pores in between.

Because of the large number of manufacturing variables, and because of the complex structures of these materials, the interaction between the different factors governing their properties is complicated. For example, the compressive strength of bricks generally decreases with increasing porosity but is also influenced by the raw materials and by the firing conditions.

Concrete is a complex material consisting of particles of aggregate that are bound together in a matrix of hardened cement. Figure 23.6 shows a simplified model of the internal structure of ordinary concrete. The gaps between the large aggregate particles are filled by smaller aggregate particles, and the gaps between those are filled by still smaller particles – and so on down to fine sand. Aggregates are often classified as coarse or fine, depending on whether their particle size is greater or less than 5 mm. Typical aggregates used in ordinary concrete consist of naturally occurring gravel and sand, or crushed rock. Generally there are voids, for example air bubbles trapped during the mixing operation. Voids can reduce the strength considerably (as much as a 30% reduction by 5% voids in some cases) so freshly mixed concrete is generally *compacted* by vibrating it to remove excessive air.

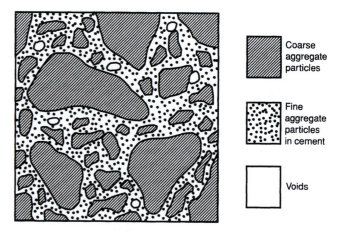

Figure 23.6

The strength of the hardened cement arises from complicated chemical hydration reactions between the cement particles and the mixing water. As these reactions proceed, their products progressively replace the water-filled spaces between the cement grains and, as they do so, they become interlocked to form a porous, rigid mass. Figure 23.7 represents the hydration process in a simplified form.

A factor of prime importance in producing good quality concrete is the proportion of water used when mixing it. If the mix is too dry then the concrete cannot be properly compacted and there will be an excessive volume of voids. On the other hand, the strength of a properly compacted concrete decreases with increasing *water/cement ratio* because a higher proportion of water produces a more porous, weaker cement matrix.

Ordinary concrete generally has a low tensile strength, typically about a tenth to a fifteenth of its compressive strength. One way round the problem of low tensile strength is to design structures in such a way that the concrete is kept in compression. Another is to use steel or

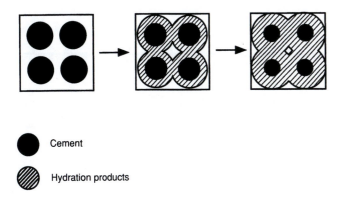

● Cement

▨ Hydration products

Figure 23.7

other suitable materials as reinforcement embedded in the concrete to carry tensile stresses, for example steel rods in the tension part of a beam. In prestressed concrete, the reinforcing rods are stressed in tension before the concrete is poured round them. When the concrete is hard the external tension is released and the rods pull the surrounding concrete into compression. The effect of this is analogous to the compressive stress in the surface of thermally tempered glass; the built-in compression must be overcome before a tensile stress can be developed.

Questions

(Use any previously tabulated data as required.)

1. Estimate (a) the theoretical tensile strength of a ceramic material and (b) the tensile strength of a test piece of the same material with a surface crack 1 μm in depth. (Assume a fracture surface energy of 5 J m^{-2} and a Young's modulus of 70 GPa and that, for a defect-free specimen, the effective crack length is the interatomic spacing which, in this case, can be assumed to be about 0.2 nm.)

2. A ceramic specimen with an internal flaw was found to have a tensile strength of 9.0 MPa, Young's modulus of 16.0 GPa and fracture surface energy of 4.9 J m^{-2}. Estimate the effective length of the flaw.

Also see the Further Questions on page 252.

TOPIC 24 THE NATURE OF METALS

COVERING:

- plastic deformation;
- yield strength and tensile strength;
- toughness;
- slip processes;
- dislocations;
- strengthening mechanisms.

We have already discussed some aspects of the nature of metals. In Topic 16 we saw that the valence electrons in the metallic bond are free to move randomly between the positive ions, thus providing an attractive force that holds the metal together. We noted that the ions do not all have to be of the same kind, which means that different elements can be combined to form alloys. In Topic 18 we saw that the freedom of movement of the valence electrons is responsible for the high thermal conductivity of metals, and in Topic 29 we shall see that it is responsible for their electrical conductivity as well. In Topic 22 we noted that the generally non-specific and non-directional nature of the metallic bond means that, as far as any particular metal atom is concerned, one neighbour is as good as any other. As a result of this, most metals have the ability to undergo plastic deformation before they break because individual metal atoms can change their neighbours without affecting the integrity of the metal as a whole (see Figure 21.6).

Topic 22 describes various metallic crystal structures in terms of three-dimensional models based on geometrical arrangements that can be formed by packing spheres together. To summarise, at normal temperatures most common engineering metals adopt either the face-centred cubic (FCC), the hexagonal close-packed (HCP) or the body-centred cubic (BCC) structure. (See Table 22.1 on page 212.) In fact, the majority of common metals are close-packed (i.e. FCC or HCP), which means that they have the maximum co-ordination number of 12 (i.e. each atom has 12 immediate neighbours) and an atomic packing factor of 74%. This is another consequence of the non-specific and non-directional nature of metallic bonding; the bond is able to pull the metal atoms into close-packed structures because there are neither the electrostatic restraints of an ionic structure nor the directional restraints of the covalent bond. The body-centred cubic metals are not so tightly packed; they have a co-ordination number of 8 and an atomic packing factor of 68%. In Topic 22 we noted that there are various possible reasons for this; in some cases the metallic bond has insufficient strength to withstand the effect of thermal vibration, in others it appears to possess some degree of covalent character.

As we shall see, these models are very useful in helping to understand the behaviour of metals under stress. But first, we need to appreciate the general effect of plastic deformation in metals.

24.1 PLASTIC DEFORMATION

Figure 24.1 compares brittle and ductile behaviour in terms of tensile load/extension properties.

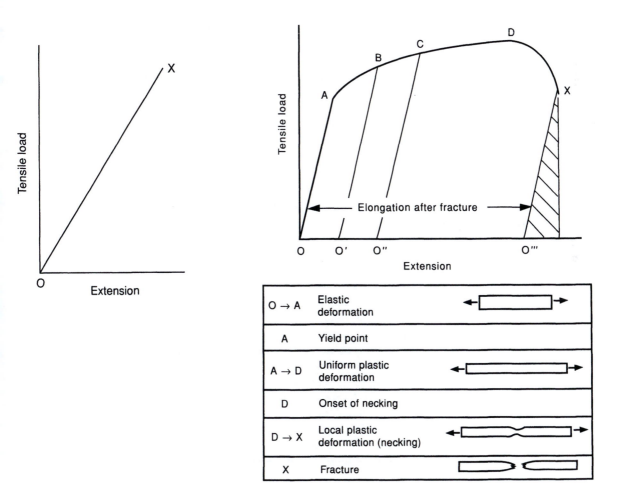

(b)

Figure 24.1

Figure 24.1(a) shows an 'ideal' brittle material which deforms elastically up to the point of fracture at X. The work done in stretching the material to any point between O and X is stored as elastic strain energy corresponding to the area under the load/extension graph up to that point (see Section 8.3, page 62). This energy can be recovered, like the energy stored in a stretched spring, by removing the load so that the material returns along the line back to the point O. If we load the material right up to the point X then it will break, allowing the individual broken pieces to recover elastically so that, in theory, they could be glued back together to reconstitute the form of the original specimen. In fact, many brittle materials undergo some plastic deformation before or during fracture, although this may be barely detectable.

Figure 24.1(b) illustrates the tensile behaviour of a typical ductile metal. After initial elastic deformation from O to A, it undergoes plastic deformation until it breaks at X. Provided that we do not stretch the specimen beyond A, we can return to O by removing the load, as with the brittle material, and provided that the material obeys Hooke's law, we can calculate the value of Young's modulus from the slope of the line (see Worked Example 2.1 on page 12). The point A corresponds to a yield point above which plastic deformation occurs in addition to the elastic deformation already experienced by the material. The *yield stress* or *yield strength* of the material can be calculated by dividing the load at the yield point by the original cross-sectional area of the test specimen. Many ductile metals do not have an obvious yield point because the transition from elastic to plastic behaviour is gradual. In such cases the *offset yield stress* or *proof stress* may be used; this is the stress required to produce a specified permanent strain, for example 0.1% or 0.2%.

If we stretch the metal beyond the point A, say to point B, it cannot return along the path BAO. Instead, it recovers elastically along BO', and the distance between O and O' represents plastic deformation which appears as a permanent increase in the length of the specimen. The figure shows that the plastic deformation is accompanied by an increase in yield stress and that the elastic behaviour of the material is now represented by the line O'B. (Note that the work done in plastic deformation is represented by the area enclosed by OABO'.)

Further plastic deformation occurs if we stretch the material beyond B, say to C. The total plastic deformation is now represented by the distance between O and O'' and the elastic behaviour by the line O''C. Again, there is an increase in the load needed to produce further plastic deformation. This progressive 'strengthening' by plastic deformation is called *work-hardening* or *strain-hardening*.

Figure 24.1(b) shows a peak at D which is a feature of many ductile metals. Between A and D, plastic deformation is uniform along the length of the specimen. However, at D the material becomes unstable at some point along its length and a neck begins to form. Further deformation is concentrated around this point and, as the neck develops, there is a large local reduction in cross-sectional area. The load needed to maintain further deformation decreases as the cross-sectional area of the neck decreases until fracture occurs at X. The broken ends of the material then recover elastically, as indicated by the line XO''' in the figure. The *elongation* of the specimen after fracture (the permanent extension measured by fitting the broken ends together) is a measure of the ductility of the material.

The area under the load/extension curve corresponds to the work done in fracture. The area to the left of O'''X corresponds to the work done in plastic deformation up to the point of fracture. The shaded area to the right corresponds to the stored elastic energy released when the test piece breaks.

The *tensile strength* of the metal is calculated by dividing the load at D by the original cross-sectional area of the test specimen. It is important to recognise that, depending on the material, the actual cross-sectional area of the specimen may be significantly reduced at point D. This means that the true stress at this point will actually be higher than the *nominal* stress (sometimes called the *conventional* or *engineering* stress) based on the original cross-sectional area. The peak at D suggests, quite wrongly, that the metal becomes weaker between D and X; the true stress at the neck, allowing for the progressive reduction in cross-sectional area as the neck develops, actually increases.

The toughness of many materials can be attributed to their capacity for plastic deformation. For example, the tough material in Figure 24.2(a) has a high degree of mobility in its internal structure and can adjust itself to the effect of the stress concentration by localised slip processes which blunt the tip of the crack. The brittle material in Figure 24.2(b) has a low degree of internal mobility, and cannot adjust itself, so the stress is concentrated in the bonds at the crack tip and the crack will tend to propagate. Furthermore, if the energy term γ in Equation 23.1 (page 224) is taken to include work done in plastic deformation during fracture, then it will generally have a high value for metals; for instance, mild steel has a value of the order of about 10^4 times greater than the value for glass. This makes mild steel a 'safer' engineering material than glass in the sense that it is very much less sensitive to damage; for example, if we scratch the surface of a mild steel sheet with a glass cutter, we would not normally expect it to break like glass.

We now need to consider the mechanism of plastic deformation in metals.

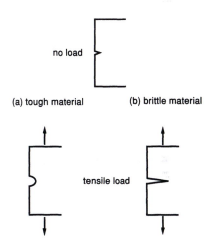

(a) tough material   (b) brittle material

no load

tensile load

Figure 24.2

24.2 SLIP PROCESSES IN METALS

Figure 21.6 (page 203) suggests that plastic deformation in a metal involves adjacent planes of atoms slipping over one another. We can see why this happens if we consider the two rows of atoms in Figure 24.3. In (a), tensile forces are tending to split the rows apart and, in (b), shearing forces are tending to make them slip over one another. It is fairly obvious that it is more difficult to tear the rows apart as in (a)

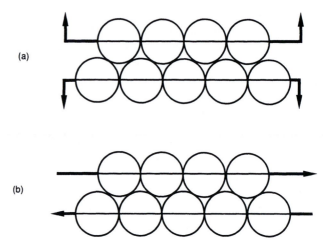

Figure 24.3

rather than just make them slip as in (b). In the first case the chemical bonds between them must be broken completely. In the second, the rows remain in contact with each other and only sufficient force is needed to pull each upper atom out of the trough between the two lower atoms and move it over into the next trough along. From this it is clear that our model is weaker in shear than in tension.

As Figure 21.6 suggests, slip occurs preferentially across particular sets of *slip planes*. Figure 24.4 is a two-dimensional model of a metal crystal which shows why this happens.

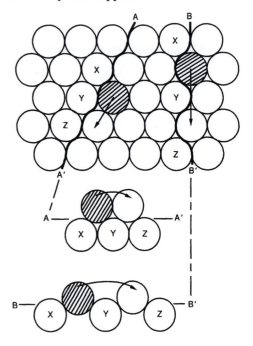

Figure 24.4

Assuming an applied force is orientated similarly in each case, let us consider whether slip would occur more readily between the rows of atoms along AA′ or those along BB′. In each case, we need to examine the movement of the shaded atom from its rest position between the atoms X and Y to a new position between atoms Y and Z. (Of course, the entire row containing the shaded atom will slip, but it is easier to compare the alternatives in terms of the movement of just one atom on its own.)

In the case of BB′, the shaded atom lies in a deeper trough than in AA′; furthermore, the distance between successive troughs is greater along BB′ than along AA′. For the shaded atom to move along BB′, more force is needed to pull it out of each trough and it must travel in longer steps. We would therefore expect slip to occur more easily along AA′ because the atoms are more closely packed in that direction.

We shall now apply this idea to three-dimensional crystals, assuming that we can decide where slip is likely to occur by identifying the most closely packed planes.

Figure 22.4 (page 208) shows how the face-centred cubic structure is built up from close-packed planes of spheres stacked in such a way that their relative positions are repeated every three layers. Figure 24.5 shows that the same face-centred cube can be constructed by stacking the close-packed planes perpendicularly to any one of the four diagonals running through the centre of the cube from corner to opposite corner. This means that there are four sets of close-packed planes across which slip can readily occur. Figure 24.6 shows that there are three *slip directions* in any close-packed plane; these correspond to the three sets of valleys, between the close-packed rows, along which atoms in the plane above can easily move. Each of the three slip directions in each of the four slip planes constitutes a *slip system*, therefore there are twelve such slip systems in the face-centred cubic structure. Whatever the orientation of a load applied to the crystal, there are always several slip systems available for plastic deformation. As a result of this, face-centred cubic crystals tend to be relatively ductile and soft.

Figure 22.3 (page 208) shows that the hexagonal close-packed structure is also constructed from close-packed planes but they are stacked in such a way that their relative positions alternate. In this case, there is only one way of stacking the planes to form the unit cell, and there are no close-packed planes in other directions. There is therefore only one set of close-packed slip planes, which all lie parallel to the base of the unit cell, so we would expect hexagonal close-packed metals to be less ductile than face-centred cubic metals. In fact, there are other slip planes but they are not close-packed and they only operate in certain cases.

The body-centred cubic structure has no close-packed planes. The densest packing occurs in the planes which pass through the diagonally opposite edges of the unit cell. Slip can still occur along these, and Figure 24.7(a) shows that there are two slip directions in each. Figure 24.7(b) shows that there are six such planes giving twelve slip systems

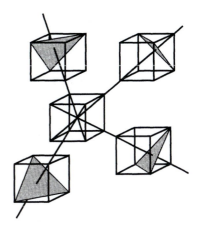

Figure 24.5

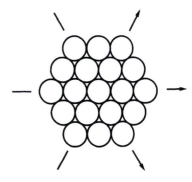

Figure 24.6

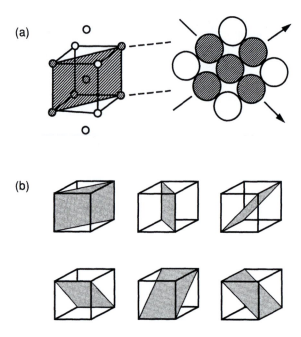

Figure 24.7

of this kind. In fact there are other less closely-packed slip planes, making a total of 48 in all. From Figure 24.4, we would expect that the lack of close-packing in body-centred cubic structures would make slip more difficult than in face-centred cubic structures. In practice, body-centred cubic metals tend to be harder and less ductile than face-centred cubic.

24.3 STRENGTH AND TOUGHNESS OF METALS

Now we shall extend our simple model of the slip mechanism to consider how strong metals should be. Figure 24.8 shows how the force and potential energy vary as we pull the shaded atom from its trough between atoms X and Y to the trough between Y and Z. The force/displacement curve shows that the force reaches a maximum as the shaded atom begins to move upwards out of the trough. (Prior to this we would expect some bond deformation to occur.) The side of the trough becomes progressively less steep towards the top and the force decreases until it becomes zero where the shaded atom is perched directly above Y. The force then becomes negative as the atom tends to run down the other side (since, in effect, it would have to be reversed to stop the atom running down). When the atom reaches the rest position at the bottom of the trough between Y and Z, the force again returns to zero. As long as the applied force remains large enough for the cycle to be repeated, then plastic deformation will continue. If we know the value

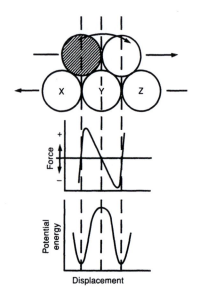

Figure 24.8

of the peak force on the force/displacement curve, then we can estimate the total force that a metal crystal will withstand before plastic
deformation occurs.

This model also tells us something about the toughness of metals.
The potential energy/displacement curve in Figure 24.8 shows two
positions of minimum potential energy, corresponding to the two troughs,
which are separated by an energy barrier which corresponds to the
atom Y. The work done in raising the shaded atom out of the trough
between X and Y is stored as potential energy. If the shaded atom is
allowed to run down the other side into the trough between Y and Z
then it will tend to oscillate about the rest position at the bottom of
the trough, rather like a pendulum. As we saw in Topic 17, oscillations
of this kind represent thermal energy on an atomic scale. Our model
therefore suggests that the work done in the plastic deformation of a
metal is dissipated as heat. That this is at least partly true is indicated
by the temperature rise that can often be detected when metals are
cold worked.

However, there is an important inconsistency. The estimated strength
of a metal crystal, based on this model, turns out to be very much
greater than the actual strength of ordinary metals as measured in the
laboratory. Clearly there is some weakening mechanism in real crystals
that we have not taken into account. This weakness is due to imperfections
in the crystal lattice called *dislocations*.

24.4 DISLOCATIONS

Figure 24.9 shows an end-on view of one type of dislocation, called
an *edge dislocation*, which is an incomplete plane of atoms (rather
like a strip of paper used as a bookmark between the pages of a book).
The figure shows that the crystal lattice is distorted around the tip of
the dislocation where the atoms around it are forced out of their stable
positions. To help understand how dislocations work, let us consider
the analogy with a wrinkle in a carpet. If we try to slide a carpet
across the floor by pulling on one edge then there will be considerable
resistance due to friction. On the other hand, if we make a wrinkle at
one edge and push the wrinkle across the carpet then we still move
the carpet, but much more easily because we do it a little at a time.

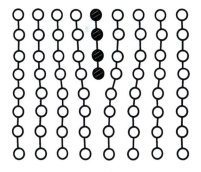

Figure 24.9

Figure 24.10 shows how the edge dislocation moves. In (a) a shearing
force applied to the crystal tends to move the dislocation to the right.
In (b) the plane of atoms in the path of the dislocation breaks where it
is already under considerable strain. In (c) the lower half of the freshly
broken plane will form a new complete plane with the original incomplete
plane; the top half now becomes the new incomplete plane and the
dislocation has moved along one step. If the shearing force is maintained
then the process is successively repeated and the dislocation runs through
the crystal like the wrinkle across the carpet. The atoms in front of the
dislocation will resist its movement because it tends to push them out

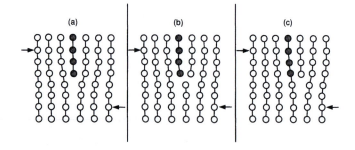

Figure 24.10

of their stable positions on the crystal lattice. However, the atoms behind it tend to drop into new stable positions, thereby pushing the dislocation forwards. In the absence of an external force these effects oppose one another and the dislocation remains stationary; but the balance is easily upset and only a small shearing force is needed to make the dislocation move.

The science of dislocations is a complicated subject, and a detailed treatment of it is beyond the scope of our discussion. But it is important to recognise that metals can be strengthened in various ways by reducing the mobility of the dislocations that tend to weaken them. We shall now briefly consider a few ways in which this can be achieved.

24.5 STRENGTHENING MECHANISMS

Dislocations can actually multiply during plastic deformation, and they can interact with one another. If a metal is deformed sufficiently, dislocations moving along intersecting slip planes may become entangled and interlocked. As a result of this, further deformation is restricted and the metal is hardened and strengthened. This leads to work- or strain-hardening, which we discussed in Section 24.1.

Metals used in engineering are generally *polycrystalline*, that is to say they consist of many small crystals, usually called *grains*, which interlock with one another like a three-dimensional jigsaw puzzle. As Figure 24.11 suggests, the crystal lattices of individual grains are orientated in different directions. Where one grain meets another there is a narrow region, called the *grain boundary*, where the transition occurs from one grain orientation to another. Because of the discontinuity in the regular crystal structure, a grain boundary effectively acts as a barrier where dislocations tend to pile up, thereby restricting further deformation. If the average grain size is small throughout a piece of metal there will be a relatively high proportion of grain boundaries, and the metal will tend to resist continued deformation to a greater extent than if the grain size were larger. The strength and ductility of a polycrystalline metal can therefore be controlled by adjustment of the grain size.

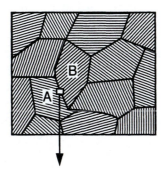

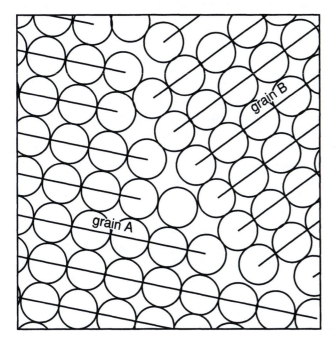

Figure 24.11

In practice, metals are generally used in the form of alloys in which they are combined with other elements to form solid solutions. By substituting atoms of different sizes in a crystalline arrangement of uniform atoms, the original crystal lattice becomes distorted. This distortion provides a resistance to slip in a way that can be regarded rather like friction. If a slip plane includes a number of substituted atoms of different sizes then these will create bumps and hollows which will make it more difficult for an adjacent plane, with its own bumps and hollows, to slip over it. Looking at this in terms of dislocation movement, the misfit of the substituted atoms leads to localised regions of strain in the crystal lattice; when a dislocation approaches, the strain around its tip will interact with the strained regions around the substituted atoms, making its progress more difficult.

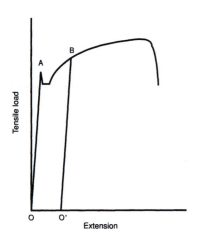

Figure 24.12

To take an example, copper and nickel are face-centred cubic metals, and both copper and nickel atoms are able to occupy positions on the same face-centred cubic lattice in any proportion. The result is a series of alloys with a continuous range of possible compositions between 100% copper and 100% nickel. However, nickel atoms are smaller than copper atoms so, when they form an alloy together, the basic face-centred cubic lattice is distorted due to their different sizes. This distortion makes it more difficult for dislocations to move through the structure, and there is a corresponding increase in tensile strength and hardness.

Brasses are essentially copper–zinc alloys. However zinc is normally a hexagonal close-packed metal; furthermore zinc atoms are larger than copper atoms. Zinc atoms are able to replace up to about 35% of the atoms in a face-centred cubic lattice of copper, and the distortion they produce leads to an increase in tensile strength and hardness.

Steels are alloys, predominantly iron with small quantities of carbon and usually other elements too. Mild steels generally contain about 0.1–0.3% carbon. Some mild steels give load/extension curves with an *upper yield point* such as that at A in Figure 24.12. This occurs as a result of carbon atoms occupying spaces between the iron atoms in the regions around dislocations. The carbon atoms have the effect of 'pinning' the dislocations. Loading the steel beyond A enables the dislocations to 'escape', allowing plastic deformation to proceed at a lower load which then rises as work-hardening occurs. Unloading at B results in elastic recovery to O', and the material then behaves elastically along O'B. If the material is immediately loaded beyond B then plastic deformation will continue as shown by the figure. However, if sufficient time is allowed, the carbon atoms migrate through the structure and re-establish themselves around the dislocations and an upper yield point reappears at B.

Finally, a very effective method of strengthening metals is to form an internal dispersion or precipitate of fine particles of a relatively hard material which act as barriers to dislocation movement. For example, aluminium can be hardened by a dispersion of alumina particles (Al_2O_3), and certain aluminium–copper alloys by precipitated particles of $CuAl_2$.

Questions

See Further Questions on page 252.

TOPIC 25 THE NATURE OF POLYMERS

COVERING:

- bond rotation and rubber-like elasticity;
- glass transition temperature;
- time-dependent effects;
- chain size and structure;
- crystallinity;
- effects due to side groups;
- cross-linking;
- polymer-based composites.

In Topic 16 we saw that the specific and directional nature of the covalent bond leads to the formation of compounds consisting of individual molecules of particular shapes and sizes. We also saw that a carbon atom, with its valency of 4, can form four covalent bonds that tend to be distributed in a tetrahedral configuration – as, for example, in methane. We noted that ethane is the first member of a series of chain-like hydrocarbon molecules. Figure 25.1 suggests that, by successively inserting —CH_2— groups into a chain, we could make it as long as we like.

Polymers are compounds consisting of large molecules formed from small repeating units called *monomers*. The process of forming polymers is called *polymerisation*, for example *polyethene* (sometimes called *polyethylene* or, more commonly, *polythene*) is made by polymerising ethene (ethylene). Figure 25.2 shows that, in effect, the polymerisation of ethene molecules involves opening their double bond so that the resulting —CH_2—CH_2— units can join end-to-end to form polymer chains. We shall not be discussing the process of polymerisation itself, other than to note that an enormous variety of polymers can be made by using different monomers and polymerising them in different ways.

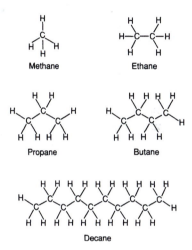

Figure 25.1

Ethene   Polymerisation   Polyethene

Figure 25.2

241

Ordinary rubber and many plastics consist of chain-like covalent molecules which are generally hundreds or even thousands of carbon atoms long. As we shall see later, some polymers are partially crystalline. Some have side chains branching out of the main backbone chain, like branches out of a tree. Most have atoms, or groups of atoms, other than just hydrogen attached to the backbone carbon atoms, for example there are chlorine atoms in PVC and $-CH_3$ groups in polypropene. Many have atoms other than just carbon in the backbone chain, for example nitrogen atoms in nylon; some have no carbon atoms in the backbone, for example *silicones* are polymers based on chains of alternating silicon and oxygen atoms. There are always van der Waals' forces tending to hold neighbouring polymer chains together but, by building polar groups onto the backbone, the interchain forces can be increased. Furthermore, covalent cross-links can be used to join adjacent chains together permanently.

The general behaviour of polymers differs markedly from that of ceramics and metals. In particular, rubber stretches very much more easily because its elasticity depends on a mechanism that is fundamentally different from the bond deformation processes that occur in ceramics and metals.

25.1 RUBBER-LIKE ELASTICITY

In Topic 16 we noted that one end of the ethane molecule can rotate relative to the other about the C—C bond, rather like a propeller. Figure 25.3(a) shows that there are potential energy barriers that tend to oppose this rotation. These are due to the slight repulsion that arises when the three hydrogen atoms at one end of the bond are exactly aligned with the three at the other so that they are at their closest together. The potential energy peaks in Figure 25.3(a) correspond to these 'eclipsed' conformations; the troughs correspond to the 'staggered' conformations where the hydrogen atoms at either end are out of alignment and are furthest apart. The figure shows that, in rotating one end of the molecule through a full circle, there are three energy barriers separated by low-energy troughs. In ethane, the energy needed to overcome these barriers is very low and, at normal temperatures, the molecule possesses sufficient thermal energy for continuous bond rotation to occur spontaneously.

We can extend this idea to explain rubber-like elasticity by considering the short segment of carbon chain shown in Figure 25.3(b). (The hydrogen atoms have been omitted for simplicity.) The carbon atoms in the chain have a zigzag arrangement because of the tetrahedral bond angles. Bond rotation about the C—C bond therefore causes the chain to change direction, as indicated by the broken lines in the figure. Continuous bond rotation occurs along the entire length of the chain, which gives it a constant wriggling motion that becomes more vigorous if the temperature is raised, and less vigorous if it is lowered.

(a)

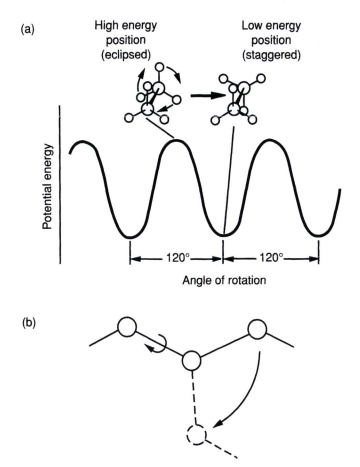

Figure 25.3

Figure 25.4 represents one possible instantaneous conformation of a single isolated chain. In fact there is an enormous number of possible conformations that it could adopt at any one moment. On statistical grounds, it can be shown that most of these tend to be fairly compact, as in Figure 25.4. However, more elongated conformations are possible but they become progressively fewer as the chain is straightened, and we reach the limit when the chain is fully extended in a straight line. The wriggling motion of the chain therefore tends to be confined to the relatively compact configurations most of the time because they are more numerous and therefore statistically more probable.

If we pull on each end of the chain we would expect it to straighten by bond rotation along its length; and if we then release the ends, we would expect it to wriggle back into more compact and statistically more probable conformations. This is the basis of rubber-like elasticity; extending rubber chains in this way, by bond rotation, requires much less force and gives much larger extensions than deforming chemical

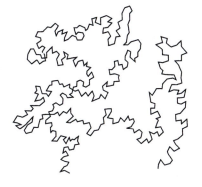

Figure 25.4

bonds by stretching them as in Figure 21.1 on page 198. In Topic 23 we saw that there is a hazard associated with the elastic strain energy stored in a material when it is converted into kinetic energy of the broken pieces after fracture. Although we discussed this in the context of brittle behaviour, similar hazards exist with other materials (for example, ropes and steel cables under stress). Even rubber is capable of storing large amounts of elastic strain energy; although the loads involved may be relatively low, the high extensions mean that the area under the load/extension curve may be considerable.

It is important to recognise that the single rubber chain in Figure 25.4 is actually entangled amongst neighbouring chains that tend to restrict its movement. In practice, the deformation of rubber depends on co-operative rotational movements of relatively small adjacent segments of neighbouring chains throughout the material. Nevertheless, the basic argument is still valid. Bond deformation processes may come into play if chain segments are extended so far that they have limited capacity for further extension by bond rotation processes.

25.2 THE GLASS TRANSITION TEMPERATURE

We now need to consider why many plastics are harder and more rigid than ordinary rubber even though they are based on similar long chain molecules.

If rubber is cooled to a sufficiently low temperature, bond rotation ceases because the energy barriers can no longer be overcome and the chain segments are frozen into fixed conformations; the polymer then becomes hard and rigid. Elasticity now depends on bond deformation (rather than bond rotation) in much the same way as we would expect in inorganic glass. Polymers in this state are often described as *glassy*,

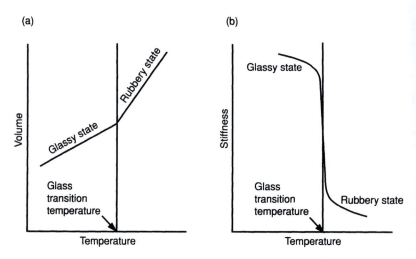

Figure 25.5

and the temperature at which the transition between rubbery and glassy behaviour occurs is called the *glass transition temperature*.

Figure 25.5(a) shows how the volume of the rubber changes as we raise the temperature from the glassy state through the glass transition temperature. In the glassy state, thermal expansion occurs as the result of increased oscillation of atoms about fixed positions (as we saw in Topic 17). When we reach the glass transition temperature, the thermal energy is sufficiently high for bond rotation to begin and for chain segments to start moving. However, the wholesale movement of a chain segment requires plenty of room, rather like a skipping rope; as the temperature is raised further, the extra volume swept out by the increasingly vigorous motion of the segments increases more rapidly, as indicated by the greater slope of the line in the rubbery region. Figure 25.5(b) shows how the stiffness of the polymer undergoes a sharp decrease corresponding to the change from the glassy to the rubbery state.

Figure 25.6 compares typical glass transition temperatures of some important polymers in everyday use. (In practice the glass transition temperature of different samples of the same type of polymer may vary considerably depending on a variety of factors.) The usefulness of a particular polymer for a specific application generally depends on whether its glass transition temperature is below or above the temperature at which it is to be used; at room temperature, for example, ordinary rubber is rubbery whereas polystyrene is hard and rigid.

25.3 TIME-DEPENDENT BEHAVIOUR

Many polymer properties are dependent on time as well as on temperature. This is illustrated by bouncing putty, which is a silicone polymer that can be moulded into shape like ordinary putty, but also bounces like rubber; furthermore, left to stand, it slowly flows under its own weight like a very viscous liquid. The reason for this strange behaviour is that, under slow loading conditions, its long chain molecules have time to disentangle themselves and to slip past one another so that the material deforms plastically. (Note that the chains in ordinary rubber are normally 'cross-linked' at intervals along their length to prevent this.) Under fast loading conditions the chains do not have enough time to slip, and the material shows rubber-like elasticity; furthermore, if it is rapidly pulled apart then it tends to fracture like a brittle material.

Although the behaviour of this silicone putty is extreme, most polymers exhibit time-dependent behaviour to a greater or lesser extent. In general, the faster they are deformed the less time they have for molecular readjustment, and this often results in increased failure load and reduced ductility. If the deformation is extremely rapid, under impact conditions for example, then there may be virtually no time for molecular readjustment. This means that, at a given temperature, some polymers that are normally tough and resilient may fracture in a brittle manner under high speed impact conditions. At the other end of the

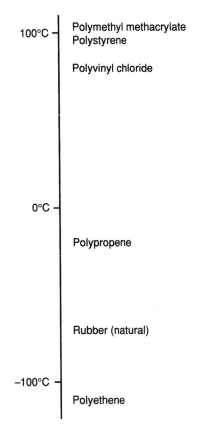

Figure 25.6

time scale, some polymers that are capable of supporting a given load in the short term may undergo time-dependent plastic deformation, called *creep*, if subjected to the same load conditions over a long period.

25.4 CHAIN SIZE

Chain size influences many polymer properties. (We normally think in terms of average size since polymers are generally mixtures of different sized chains.) For illustration, let us consider a linear (i.e. unbranched) polymer consisting of very long chains. If the material is above its glass transition temperature we would expect it to show rubber-like elasticity; even if the chains are not cross-linked there should be a sufficient number of points of entanglement between them to act, in effect, as temporary cross-links provided that the material is not deformed for too long. If the chains are shorter then they can slip past one another more readily and the polymer will tend to behave more like a liquid; in fact there may be an intermediate stage where they are short enough to allow the polymer to flow like a very viscous liquid but still long enough for some entanglement to cause rather weak rubbery retraction if it is stretched and quickly released.

Figure 25.7 shows how we might expect the glass transition temperature to vary with the chain length. We can see that, at any one temperature in the rubbery region, short chains occupy a greater volume than long chains. This is because the motion of a 'terminal segment' (at the end of a chain) is less restricted than the motion of a segment in the middle; a terminal segment is only attached to the rest of the chain by one of its ends whereas a segment in the middle is attached by both ends and has less freedom of movement. The wriggling motion of terminal segments is therefore less restricted, so they sweep out more space. A given mass of short chains will obviously contain a greater number of terminal segments than the same mass of long chains; this means that, in the rubbery region, the short chain material will occupy a greater volume. In the glassy state there is no difference between the volume of long chains and short chains because the movement of segments is frozen. Figure 25.7 shows that the rubber line for the short chains intersects the glass line at a lower temperature than the rubber line for the long chains. The glass transition temperature is therefore lower for the short chain material.

25.5 CRYSTALLINITY

So far we have considered the structure of polymers in terms of a mass of chains, in random conformations, haphazardly entangled amongst each other rather like spaghetti. However many polymers are semi-crystalline. Figure 25.8 shows, in a simplified way, how neighbouring chains can pack parallel to one another under the influence of inter-

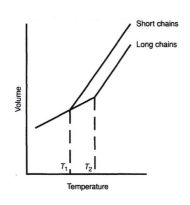

Glass transition temperatures: T_1 for short chains
T_2 for long chains

Figure 25.7

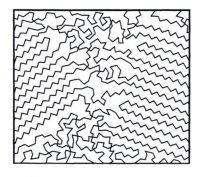

Figure 25.8

molecular forces. Crystalline order exists within these regions and we can regard polymers of this kind as consisting of crystalline regions dispersed in a matrix of amorphous material.

Because of its simple chain structure, polyethene is able to crystallise very easily. As Figure 25.6 shows, polyethene is well above its glass transition temperature under normal conditions and we might therefore expect it to be a rubber-like material. However, the crystalline regions do not melt until they are above room temperature so they reinforce the polymer structure, restricting its mobility as a whole and giving it extra rigidity. Polypropene behaves in a similar way.

The extent to which polyethene crystallises is affected by side chains branching out of the main backbone chain; these interfere with the ability of the chains to pack together, therefore highly branched polyethene tends to be less crystalline than the more linear material. The crystalline regions tend to be denser than the amorphous matrix, therefore an increased degree of crystallinity leads to greater density: so-called *high density polyethene* (HDPE) and *low density polyethene* (LDPE) have typical densities around 960–970 and 920–930 kg m^{-3} respectively.

Figure 25.9 compares the tensile load/extension curves for samples of high and low density polyethene, which illustrates the considerable property differences that may be observed.

The initial part of the curve for the high density material shows the stiffening and strengthening effect due to the higher proportion of crystalline material. The load rises to a yield point where a neck begins to form. Once formed, the neck propagates along the length of the specimen, as can be demonstrated by stretching the polyethene rings used to hold packs of beer cans together. (This is in contrast with the behaviour of ductile metals discussed in Topic 24, where the neck deepens rather than lengthens.) This process, called *cold drawing*, occurs because the linear chains slip past one another and become aligned parallel to the loading direction. The cold drawn material is generally stiffer and stronger than the original because chain alignment limits its capacity for bond rotation; this means that a higher proportion of the load is carried by bond deformation. Eventually fracture occurs, often accompanied by an increase in load, when the neck has travelled the length of the specimen.

The low density polyethene in Figure 25.9 is weaker and less stiff than the high density material because of its lower crystallinity. Furthermore the specimen does not exhibit cold drawing, and it breaks at a much lower extension, because the presence of the chain branches inhibits the parallel alignment of neighbouring chains and their ability to slip past one another.

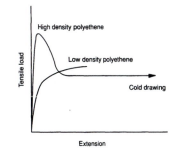

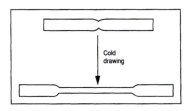

Figure 25.9

25.6 SIDE GROUPS

Figure 25.10 shows a series of important polymers where, in effect, hydrogen atoms on alternating carbon atoms in the polyethene structure

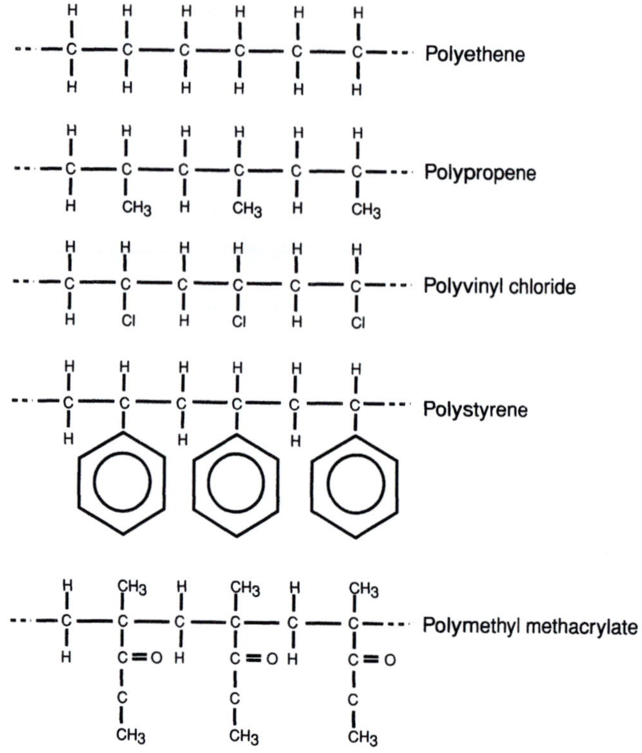

Polyethene

Polypropene

Polyvinyl chloride

Polystyrene

Polymethyl methacrylate

Figure 25.10

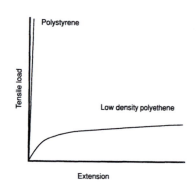

Figure 25.11

have been replaced by a variety of other atoms or groups of atoms. In polypropene each alternate carbon carries a —CH_3 group and in polyvinyl chloride (PVC) it carries a chlorine atom. In polystyrene it carries a benzene ring, which is a bulky group based on a six-membered ring of carbon atoms represented by the hexagon in the figure. Finally, in the case of polymethyl methacrylate, it carries a —CH_3 group and a —$COOCH_3$ group.

Figure 25.6 shows that the glass transition temperature increases along this series of polymers. This is because the bulkier and more complex side groups provide greater barriers to bond rotation and increase mutual restraint between neighbouring chains. There is also a corresponding overall increase in strength and stiffness and a tendency towards brittleness because of the increasing restriction on chain movement. This is illustrated by Figure 25.11, which compares tensile load/extension curves for polystyrene and low density polyethene. (The low density polyethene curve looks different in Figure 25.9 because it is plotted on a different scale.)

Polymethyl methacrylate and polystyrene both have relatively high glass transition temperatures because of their bulky side groups; but Figure 25.6 shows that polyvinyl chloride has a similar glass transition temperature, even though the single chlorine atoms might be expected to have a much smaller effect on chain mobility. To understand this we need to appreciate that there is polarisation of the covalent bonds across the chain where the chlorine atom has a negative bias and the hydrogen on the same carbon atom has a positive bias (see Section 16.3); this leads to relatively strong attractive forces between adjacent chains, giving polyvinyl chloride a higher glass transition temperature, and making it a stronger and stiffer material than we would expect from considering van der Waals' forces alone (see Section 16.5).

PVC can be made soft and flexible by the addition of *plasticisers*. In practice these are often relatively small polar molecules which tend to attach themselves to the polar parts of polymer chains, in effect neutralising them and reducing the interchain forces. Plasticisers may also act as lubricants, separating the polymer chains and reducing the attractive forces between them.

25.7 CROSS-LINKING

In the previous two sections we saw that chain entanglement, and the presence of intermolecular forces and crystalline regions have the effect of linking adjacent polymer chains together. In this section we shall consider permanent cross-links formed by covalent bonding between the chains.

Rubber is *vulcanised* using sulphur to form cross-links at intervals along the chains to stop them slipping past one another. Varying the amount of sulphur, hence varying the number of cross-links, gives a wide range of behaviour; for example, rubber bands are soft and flexible because they are lightly cross-linked whereas 'ebonite' is a hard, rigid material because it is highly cross-linked and chain mobility is greatly restricted.

Thermosettting plastics are polymers with cross-linked structures. Examples include phenolic, polyester and epoxy resins which are used as matrix materials in composites with applications as diverse as kitchen work surfaces, boat hulls and racing car bodies. (We shall discuss composites in the next section.) As their name implies, thermosetting plastics 'set' or harden when heat is applied to them; the elevated temperature brings about the cross-linking reactions (although in some cases room temperature may be enough to do this). Once they are cross-linked, thermosetting plastics cannot be remoulded. It is important to distinguish between them and *thermoplastics*, such as polyethene and polystyrene, which we discussed in earlier sections. Thermoplastics are not covalently cross-linked; as their name implies, they become plastic and mouldable at elevated temperatures (because the interchain attractive forces are opposed by the effect of heat) – they then reharden on cooling. The cross-linked structure of thermosetting resins generally tends to give them relatively high strength, stiffness and temperature resistance compared with many thermoplastics, but they tend to be brittle.

25.8 POLYMER-BASED COMPOSITES

The mechanical performance of most polymers tends to be rather poor compared with many other materials. For example, nylon has an approximate Young's modulus value of 3 GPa compared with 70 GPa for glass and 200 GPa for steel (see Table 21.1 on page 199). Table 25.1 gives a rough comparison between some typical tensile strength values.

Table 25.1

Material	Typical Tensile Strength/MPa
Concrete	< 5
Polymers	10–100
Glass (bulk)	100*
Mild steel	400
Glass fibres	2000–4000

* Indirect value.

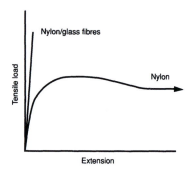

Figure 25.12

The mechanical properties of polymers can be improved by combining them with other materials to form *composites*. (Composites are materials containing more than one distinct constituent, for instance concrete is a composite formed from aggregate particles bound together in a matrix of hardened cement.) Sometimes a penalty must be paid for the improvement. For example, Figure 25.12 shows the general stiffening and strengthening effect of reinforcing nylon with glass fibres; however there is a dramatic reduction in elongation and a corresponding reduction in the area under the load/extension curve. To take another example, Figure 25.13 shows how polystyrene may be toughened by incorporating into it a fine dispersion of rubber particles; these interfere with crack propagation, and the area under the load/extension curve is greatly increased, but the strength and stiffness are reduced.

Fibre reinforced thermosets, such as glass reinforced polyester resin, form one of the most important groups of polymer-based composites. Glass has useful stiffness but it is very brittle and susceptible to the weakening effect of surface defects; freshly drawn glass fibres are very smooth and have extremely high tensile strength values compared with bulk glass. Thermosetting resins used to bind glass fibres together tend to be rather brittle themselves, and are generally much weaker and less stiff than the fibres. Nevertheless, composites of this kind have useful strength and stiffness, and can have comparable toughness with metals.

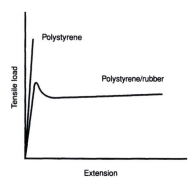

Figure 25.13

Figure 25.14 represents a fibre reinforced composite under load. For simplicity the fibres are shown parallel to one another, and the stiffness and strength would be concentrated in that direction; in practice, the alignment of the fibres is usually varied to give the composite more uniform properties in different directions.

Figure 25.14

The resin acts as a glue, bonding the fibres together and protecting them from surface damage which would weaken them. The resin transmits stress between the fibres so that they act collectively in resisting the applied load. Under the loading conditions in the figure the strain in the resin would be about the same, on average, as the strain in the fibres. But, because the glass fibres are much stiffer than the resin, they carry most of the load; glass, with its higher Young's modulus, experiences a higher stress than the resin when they are strained by the same amount.

Figure 25.15 helps us to understand the toughness of this type of material in terms of a crack growing through the resin matrix. Provided that the adhesion between the resin and the glass is not too great, then fibres can act as 'crack stoppers'. The figure shows that, as the crack approaches a fibre, the adhesion at the interface breaks down and the resin splits away from the glass; in effect, the crack is blunted and further propagation is inhibited. Even if the crack is able to find its way past the fibre, as in Figure 25.16, further progress is made difficult because the fibre forms a bridge that tends to hold the opposite faces of the crack together. For the crack to grow any further, the faces must be moved apart and, for this to happen, the fibre must be pulled out of the resin against the resistance due to adhesion at the resin/glass interface. So although the resin and the glass themselves have very little resistance to crack propagation on their own, the fibres throughout the resin matrix provide barriers to crack propagation that toughen the composite as a whole.

Wood is a natural polymer-based composite based on cellulose bonded with lignin. The structure of wood is very much more complex than the model composite we have just discussed above, but it can be viewed in terms of tubular cells packed parallel to one another to give an arrangement rather like a bundle of drinking straws. The parallel arrangement of these cells can lead to considerable tensile strength in the direction along which they are aligned (i.e. along the grain); furthermore, the fibrous nature of wood gives it toughness.

(a)

(b)

(c)

Figure 25.15

Figure 25.16

Questions

See Further Questions on page 252.

FURTHER QUESTIONS

The following questions are for the benefit of readers who are particularly interested in the descriptive aspects of materials science covered in Topics 14, 16 and 21–25. Maximum benefit will be achieved by using them as the basis for tutorial group discussions, but they are also a useful aid for individual study.

1. Discuss the role of electrons, protons and neutrons in determining the structure and properties of atoms.

2. Discuss the significance of ionisation energy and the electronic configuration of the inert gases in describing the structure and properties of atoms.

3. With reference to the nature and origin of the component attractive and repulsive forces, explain how the net force/separation curve arises for (a) the ionic bond, (b) the covalent bond and (c) the metallic bond. Give an example in each case.

4. Explain how each of the following arises: (a) intermediate (ionic–covalent) bonding; (b) hydrogen bonding; (c) van der Waals' forces. Give an example in each case.

5. Show, with the aid of diagrams, the general way in which the forces between chemically bonded atoms vary with separation/displacement and describe how these relationships might be expected to be reflected ultimately in the mechanical properties of (a) an 'ideal' brittle material, and (b) an 'ideal' metal.

6. Identify the elements in the table on page 253 and deduce the type of chemical bond between the atoms in each pair, hence discuss the general structure and properties of the resulting materials.

7. State the types of chemical bond present in the following substances and, in each case, briefly explain how the nature of the bonding determines their general physical characteristics at room temperature: (a) sodium chloride; (b) diamond; (c) glass; (d) copper; (e) helium; (f) polyvinyl chloride.

8. (a) With reference to their electronic structure, explain how hydrogen bonding arises between water molecules.
(b) Account for the progressively lower temperatures at which magnesium oxide, sodium chloride, water and argon liquefy.

	Atomic number	
	Element A	Element B
Pair 1	8	1
Pair 2	8	12
Pair 3	8	8
Pair 4	6	17
Pair 5	55	17
Pair 6	28	29
Pair 7	54	54

9. Briefly explain the following in terms of the relevant scientific principles:
   (a) a charged plastic rod will attract a thin stream of water;
   (b) a charged plastic rod will attract a small piece of aluminium foil;
   (c) work must be done to stretch a film of liquid;
   (d) the fundamental basis of Hooke's law;
   (e) the thermal expansion of crystalline solids;
   (f) the thermal expansion of rubber;
   (g) ice floats on water;
   (h) the boiling point of water is much higher than that of methane;
   (i) the electrical conductivity of metals;
   (j) glass is 'strengthened' by thermal tempering;
   (k) the formation of metallic alloys;
   (l) crystallinity in polymers;
   (m) the brittleness of polystyrene;
   (n) the mechanical behaviour of silicone putty;
   (o) the apparently high glass transition temperature of PVC;
   (p) the effect of polymer chain length on glass transition temperature;
   (q) toughening polystyrene by rubber addition;
   (r) the effect of grain size on the strength of metals;
   (s) reinforcement of nylon with glass fibres;
   (t) the upper yield point of mild steel;
   (u) the electrical conductivity of graphite;
   (v) the tensile strength of glass fibres;
   (w) the effect of side groups on the properties of polymers;
   (x) the principle of the glass cutter;
   (y) work-hardening;
   (z) stress concentration.

10. With special reference to chemical bonding and/or internal structure, explain the differences between:
    (a) the plastic deformation of aluminium and sodium chloride;
    (b) the hardness of diamond and graphite;
    (c) the density of ice and water;
    (d) the tensile properties of high and low density polyethene;
    (e) the toughness of ceramics and metals;

(f) the electrical conductivity of copper and polymethyl methacrylate;

(g) the ductility of body-centred cubic and face-centred cubic metals;

(h) the heat resistance of thermosets and thermoplastics.

11. (a) Describe the mechanism of rubber elasticity, carefully explaining how it leads to low stiffness and high extension.
(b) Explain why rubber becomes brittle below its glass transition temperature.

12. Outline the main structural and property characteristics that distinguish ceramics from metals.

13. (a) With the aid of sketches, explain the similarities and differences between the hexagonal close-packed and the face-centred cubic crystal structures.
(b) Explain the significance of the radius ratio in determining the crystal structure that a particular cation–anion pair will adopt.

14. (a) Discuss the factors which reduce the tensile strength of (i) ordinary sheet glass, and (ii) concrete.
(b) Explain the basis of the technique, applicable in principle to both sheet glass and to concrete beams, which may be employed to offset their weakness in tension and show how it can be applied in both cases.

15. The tensile strength and Young's modulus of thermosetting resins tend to be relatively low, and such materials tend to be brittle. Explain how the combination of thermosetting resin and glass fibres, which are also brittle, gives composite materials with useful strength, stiffness and toughness.

16. Draw the tensile load/extension curve for a typical ductile metal and use it to show what is understood by (a) Young's modulus, (b) yield stress, (c) 0.1% proof stress, (d) tensile strength, (e) work-hardening, (f) elastic deformation, (g) uniform plastic deformation, and (h) localised plastic deformation (necking).

17. Explain why mild steel may be regarded as a safer engineering material than glass.

18. (a) Describe the simple four-atom model for the plastic deformation of metals and use it to show the origin of the theoretical yield strength and the toughness of metals.
(b) Explain the role of dislocations in reducing the actual yield strength below the theoretical value and briefly describe three methods whereby dislocation movement can be impeded in order to increase the yield strength.

19. Explain how and why you would expect body-centred cubic metals and face-centred cubic metals to differ from one another in their mechanical behaviour.

20. A number of materials were compared by placing sheets of identical size on two parallel supports and allowing a weight to fall from a fixed height onto the centre line between the supports. The materials used were glass, aluminium and polyethene. The glass shattered, the aluminium bent and the polyethene remained undamaged. An observer was heard to conclude that the polyethene was stronger than the aluminium which, in turn, was stronger than the glass. Explain the nature of the observer's misunderstanding.

Part 3: Electricity and Magnetism

TOPIC 26 ELECTRIC CHARGE

COVERING:

- the nature of electric charge;
- force between electric charges (Coulomb's law);
- electrostatic induction.

Electricity and magnetism both stem from electric charge.

Charge, like mass, is a fundamental concept that lies at the limit of our absolute understanding of the physical world. We do not know precisely what it is, but we can describe its properties in terms of the effects it produces.

We can deliberately charge certain objects, such as plastic combs or pens, by rubbing them with a cloth so that they attract scraps of paper or even a thin stream of water running from a tap. This demonstrates a fundamental similarity between charge and mass in that they both give rise to forces.

In Topic 14 we saw that the electron and the proton carry equal but opposite charges (negative and positive, respectively) of 1.60×10^{-19} C. It is helpful to bear in mind that, no matter how we choose to define charge, these two particles form its natural units. Matter normally contains electrons and protons in more or less equal numbers and is therefore generally electrically neutral. Charge only becomes apparent when electrons and protons become separated from one another – by friction, for example.

If we rub certain objects with a cloth, then they become charged, either positively or negatively, depending on the materials involved. This can be explained in terms of the transfer of electrons from the cloth to the object, or vice versa, when they are rubbed together. The loss of electrons from one to the other leaves a positive charge due to the unbalanced protons left behind.

It is important to note the *principle of conservation of charge*. Although positive and negative charges may be separated as above, or indeed combined to neutralise each other, the net charge within any particular system remains the same.

We should also note that electric current is simply a flow of charge and that it is the freedom of movement of valence electrons through metals (see Topic 16) that makes them good electrical conductors. By contrast, ideal electrical insulators do not conduct electricity, because they possess no free charge-carriers. For example, the valence electrons in solid covalent and ionic substances are localised within their respective bonds or ions and are not free to carry current. It should be noted, however, that *electrolytes* contain ionic substances, either molten or in solution, that enable them to conduct electricity because of the freedom of movement of the ions. We shall see later (Topic 29) that *semiconductors* are intermediate between insulators and conductors in their electrical conductivity. (In practice, real insulators do allow some very slight movement of charge.)

In this topic we shall confine our discussion to *electrostatics* – that is to say, the study of electric charge at rest.

26.1 COULOMB'S LAW

The force between electric charges was investigated in the eighteenth century by the French scientist Coulomb. In Topic 14 we briefly met the law named after him. This is expressed by the equation

$$F = \frac{1}{4\pi \, \varepsilon_0} \times \frac{Q_1 Q_2}{r^2} \tag{26.1}$$

F is the magnitude of the force in newtons between two charges and is repulsive or attractive depending on whether they are of like or opposite sign. Q_1 and Q_2 are the magnitudes of the charges in coulombs, r is the distance in metres between them and ε_0 is a constant with the value $8.85 \times 10^{-12} \text{ C}^2 \text{ N}^{-1} \text{ m}^{-2}$. (The units for ε_0 are obtained here by consideration of Equation 26.1 but it is more usual to express them rather differently, as we shall see in Topic 28.) ε_0 is called the *absolute permittivity of free space* or the *electric constant*.

Equation (26.1) only applies under vacuum conditions; the presence of a *dielectric* or non-conducting substance between the charges reduces the magnitude of the force and the equation becomes

$$F = \frac{1}{4\pi \, \varepsilon} \times \frac{Q_1 \, Q_2}{r^2} \tag{26.2}$$

where ε is the *absolute permittivity* of the substance (often simply called the *permittivity*). The *relative permittivity* ε_r, sometimes called the *dielectric constant*, is given by the ratio between the absolute permittivity of the substance and that of free space, so that $\varepsilon_r = \varepsilon/\varepsilon_0$. Table 26.1 gives some typical approximate relative permittivity values. For many practical purposes, air may be considered to have the same

Table 26.1

Substance	Relative permittivity
Air	1.00
Polythene	2.3
Glass	4–7
Water	80

permittivity as that of free space. We shall discuss dielectrics in more detail in Topic 28.

Coulomb's law is an inverse square law, like Newton's law of gravitation. The electric force acts along the straight line between the two charges and is a vector quantity like gravitational force. However, there is an important difference in that gravitational force is always attractive, whereas electric force can be either attractive or repulsive.

Before examining electric forces any further, let us put the magnitude of the coulomb into perspective. At the atomic level we know that an electron carries a charge of 1.60×10^{-19} C. This means that we need 6.25×10^{18} electrons (i.e. about 1×10^{-5} mol) to provide a total negative charge of one coulomb. In due course we shall see that the ampere is the basic unit used to measure electric current (Topic 29). One coulomb is the amount of charge that is transported past any point in a conductor in one second by a current of one ampere. In everyday terms, one ampere would be the current drawn by a 240 volt lighting circuit with four 60 watt light bulbs. On this basis a coulomb seems to be a fairly modest amount of charge.

Now let us consider the repulsive force between two like charges of one coulomb each if, for instance, we placed them 25 mm apart in air. Letting $Q_1 = Q_2 = 1$ C and $r = 0.025$ m in Equation (26.1),

$$F = \frac{1}{4\pi \times 8.85 \times 10^{-12}} \times \frac{1 \times 1}{(0.025)^2} = 1.44 \times 10^{13} \text{ N}$$

A medium-sized apple weighs about 1 N, so the force between the charges is equivalent to the weight of over fourteen million million apples. This extraordinary figure tells us that, although the coulomb represents a modest charge in terms of electric current, it is a very large charge in terms of electrostatics. In practice, the magnitude of ordinary electrostatic charges is rarely more than a tiny fraction of a coulomb, although thousands of volts may be involved.

You will find that Question 3 makes the interesting point that, on an atomic scale, the gravitational force between electrons and protons is negligible compared with the electrical force. Gravitational forces become important where very large masses, such as bodies on an astronomical scale, are involved.

26.2 ELECTROSTATIC INDUCTION

So far we have tended to think in terms of point charges of negligible volume, as opposed to charged objects on the human scale. The charge on an insulator rubbed with a cloth would tend to be distributed over the area covered by the cloth.

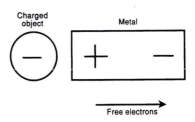

Figure 26.1

In the case of metals, with their free valence electrons, it is possible to 'induce' a temporary unevenness in the charge distribution. If a piece of metal is brought close to a charged object, then an attractive force will arise between them. The explanation is simple. If the object is negatively charged, then it will repel the free electrons towards the far side of the metal, as in Figure 26.1. This will leave an equivalent surplus of protons behind. Thus, the metal nearest the charged object has a net positive charge, resulting in an attractive force between the two. If the object had been positively charged, then a surplus of delocalised electrons would have been attracted towards it, making the nearer side of the metal negative – again leading to an attractive force.

Electrostatic induction, as this way of inducing charge separation is called, is also responsible for the attractive force that pulls a thin stream of water towards a charged object. The water molecule is polar, as we saw in Topic 16, with two positively and two negatively biased orbitals. Any charged object nearby will tend to attract the orbitals of opposite charge, and repel those of like charge, so that the molecules will tend to orientate themselves accordingly, as in Figure 26.2. This, in turn, will give rise to an attractive force between the charged object and the water. Similar electrostatic induction effects cause scraps of paper and hair to be attracted by charged objects.

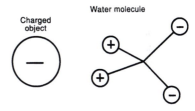

Figure 26.2

Questions

(Use any previously tabulated data as required. Assume that $\varepsilon_0 = 8.85 \times 10^{-12} \text{ C}^2 \text{ N}^{-1} \text{ m}^{-2}$.)

1. What happens to the force between two separate charges if the distance between them is (a) halved, (b) trebled?

2. Two positive charges of equal magnitude experience a repulsive force of 0.133 N between them when they are separated by a distance of 13 mm in air.

   (a) What is the magnitude of each charge?
   (b) What is the magnitude of the force if the charges are moved to a distance of 65 mm apart?
   (c) What is the magnitude of the force if the charges are separated by a 14 mm thickness of polythene?

3. Assuming that the hydrogen atom consists of an electron orbiting around a proton at a distance of 5.29×10^{-11} m, find (a) the electric force and (b) the gravitational force between them. (See Table 14.1 on page 118

and Equation 2.1 on page 9, and assume that $G = 6.7 \times 10^{-11}$ N m^2 kg^{-2}.)

4. Two identical charges, each of 15 g mass and negligible volume, are both suspended from the same point by non-conducting weightless strings 1500 mm in length. Find the magnitude of each charge if the angle between the strings is 5°. ($g = 9.8$ m s^{-2}.)

TOPIC 27 ELECTRIC FIELD

COVERING:

- electric field strength;
- the analogy with gravitational field;
- field lines;
- uniform and non-uniform fields;
- potential and potential difference.

In the previous topic we saw that forces exist between electric charges. It follows that a charge must in some way influence the space around itself. This property can be described in terms of an *electric field* surrounding the charge. Where two or more charges are involved, their fields interact with one another to produce forces.

The idea of an electric field is nothing more than an imaginary device to help us picture how charges behave. The same idea is used where other types of force operate at a distance – for instance, gravitational force and, as we shall see later, magnetic force. Nuclear scientists think in terms of fields within the nucleus which are responsible for the forces that hold protons and neutrons together. Thus, the effects of field forces range in scale from the nuclear to the astronomical.

27.1 FIELD STRENGTH

Newton's law of gravitation (Topic 2) tells us that there is an attractive force between any two objects by virtue of their mass – between the earth and an apple, for example. For convenience we tend to think of the earth as having a gravitational field which causes any other mass close to its surface to experience a weight of 9.8 N kg^{-1}. We can therefore say that the earth has a gravitational field strength of 9.8 N kg^{-1} close to its surface. By contrast, the moon has a corresponding gravitational field strength of about 1.6 N kg^{-1}. As this suggests, we could determine the strength of an unknown gravitational field by measuring the force acting on a test mass.

In a similar way we can determine the strength of an electric field by measuring the force acting on a test charge. The electric force F is given by

$$F = qE \tag{27.1}$$

where E represents the electric field strength and q represents the magnitude of the charge used to measure it. (Note the close analogy with the gravitational force $F = mg$.) The electric field strength can be expressed as force per unit charge, N C^{-1} (since $E = F/q$), in just the same way that gravitational field strength can be expressed as force per unit mass. (The magnitude of our test charge would have to be very small to avoid the possibility of inducing charge separation in nearby objects, and, hence, altering the field that we are trying to measure.)

In the previous topic we noted that electric force is a vector quantity. By convention, the field direction is taken to be that of the force acting on a positive charge within the field.

Now let us use Coulomb's law to find an expression for the strength of the field due to a point charge of magnitude Q. We shall use a very small positive test charge q and place it at a distance r from Q. The force acting between the charges is given by Equation (26.1) (page 257) as follows:

$$F = \frac{1}{4\pi\varepsilon_0} \times \frac{Qq}{r^2}$$

If we substitute the right-hand side of this expression for F in Equation (27.1), we get

$$E = \frac{F}{q} = \frac{Q}{4\pi\varepsilon_0 r^2} \tag{27.2}$$

Remember that, since q is positive, the direction of the field is towards Q if Q is negative, and away if it is positive. Also remember that the force acting on a negative charge will be in the opposite direction to that of the field.

27.2 FIELD LINES

At this stage it is useful to introduce another imaginary thinking aid – namely the *electric field line*. This helps us to picture the strength and direction of an electric field. A field line represents the direction of the force acting on a positive test charge at any given point; hence, it represents the field direction there. Figure 27.1 shows field lines around (a) a negative charge, (b) a positive charge and (c) a stronger positive charge. Field strength is indicated by the concentration of field lines; where they are closer together, the field is stronger. The greater the magnitude of the charge the greater the number of field lines entering it or emerging from it. And as the distance from the charge increases, the field lines diverge and the field becomes weaker.

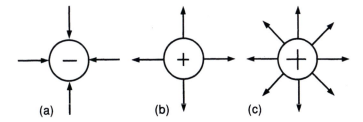

Figure 27.1

Figure 27.2(a) represents the pattern of field lines around two equal and opposite charges. Some readers will recognise the similarity with the pattern that iron filings make in the magnetic field around a bar magnet (see Figure 32.1 on page 317). If the charges are alike and are of equal magnitude, as in Figure 27.2(b), then their overall pattern is similar to that around a single charge, particularly when they are close together.

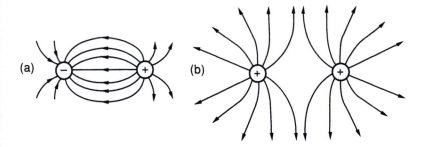

Figure 27.2

Since field lines reflect the force acting on a test charge, they are sometimes called lines of force. The nineteenth century scientist Michael Faraday pictured field lines as representing a state of strain, imagining them to be in tension (hence, tending to shorten), while at the same time repelling each other laterally. Thought of in those terms, the patterns in Figure 27.2 help us to picture the way in which force fields arise between charged objects.

The fields that we have considered so far vary in magnitude and direction from point to point, and are therefore described as being non-uniform. By contrast, a uniform field has constant strength and direction throughout. Figure 27.3 represents the uniform electric field between two parallel, oppositely charged metal plates. The field lines are parallel and uniformly spaced, apart from at the edges, where they tend to escape to some extent and bulge outwards. We shall ignore the edge effect for the purposes of our discussion.

Returning to the gravitational analogy, because the earth is so large the gravitational field over a small area of its surface is virtually uniform.

Locally, the gravitational field lines, along which test masses would fall under gravity, can be regarded as parallel for most practical purposes.

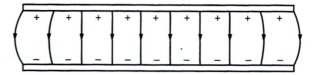

Figure 27.3

27.3 POTENTIAL AND POTENTIAL DIFFERENCE

Another aspect of the analogy between gravitational and electric fields is potential energy. In Topic 8 we saw that the work W needed to raise a mass m to a height h above the ground is equal to mgh joules (Equation 8.3 on page 61). This work is stored as mgh joules worth of potential energy so long as the mass remains at that height. Thus, the *potential* at any point in a gravitational field can be expressed in terms of $(mgh)/m = gh$ joules of potential energy per kilogram mass. Furthermore, we can define an *equipotential* as a surface over which the potential has a constant value – for example, at a particular height above ground level. Obviously an object can move anywhere across an equipotential surface without its potential energy changing. (No component of the force due to the field can lie within the plane of the equipotential; therefore, any field line must cut an equipotential perpendicularly to its surface.)

For practical purposes, we are often interested in the *potential difference* between two levels, as in lifting a mass from one height to another. It is important to remember that the potential difference between two levels corresponds to the distance between them in the field direction (e.g. vertically at the surface of the earth), irrespective of the actual route taken.

Similar considerations apply to a charge in an electric field. Figure 27.4 shows two oppositely charged parallel plates. A positive charge Q placed between the plates will be repelled by the positive plate and attracted towards the negative. To move the charge towards the positive plate, say from the equipotential at X to that at Y, we would need to apply a force F of QE newtons (Equation 27.1) in the opposite direction to that of the field. To move it a distance d in that direction, we would have to do QEd joules of work (just as we have to do mgh joules of work to lift a mass through a height h). Thus, we can express electric potential and define electric equipotentials in joules per coulomb ($J\ C^{-1}$), just as we can express their gravitational counterparts in joules per kilogram. (Remember that electric potential decreases in the field direction.)

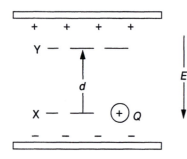

Figure 27.4

Because of its size, it is convenient to regard the earth as representing the practical zero of electrical potential, since it acts as an enormous reservoir of charge whose level is virtually constant. By *earthing* a charged object any excess or deficiency of electrons would flow to or from the earth to give the object zero potential relative to the earth.

Later we shall find it necessary to consider potential differences in electrical circuits. For example, we shall regard the function of a battery as raising a charge through a potential difference in much the same way as a hoist raises a mass.

27.4 THE VOLT

Potential difference V is defined as the energy change involved in raising or lowering unit charge from one point to another; obviously, it could be expressed as joules per coulomb but potential difference and potential are given a special unit of their own – namely the *volt* (V). If the transfer of 1 coulomb of charge between two points involves an energy change of 1 joule, then the potential difference between them is 1 volt. For instance, a 1.5 volt battery will give 1.5 joules of energy to each coulomb of charge that passes through it.

Figure 27.4 helps to make this clear. If the work W required to move the charge Q from X to Y is given by $W = QEd$ joules, then V, the potential difference between X and Y (i.e. the energy change per coulomb moved), is given by

$$V = W/Q = (QEd)/Q = Ed \ (\text{V or J C}^{-1})$$

This provides us with two useful relationships:

$$W = QV \tag{27.3}$$

and

$$V = Ed \tag{27.4}$$

Equation (27.3) gives us the energy change involved when a charge moves through a potential difference without our needing to know the field strength or the route taken by the charge. Later on we shall see that this equation can be put to use in various contexts – for example, in converting an electric current to heat, or in storing energy in an electric field. For the time being, as an illustration, let us use it to consider what happens if we release an electron in an electric field between two parallel plates, as in Figure 27.5(a).

The electron will experience a force (hence, an acceleration) in a similar way to a mass in a gravitational field. (We assume that there is a vacuum between the plates; otherwise air molecules will get in the

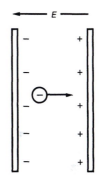

(a)

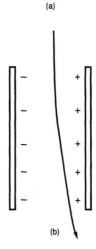

(b)

Figure 27.5

way of the electron and interfere with its acceleration.) As the electron falls through a given potential difference, it trades potential energy for kinetic energy, like a mass falling in a gravitational field. (Remember that an electron 'falls' in the opposite direction to a positive charge.) The kinetic energy ($\frac{1}{2}mv^2$) acquired by the electron is equal to the work done on it (QV) by the electrostatic force due to the field. (Note that if a moving electron enters the field between the plates, then its path will be deflected accordingly, as in Figure 27.5b, for example.)

The *electronvolt* (eV) is a non-SI unit of energy which is very convenient for dealing with events on an atomic scale. It can be defined as the kinetic energy an electron acquires as it falls freely through a potential difference of 1 volt. Its value in joules can be obtained from Equation (27.3) as follows:

$$W = QV = 1.6 \times 10^{-19} \times 1.0 = 1.6 \times 10^{-19} \text{ J}$$

(Since 1 V is equivalent to 1 J C^{-1}, the units of $Q \times V$ are C $\times$ J C^{-1} = J.)

Let us now consider Equation (27.4) ($V = Ed$). This relates the potential difference between two points in a uniform electric field to the distance between them. As an illustration, Figure 27.6 shows a uniform field (neglecting edge effects) between two parallel plates that are 0.25 m apart with a potential difference of 500 V between them. Equipotentials are drawn at 100 V (i.e. 0.05 m) intervals, and potential is plotted against distance in the lower half of the figure.

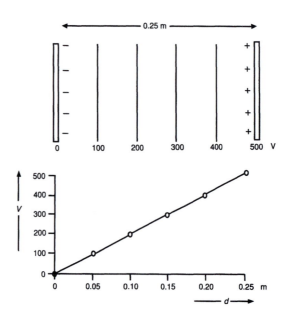

Figure 27.6

The plot is a straight line and the slope represents the electric field strength; thus,

$$E = \frac{V}{d} = \frac{500}{0.25} = 2000 \text{ V m}^{-1}$$

We now have two ways of regarding electric field strength: in terms of either force (N C^{-1}) or energy (V m^{-1}). We can readily show that N C^{-1} and V m^{-1} are equivalent to one another, as follows:

$$1 \text{ V} = 1 \text{ J C}^{-1}$$

and because 1 J = 1 N m,

$$1 \text{ V} = 1 \text{ N m C}^{-1}$$

Dividing both sides by m,

$$1 \text{ V m}^{-1} = 1 \text{ N m C}^{-1} \text{ m}^{-1} = 1 \text{ N C}^{-1}$$

The following worked example compares both viewpoints.

Worked Example 27.1

If an electron is released at the inner surface of the left-hand plate in Figure 27.6, find its velocity when it reaches the other plate (a) by force considerations and (b) by energy considerations.

(a) The electron would experience a force of magnitude F towards the right-hand plate, given by

$$F = QE = 1.6 \times 10^{-19} \times 2000 = 3.2 \times 10^{-16} \text{ N}$$

The resulting acceleration a is, from $F = ma$ (Equation 6.1 on page 47), given by

$$a = \frac{F}{m} = \frac{3.2 \times 10^{-16}}{9.1 \times 10^{-31}} = 3.5 \times 10^{14} \text{ m s}^{-2}$$

(The acceleration due to this tiny force is enormous because of the extremely low mass of the electron.)

The final velocity v of the electron is given by Equation (5.4) (page 39):

$$v^2 = u^2 + 2as$$

Since u (the initial velocity) is zero and s (the distance between the plates) is 0.25 m, the magnitude of v is given by

$$v = \sqrt{2as} = \sqrt{2 \times 3.5 \times 10^{14} \times 0.25} = 13 \times 10^6 \text{ m s}^{-1}$$

(b) If we assume that all the work done on the electron (QV) is converted to kinetic energy ($\frac{1}{2}mv^2$), then

$$\frac{1}{2} mv^2 = QV$$

and

$$v = \sqrt{\frac{2QV}{m}} = \sqrt{\frac{2 \times 1.6 \times 10^{-19} \times 500}{9.1 \times 10^{-31}}} = 13 \times 10^6 \text{ m s}^{-1}$$

(Note that we have to be careful with classical dynamic calculations like this, since the mass of an object increases when it is moving close to the speed of light. This is a consequence of the theory of relativity, which we shall not consider here. In this particular case the speed is less than 5% of that of light and the mass increase is less than 0.1% of the electron's mass at rest.)

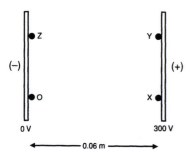

Figure 27.7

Questions

(Use any previously tabulated data as required. Where appropriate, assume vacuum conditions unless otherwise stated. Assume that $g = 9.8$ m s^{-2} and that the speeds involved are too low to have a significant effect upon mass.)

1. Figure 27.7 shows two parallel metal plates 60 mm apart with a potential difference of 300 V between them.

   (a) What is the direction of the electric field between the plates?
   (b) If a proton at rest is released from the mid-point between the plates, in which direction will it move?
   (c) What is the field strength between the plates (i) in N C^{-1} and (ii) in V m^{-1}?
   (d) Find the energy required to move a proton from (i) O to X, (ii) O to Y and (iii) O to Z.
   (e) Find the energy required to move an electron from X to O.
   (f) If a proton is released from X, find its speed when it reaches O.
   (g) If an electron is released from O, find its speed when it reaches X.

2. A potential difference of 200 V exists between two parallel plates that are 50 mm apart. Find (a) the force and (b) the acceleration experienced by an electron in the electric field between the plates.

3. What strength of electric field would just support the weight of a Ca^{2+} ion?

4. A negatively charged droplet of 4.9×10^{-15} kg mass is suspended in the electric field between two parallel horizontal plates 6 mm apart with a potential difference of 450 V between them. Find the number of surplus electrons carried by the droplet.

5. A 10 nC charge is moved 70 mm perpendicularly and then 50 mm parallel to the direction of a uniform electric field of 2000 V m^{-1} strength. Find the overall change in potential energy of the charge.

6. An electron, initially at rest, is released in a uniform electric field 500 V m^{-1} in strength. Estimate (a) its speed after it has travelled 400 mm and (b) how long it takes to travel this distance.

7. Repeat Question 6, replacing the electron with a proton.

8. An electrically charged droplet of 4×10^{-15} kg mass falls vertically at a steady speed through air between two vertical parallel plates set 10 mm apart. A potential difference of 500 V is applied to the plates, whereupon the droplet falls at an angle of 31.5° to the vertical. Estimate the magnitude of the charge carried by the droplet.

9. An electron, initially at rest, is accelerated through a potential difference of 285 V. It then passes midway between two parallel plates providing a uniform electric field perpendicular to the direction in which it is travelling. The plates are 50 mm long and 25 mm apart and there is a potential difference of 71 V between them. Find (a) the speed of the electron after its initial acceleration and (b) the transverse deflection experienced by the electron as it emerges from between the plates.

TOPIC 28 CAPACITANCE

COVERING:

- the parallel-plate capacitor;
- dielectrics;
- energy stored in a capacitor;
- capacitors in parallel and in series.

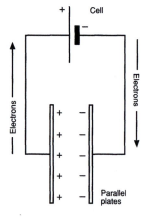

Figure 28.1

The creation of an electric field involves separating positive and negative charge. Work has to be done which is then stored in the field as potential energy. Figure 28.1 illustrates this in terms of an electric cell or battery connected across two parallel metal plates. The figure shows the conventional symbol for an electric cell (although the signs are usually omitted). A battery is simply a number of cells connected together to form a single unit.

Electrons will flow towards the positive terminal of the cell from the plate to which it is connected. At the same time electrons will flow from the negative terminal of the cell onto the other plate. In effect, the cell transfers electrons from one plate to the other. The build-up of charge on the plates increasingly opposes the flow of electrons, which therefore decreases and eventually stops when the potential difference across the plates is equal to the voltage of the cell. If the cell is disconnected and the plates connected directly to one another, then electrons will flow from the negative plate to the positive until the potential difference between them is zero.

A device such as this, capable of storing electric charge, is called a *capacitor*. Its ability to store charge is measured in terms of its *capacitance C*, given by

$$C = Q/V, \text{ or } Q = CV \tag{28.1}$$

where Q is the charge stored on either plate and V is the potential difference between them. The unit of capacitance is called the *farad* (symbol F) but, as it stands, this is much too large a quantity for many purposes, so the microfarad ($1 \mu F = 1 \times 10^{-6}$ F) and the picofarad ($1 pF = 1 \times 10^{-12}$ F) are in common use.

In this topic we shall confine ourselves to the so-called *parallel-plate capacitor* of the type shown in Figure 28.1.

Note that capacitance is dependent on the area of the plates and on the separation between them. Assuming that they are each of area A

and separated by a distance d, and that the non-uniformity of the field at the edges can be ignored, then C is proportional to A/d. However, as we shall now see, capacitance is also increased by the presence of a *dielectric* material between the plates.

Practical capacitors are normally of the basic parallel-plate type, although a variety of geometrical arrangements are used. For example, metal foil and thin sheets of dielectric material are stacked in layers or rolled up together like a swiss roll to provide high capacitance in a small volume. Some capacitors have movable plates with adjustable overlap so that their effective area and, hence, their capacitance, can be varied.

28.1 DIELECTRICS

A dielectric is a non-conductor of electricity and can be used simply as an insulator. For this purpose it must have adequate *dielectric strength* – that is to say, it must not break down electrically and lose its insulating properties. Dielectric strength is generally expressed in terms of the maximum electric field the material will withstand. (Note that air normally has a dielectric strength of about 3 kV mm^{-1}. Some ceramics and polymers used as insulators have values well in excess of 10 kV mm^{-1}.)

Although dielectrics are insulators, they do respond to electric fields. In Topic 26 we saw that the force between two charges is reduced by the presence of a dielectric between them. Equation (26.2) (page 257) shows that this force is inversely proportional to the permittivity ε of the dielectric. (Remember that $\varepsilon = \varepsilon_r \varepsilon_0$, where ε_r is the relative permittivity, or dielectric constant, and ε_0 is the permittivity of free space.)

The presence of a dielectric between the plates of a capacitor will increase its capacitance C in accordance with the expression

$$C = \varepsilon_r C_0 \tag{28.2}$$

where C_0 is the capacitance when there is a vacuum between the plates. From this it follows that relative permittivity can be defined as the ratio C/C_0.

Now let us consider what actually happens when a dielectric is inserted into the space between a pair of charged plates separated by vacuum. If the capacitor is electrically isolated, then the charge on the plates remains fixed and the potential difference between them will fall. This is what we would expect from Equation (28.1). Since $V = Q/C$, and Q is fixed, then an increase in C by a factor of ε_r (from Equation 28.2) will lead to a corresponding reduction in V. If the capacitor is connected to a cell or a battery, as in Figure 28.1, then the insertion of a dielectric will cause a further flow of electrons, so that the potential difference across the plates remains equal to the voltage of the cell. In this case V is fixed, where $Q = CV$, and the presence of the dielectric raises C by a factor of ε_r, so the charge stored by the capacitor is increased accordingly. These effects suggest that the presence of the dielectric reduces the electric field between the plates, so that more charge is required to restore the potential difference between them to its original value. The reason for this is polarisation.

As Figure 28.2 suggests, the nucleus of an atom in an electric field tends to be shifted in the field direction, whereas the outer electrons tend to be shifted in the opposite direction. The centres of positive and negative charge distributions within the atom are thus displaced from one another, producing a dipole orientated in opposition to the field. In effect, this gives a resultant field smaller than the original. Where a substance consists of polar molecules with permanently polarised structures (Topic 16), their tendency to orientation can lead to marked dielectric behaviour. For example, as we saw in Topic 26, water is a highly polar molecule which, in the liquid state, has considerable freedom to align itself in an electric field. Water therefore has a high relative permittivity value (see Table 26.1 on page 258). The bending and stretching of polar bonds can also contribute to polarisation.

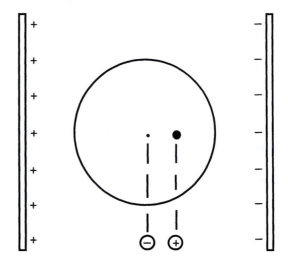

Figure 28.2

Earlier we noted that capacitance C is proportional to A/d, where A is the area of each plate and d is the distance between the plates. We can now make this relationship into an equation, because the permittivity ε of the dielectric (where $\varepsilon = \varepsilon_r \varepsilon_0$) gives us the constant of proportionality as follows:

$$C \left(= \frac{\varepsilon_r \varepsilon_0 A}{d} \right) = \frac{\varepsilon A}{d} \tag{28.3}$$

Note that the normal unit of permittivity (given by rearrangement of this equation) is farads per metre, F m^{-1}. Also note that ε_r is a dimensionless quantity, since it is given by the ratio $\varepsilon/\varepsilon_0$.

28.2 ENERGY STORED IN A CHARGED CAPACITOR

The energy stored in a charged capacitor is the total work done in transferring the electrons from one plate to the other. The work done in transferring the first electron is virtually zero, because the initial potential difference between the plates is zero. Since the relationship between Q and V is a straight line (see Figure 28.3), the work done in transferring successive electrons increases linearly as the charge builds up on the plates. If, at a given moment, the total charge transferred is Q and the potential difference is V, then we can say that the electrons have been transferred through an average potential difference equal to $(0 + V)/2$. Thus, W, the total work done, is given by the total charge multiplied by the average potential difference, as follows:

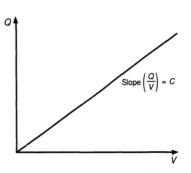

Figure 28.3

$$W = Q \times \frac{(0 + V)}{2} = \frac{1}{2} QV \qquad (28.4)$$

(We can regard the energy stored in a charged capacitor as analogous to the strain energy stored in a stretched wire. Strain energy is given by the area of the triangle under the force/extension line (Figure 8.2b on page 62). Similarly, the energy stored in the capacitor is given by the area of the triangle under the Q/V line.)

Note that, since $Q = CV$, W can also be expressed in the forms

$$W \left(= \frac{1}{2} QV \right) = \frac{1}{2} CV^2 \qquad (28.5)$$

and

$$W = \frac{1}{2} \frac{Q^2}{C} \qquad (28.6)$$

Worked Example 28.1

A capacitor of unknown value was charged using a 10 V battery. It was then disconnected from the battery and discharged through a small electric motor, which raised a 100 g mass to a height of 1 m. Estimate the unknown capacitance.

Assuming that all the stored electrical energy was converted to gravitational potential energy, then

$$\frac{1}{2}\,CV^2 = mgh$$

which can be rearranged to

$$C = \frac{2mgh}{V^2}$$

Substituting the values given above, and assuming that $g = 9.8$ m s^{-2},

$$C = \frac{2mgh}{V^2} = \frac{2 \times 0.1 \times 9.8 \times 1}{10^2} = 0.02 \text{ F}$$

28.3 CAPACITORS COMBINED IN PARALLEL AND IN SERIES

Where a number of capacitors are combined in an electrical circuit, their resultant capacitance can be found by considering them as either *parallel* or *series* combinations.

Figure 28.4 shows a parallel combination of two capacitors with capacitances C_1 and C_2, respectively. (The figure shows the conventional symbol for a capacitor.) Both capacitors are connected across the same battery, so the potential difference V across each is the same. The charge carried by each is therefore given by $Q_1 = C_1V$ and $Q_2 = C_2V$, respectively (from Equation 28.1). The combined charge Q is given by

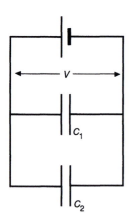

Figure 28.4

$$Q = Q_1 + Q_2 = V(C_1 + C_2)$$

Therefore, the combined capacitance C is

$$C = C_1 + C_2$$

This argument can be extended to give the general equation for any number of capacitances in parallel, as follows:

$$C = C_1 + C_2 + C_3 + \ldots C_n \tag{28.7}$$

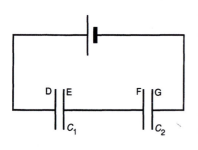

Figure 28.5

Figure 28.5 shows a series combination. In this case the battery produces a charge of $+Q$ on plate D and of $-Q$ on plate G. Provided that the plates in each capacitor are sufficiently large and closely spaced together, E and F carry charges of $-Q$ and $+Q$ induced by their proximity to plates D and G, respectively. In this case the charge stored by each capacitor is equal and the potential difference across each is given by $V_1 = Q/C_1$ and $V_2 = Q/C_2$, respectively. The overall potential difference V, governed by the battery, is divided across the capacitors, so that

$$V = (V_1 + V_2) = Q \left(\frac{1}{C_1} + \frac{1}{C_2} \right)$$

The combined capacitance C is therefore given by

$$\frac{1}{C} = \frac{1}{C_1} + \frac{1}{C_2}$$

Again the argument can be extended to give a general equation for any number of capacitances in series, as follows:

$$\frac{1}{C} = \frac{1}{C_1} + \frac{1}{C_2} + \frac{1}{C_3} + \cdots \frac{1}{C_n} \qquad (28.8)$$

Worked Example 28.2 illustrates more complex networks involving both parallel and series combinations.

Worked Example 28.2

Find the resultant capacitance of each of the combinations shown in Figure 28.6.

(a) First, considering the pair in series,

$$\frac{1}{C} = \frac{1}{6} + \frac{1}{6}$$

which gives $C = 3$ μF.
  Then, considering the parallel combination of this with the third capacitor,

$$C = 3 + 6 = 9 \text{ μF}$$

(b) First, considering the pair in parallel.

$$C = 6 + 6 = 12 \text{ μF}$$

Then, considering the series combination of this with the third capacitor,

$$\frac{1}{C} = \frac{1}{12} + \frac{1}{6}$$

which gives $C = 4$ μF.

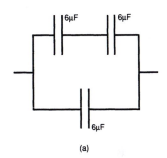

(a)

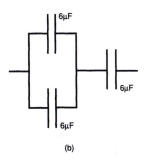

(b)

Figure 28.6

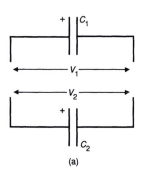

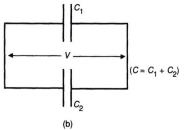

Figure 28.7

Worked Example 28.3

A 3 µF capacitor is charged to 60 V and a 6 µF capacitor is charged to 120 V. The capacitors are then connected with their like-charged plates together. Find the total stored energy before and after connection.

Figure 28.7 shows the capacitors (a) before, and (b) after connection.

(a) Before connection:

The total scored energy is given by

$$\frac{1}{2} C_1 V_1^2 + \frac{1}{2} C_2 V_2^2$$

and, substituting the given values,

$$\left(\frac{1}{2} \times 3 \times 10^{-6} \times 60^2 \right) + \left(\frac{1}{2} \times 6 \times 10^{-6} \times 120^2 \right)$$

$$= 0.0486 \text{ J}$$

The total stored charge is given by

$$C_1 V_1 + C_2 V_2$$

and, substituting the given values,

$$(3 \times 10^{-6} \times 60) + (6 \times 10^{-6} \times 120)$$

$$= 9 \times 10^{-4} \text{ C}$$

(b) After connection:

Initially the capacitors have different potentials. On connection, there will be a redistribution of the total charge Q to equalise the potential difference across the parallel combination, which has a combined capacitance $(C_1 + C_2 =)$ 9 µF. From Equation (28.6), the stored energy is given by

$$\frac{1}{2} \frac{Q^2}{C} = \frac{1}{2} \frac{(9 \times 10^{-4})^2}{9 \times 10^{-6}} = 0.045 \text{ J}$$

Note that the stored energy lost on connection (i.e. $0.0486 - 0.045 = 0.0036$ J) is converted to heat in the connecting wires.

Also note that, after connection, the potential difference V is the same across both capacitors and can be found from

$$V = \frac{Q}{C} = \frac{9 \times 10^{-4}}{9 \times 10^{-6}} = 100 \text{ V}$$

Using this approach, the stored energy after connection can be obtained from

$$\frac{1}{2} QV = \frac{1}{2} \times 9 \times 10^{-4} \times 100 = 0.045 \text{ J}$$

or from

$$\frac{1}{2} CV^2 = \frac{1}{2} \times 9 \times 10^{-6} \times 100^2 = 0.045 \text{ J}$$

Questions

(Use any previously tabulated data as required.)

1. Find the charge on a 5 µF capacitor with a potential difference of 80 V between its plates.

2. Two parallel plates, each 0.2 m square, are positioned 2 mm apart in air. How many electrons must be transferred from one plate to the other to give a potential difference of 190 V between them?

3. A parallel-plate capacitor, with air between its plates, is charged by using a 60 V battery. Find the potential difference between the plates if, after disconnecting the battery, the air between them is replaced with polythene.

4. What is the total charge stored by one 2 µF, one 3 µF, one 5 µF and two 10 µF capacitors all connected in parallel across a 24 V battery?

5. Find the various capacitances obtainable from three 3 µF capacitors in combination.

6. Two parallel-plate capacitors are identical except that the plates of one are spaced twice as far apart as those of the other. If the capacitors are connected in series across a 12 V battery, estimate the potential difference across each.

7. A capacitor holds a charge of 0.1 mC after being charged to 50 V. Find (a) its capacitance, and (b) the amount of energy stored.

8. A 20 V battery is used to charge a 0.021 F capacitor. After disconnection from the battery, the energy stored in the capacitor is used to heat 10 g of water that is initially at a temperature of 15.0 °C. Find the final temperature of the water.

9. A 150 V battery is connected across a 4 µF and an 8 µF capacitor combined (a) in parallel, (b) in series. Find the charge stored by each capacitor in both cases.

10. A 3 µF capacitor, previously charged to 120 V, is connected in parallel with an uncharged 5 µF capacitor. Find (a) the final voltage across the capacitors and (b) the energy lost in heating the connecting wires.

TOPIC 29 ELECTRIC CURRENT

COVERING:

- electric current as a flow of charge;
- current in metal conductors;
- conventional current;
- power;
- current in semiconductors.

Electric current provides a very convenient means of transporting energy from place to place. Metal conductors are used to carry it, and all sorts of electrical devices are available to convert it into heat, light or whatever other form of energy is required.

As we noted in Topic 26, electric current is simply a flow of electric charge. It may be the flow of electrons through a vacuum, or ions through a gas or an electrolyte; more importantly from our point of view, it may be the flow of electrons through a metal or, as we shall see later in this topic, electrons and positive holes through semiconducting materials.

The magnitude of an electric current could be defined as the rate of flow of charge – say the number of coulombs passing a given point in a conductor in one second. But electric current (like mass, length and time) is one of the seven base SI units (see Topic 1) and, as we shall see later, the base unit of current, the ampere (A), is defined in quite a different way. The coulomb is formally derived from the ampere, rather than the other way round, and is defined as the quantity of charge which flows in one second past a point in a conductor which is carrying a steady current of one ampere. Thus,

$$Q = It, \text{ or } I = \frac{Q}{t} \tag{29.1}$$

where I amperes is the steady current that flows when a total charge Q coulombs passes a given point at a uniform rate over a period of t seconds. To take a very simple example, the uniform flow of a total of 135 C over a period of 45 s is equivalent to a current of 3 A.

29.1 ELECTRIC CURRENT IN METAL CONDUCTORS

First let us estimate the velocity of the free electrons that constitute an electric current through a metal conductor. Figure 29.1 shows a conductor with a cross-sectional area of A square metres. If the average velocity of the free electrons flowing along the length of the conductor is v metres per second, and if X and Y are v metres apart, then all the free electrons between X and Y at a given moment will pass Y in 1 s. The volume of metal between X and Y is given by $v \times A$ (i.e. length $\times$ cross-sectional area), so, if the metal contains n free electrons per unit volume, then nvA electrons will pass Y in 1 s. This corresponds to a total charge of $nvAe$, where e represents the charge on each electron. Putting this into the form of an equation, the current I is given by

$$I = nvAe \tag{29.2}$$

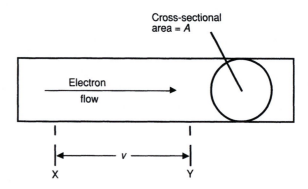

Figure 29.1

To estimate the velocity of the electrons, let us consider a current of 1 A flowing in a copper wire of 1 mm^2 cross-sectional area. In Question 7 at the end of Topic 14 you should have found that a 1 mm cube of copper contains 8.4×10^{19} atoms. A copper atom has one outer electron (Table 14.2 on page 122), so it is not unreasonable to assume a value for n of 8.4×10^{28} electrons per m^3. Rearranging Equation (29.2) and substituting these values,

$$v = \frac{I}{nAe}$$

Therefore,

$$v = \frac{1}{8.4 \times 10^{28} \times 1 \times 10^{-6} \times 1.6 \times 10^{-19}}$$

$$= 0.074 \times 10^{-3} \text{ m s}^{-1}$$

This is equivalent to about a quarter of a metre per hour, which bears no comparison with the speeds of electrons in vacuum that we met in Topic 27. Clearly the electrons carrying a current through a metal conductor encounter considerable resistance to their movement.

To understand the reason for this, we need to remember that our atomic model of a metal is an ordered crystal structure of positive ions within which the valence electrons are free to move. If the crystal was perfectly regular and the ions stationary, then the electrons would be able to move through the 'corridors' between the rows of ions without any resistance. However, because of their thermal energy, the ions continuously vibrate about their mean positions on the crystal lattice, partially blocking the corridors as they move to and fro, so that the electrons collide with them from time to time. The collision frequency is expressed in terms of the *mean free path*, which is the average length of free flight between collisions. In the case of copper at ordinary temperatures, the mean free path is about 40×10^{-9} m (40 nm). This is equivalent to a row of about 160 copper atoms. At higher temperatures, thermal vibration is more vigorous and the collisions more frequent; hence, the mean free path becomes correspondingly shorter.

Normally the free valence electrons in a metal move at very high speeds in an entirely random fashion, rather like gas molecules. They have wide ranges of velocity and kinetic energy, and each electron continually changes speed and direction as it collides with metal ions in the crystal lattice. However, because there are so many electrons moving at random, their overall distribution is uniform and there is no net flow of charge in any particular direction.

If a potential difference is applied across the ends of a metal conductor, the electrons will tend to move towards the positive end. Their movement will still be essentially random, but during each flight between collisions they will respond to the force due to the applied electric field. Electrons that are moving towards the positive end will be accelerated, and those moving away from it will be decelerated. The result will be an overall drift superimposed on their random motion.

The drifting electrons experience a gain in kinetic energy associated with their net acceleration towards the positive end of the conductor. They share this energy with the metal ions as they collide with them, then they move on, providing a continual transfer of energy to the ions which ultimately appears as a temperature rise in the metal. By its very nature, the temperature rise will itself cause increased resistance to the passage of electrons which, in turn, will lead to a further increase in temperature – and so on. Sooner or later an equilibrium may be established where the conductor loses heat to its surroundings at the same rate as that at which it gains heat from the passage of the electrons. Obviously, the conversion of electricity to heat is wasteful and potentially dangerous in an electric cable but essential in a heating element (or in a piece of fuse wire, which melts when the current exceeds a particular value). Temperature rise is therefore an important factor in the design of a conductor.

Apart from temperature, the ease with which electrons pass through a metal depends upon structural features. For instance, the electrical conductivity of pure copper is roughly halved by substituting 10% of the atoms with zinc. Because of their different size, the substituent atoms distort the copper crystal lattice and, hence, the electron pathways between the ions. This will have the effect of reducing the mean free path and increasing the frequency of collisions. The mass of a substituent atom is also important, because this will affect its response to a collision and the extent to which it vibrates afterwards, and, hence, the transfer of heat into the metal structure. The charge on the ion also has an effect: zinc ions carry a double charge, in contrast to the single charge on copper, and this will create electrical irregularities in the electron pathways.

So now we can see that metals show an almost friction-like resistance to the passage of electrons which is due to structural irregularities of one kind or another.

29.2 CONVENTIONAL CURRENT DIRECTION

Before we go any further, there is a very important convention that we must remember. In practice, the direction of an electric current is taken to be that in which positive charge carriers would move (i.e. opposite to that of the electron drift). This *conventional current* direction was established before the discovery of the electron's role as the charge carrier in metals, and it remains to this day. For many practical purposes it makes no difference whether we think about negative charge moving one way or positive charge moving the other.

29.3 ENERGY CONVERSION AND POWER

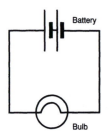

Figure 29.2

Figure 29.2 represents a lamp powered by a battery. We shall assume that the wires connecting the bulb to the battery offer negligible resistance to the current. The filament in the bulb, however, offers so much resistance that it becomes hot enough to emit light.

Thinking in terms of conventional current, the positive terminal of the battery represents a point of high potential and the negative terminal a point of low potential. Our notional positive charge will leave the positive terminal and pass through the filament, providing energy to raise its temperature. The charge, depleted of its energy, will return to the battery via the negative terminal. The potential difference across the filament, and the total energy per coulomb converted by it (to heat and light), are related by Equation (27.3) ($V = W/Q$). Thus, a potential difference of 6 V means that 6 J are extracted from each coulomb passing through the filament, 12 J from 2 C, 18 J from 3 C, and so on. This argument applies generally to devices which convert electrical energy to other forms – for example, electric motors which convert it to mechanical energy, and batteries on charge which convert it to chemical energy. We should note that most devices convert some of the electrical energy into unwanted by-products, such as heat, which reduce their efficiency.

In Topic 8 we noted that power is the quantity used to measure the rate at which energy is converted from one form to another. Thus, the power of an electrical device is the rate at which it converts electrical energy into other forms. If W joules of electrical energy are converted in t seconds, then the power P is given by W/t. But $W = QV$; therefore, W/t is equal to QV/t. Since $Q = It$ (Equation 29.1),

$$P \left(= \frac{W}{t} = \frac{QV}{t} \right) = \frac{ItV}{t}$$

and

$$P = IV \tag{29.3}$$

where P is in watts (W). This equation gives the rate of conversion to all forms of energy – for example, light plus heat in a light bulb, or mechanical energy plus heat in an electric motor.

29.4 ELECTRIC CURRENT IN SEMICONDUCTORS

In view of the great technological importance of semiconductors, we need to have some understanding of how they work. As their name implies, they are neither good conductors nor good insulators but lie somewhere in between. There are many semiconducting materials. Silicon and germanium are well-known examples which we shall use as the basis of our discussion.

As their position in group IV of the periodic table suggests, atoms of silicon and germanium form four covalent bonds with their neighbours in a similar way to carbon atoms in diamond. At absolute zero both silicon and germanium behave as insulators. However, their valence electrons are rather loosely held, so that, at higher temperatures, some of them have sufficient thermal energy to become detached from the bonds and turn into conduction electrons that are able to transport charge.

To detach a valence electron in diamond involves overcoming an energy barrier of about 6 eV, whereas it only requires about 1.1 eV and 0.7 eV for silicon and germanium, respectively. The effect of this is that, at room temperature, silicon and germanium possess enough conduction electrons to make them semiconductors, whereas diamond remains an insulator. Diamond becomes a semiconductor if its temperature is raised sufficiently and, in general, the conductivity of semiconductors increases with increasing temperature as more electrons have sufficient thermal energy to jump the barrier.

When a valence electron becomes detached, it leaves behind it a positively charged vacant site called a *hole*. A second valence electron from a nearby covalent bond may then move into the hole, in which case it will leave a new hole behind it. In effect, this electron and the original hole will exchange places, as indicated by the arrows in Figure 29.3, the electron moving one way and the positive hole in the opposite direction. This means that holes can act as charge carriers which contribute towards the electrical conductivity of the semiconductor.

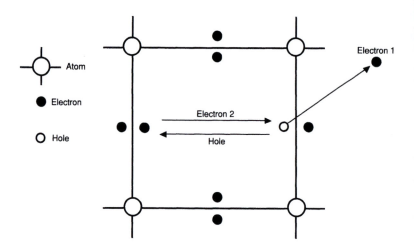

Figure 29.3

Intrinsic semiconductors are so called because their semiconductivity is an inherent property of the material. Intrinsic semiconductors may be *doped* with minute traces of impurities in order to control the flow of current by introducing additional charge carriers. These doped materials are called *extrinsic semiconductors*. They are classified as *n-type* if the additional charge carriers are electrons and *p-type* if they are holes. (p- and n- signify positive and negative charge carriers, respectively.)

n-Type semiconductors can be made from silicon and germanium by doping them with group V elements (i.e. elements with five outer electrons) such as phosphorus, arsenic and antimony. Each impurity atom takes up a position on the crystal lattice, forming four covalent bonds with its neighbours, leaving the fifth electron free to act as a charge carrier. Electrons are the *majority carriers* (i.e. the predominant type), although intrinsic *minority carriers* (holes) will still be present. In this case the impurity atoms donate electrons and are known as *donors*. It is important to note that the crystal as a whole remains electrically neutral, because all of its constituent atoms are neutral.

If silicon or germanium is doped with a group III element (such as boron, aluminium or gallium), with three outer electrons, the impurity atoms form three covalent bonds with their neighbours, leaving a hole which can accept electrons. In this case the majority charge carriers are holes and we have a p-type semiconductor with the impurity atoms described as *acceptors*.

29.5 p–n JUNCTIONS

Many semiconductor devices make use of *p–n junctions* specially formed between p- and n-type materials.

A single junction forms the basis of the *semiconductor diode*, which has the property of allowing conventional current to flow from the p side to the n side but not the other way. Figure 29.4 illustrates how this works in principle.

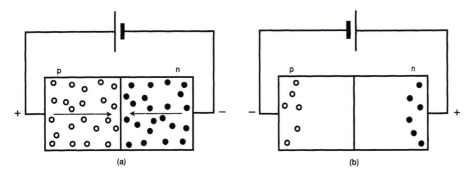

Figure 29.4

Figure 29.4(a) shows a cell connected across the junction so that the p end is positive and the n end is negative. The majority carriers on each side (holes and electrons, respectively) move towards the junction, where they meet and combine. Fresh electrons enter the n-type material from the negative terminal of the cell and fresh holes are formed at the positive terminal where electrons are withdrawn. Current will flow as these processes continue and the junction is said to be *forward-biased*.

The recombination of electrons and holes at a forward-biased p–n junction is accompanied by the liberation of energy, usually in the form of heat. Some types of junction emit useful amounts of light and can be used as the basis of *light-emitting diodes*, such as those used for digital displays.

If the applied voltage is reversed, as in Figure 29.4(b), the electrons and holes tend to be drawn away from the junction and no current flows (apart from a very small *leakage current* due to the thermal formation of a few intrinsic carriers). In this case the junction is said to be *reverse-biased*.

Questions

(Use any previously tabulated data as required.)

1. In 1 min how many electrons pass a given point in a wire carrying a current of 0.2 A?

2. What is the power consumed by an electrical device carrying a current of 500 mA and across which there is a potential difference of 10 V?

3. Find the energy lost by an electron in passing through the filament of a torch bulb across which there is a potential difference of 2.5 V.

4. In 1 min a charge of 3 C passes through an electrical device across which there is a potential difference of 120 V. Find the power consumed by the device.

5. An electrical device consumes a total of 120 J of electrical energy when 8 C passes through it at a steady rate over a period of 20 s. Find (a) the power consumption of the device, (b) the potential difference across it, and (c) the magnitude of the current passing through it.

6. Assuming that the current and duration of a lightning flash are of the order of 1×10^4 A and 1×10^{-3} s, respectively, estimate the quantity of charge involved. Assuming a potential difference of 200 MV between the cloud and the ground, estimate the energy dissipated.

7. Find the potential difference across an electrical device which consumes 1200 J of electrical energy when a steady current of 2 A flows through it for a period of 10 s.

8. Water at 10 °C passes into a continuous-flow electric water heater at a rate of 100 g per minute and emerges at 15 °C. Assuming there are no heat losses, find the current in the heater element if the potential difference across it is 14 V.

9. An electrical machine operates with an efficiency of 80% while drawing a current of 6 A at 160 V. How much power is wasted?

TOPIC 30 RESISTANCE

COVERING:

- simple measurement of resistance;
- resistance and resistivity;
- I–V characteristics;
- resistors in series and in parallel;
- e.m.f.;
- internal resistance;
- power.

In the previous topic we saw how metal conductors tend to resist the flow of electric charge that constitutes an electric current. We have now reached the point where we need to be able to quantity this electrical resistance.

We already know that a potential difference is needed to make a current flow through a conductor. Extending this idea, a bad conductor offers greater electrical resistance than a good one; it therefore requires a correspondingly greater potential difference to give the same current under the same conditions. Resistance R is defined as the ratio between the potential difference V across the conductor and the current I that is passing through it. Thus,

$$R = \frac{V}{I} \tag{30.1}$$

The unit of resistance is called the *ohm* (symbol Ω).

To take a simple example, if a bulb operating at 3 V draws a current of 0.25 A, then it has a resistance R, given by

$R = V/I = 3/0.25 = 12 \ \Omega$

Looking at this another way ($I = V/R$), a conductor with a small resistance will allow a large current to pass for a given potential difference. To illustrate this, let us compare the currents drawn by two different bulbs, say with resistances of 12 Ω and 15 Ω, respectively, both operating at 3 V.

For the 12 Ω bulb,

$I = V/R = 3/12 = 0.25$ A

and for the 15 Ω bulb,

$$I = V/R = 3/15 = 0.2 \text{ A}$$

30.1 MEASURING RESISTANCE

Figure 30.1 shows a simple way of measuring the resistance of a bulb, or indeed any other electrical component, using Equation (30.1).

The voltmeter is connected in parallel with the bulb in order to measure the potential difference across it. The current will divide at the point X, some passing through the bulb and the rest through the voltmeter; then it will recombine at the point Y. Ideally a voltmeter should have very high resistance so that it diverts an insignificant proportion of the current from the component across which it is connected. (In terms of $I = V/R$, the larger the value of R the smaller the value of I.) The voltmeter will then cause very little disturbance to the conditions in the main circuit.

The ammeter measures the current and is connected in series with the bulb, so that the same current flows through them both (assuming a negligible current through the voltmeter). Ideally, an ammeter should have very low resistance so that it disturbs the conditions in the circuit as little as possible.

The resistance of the component is then calculated from the voltmeter and ammeter readings by use of Equation (30.1).

We shall look at another method of measuring resistance in the next topic.

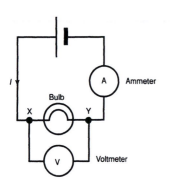

Figure 30.1

30.2 RESISTANCE AND RESISTIVITY

The resistance of a conductor (e.g. a wire) depends on its dimensions. If we double its cross-sectional area without changing anything else, then twice as much charge can pass and the current will be doubled. (The same effect would be achieved by connecting a second identical conductor in parallel with the first.) On the other hand, if we double the length (in effect, adding a second identical conductor in series with the first), then the current has to travel twice as far and will therefore experience twice the resistance. If nothing else is changed, the current will be halved (because $I = V/R$).

These observations are summarised by the equation

$$R = \rho \times \frac{L}{A} \tag{30.2}$$

where L and A are the length and cross-sectional area of the conductor in m and m^2, respectively. ρ, the constant of proportionality, represents the *resistivity* of the material from which the conductor is made.

Rearrangement of Equation (30.2) tells us that the unit of ρ is Ω m. Resistivity is a property of enormous variation, ranging, for example, from 1.6×10^{-8} Ω m for silver to around 1×10^{18} Ω m for silica glass. (The reciprocal of resistivity $(1/\rho)$ is called *conductivity* (σ). This is measured in Ω^{-1} m^{-1} or, in SI units, siemens per metre, where siemens (S) is the SI unit of electrical *conductance*. $1\ S = 1\ \Omega^{-1}$; thus, conductance is the reciprocal of resistance.)

Table 30.1

Substance	Resistivity/Ω m
Silver	1.6×10^{-8}
Copper	1.7×10^{-8}
Aluminium	2.8×10^{-8}
Tungsten	5.5×10^{-8}
Nichrome (Ni/Cr alloy)	100×10^{-8}
Germanium	0.5
Silicon	2.5×10^{3}
Silica glass	1×10^{18}

Table 30.1 shows the approximate resistivities of various materials at ordinary temperatures.

Copper has slightly higher resistivity than silver but is much more widely used as an electrical conductor, because of the cost factor. Aluminium is used in overhead power lines, because, although its resistivity is around one and a half times that of copper, its density is about three times less. Thus, on a weight basis (important for suspended cables), aluminium is an appropriate choice. Tungsten is used for light bulb filaments, which get very hot, because it has a very high melting point (about 3400 °C). Nichrome, which is a nickel/chromium alloy with fairly high resistivity, is used for making heating elements.

Knowing the dimensions of a conductor, Equation (30.2) enables us to calculate its resistance from its resistivity, or vice versa. It also suggests that, if we are simply interested in transporting electricity, we should make conductors as thick and as short as possible in order to minimise their resistance. If the resistance is high, then electrical energy will be wasted in heating the conductor and there will be a voltage drop along its length (given by $V = IR$).

30.3 *I–V* CHARACTERISTICS OF A METALLIC CONDUCTOR

We would normally expect the current flowing through a metallic conductor to increase if we increase the potential difference across its ends. Figure 30.2 illustrates several general points about the relationship between current and voltage.

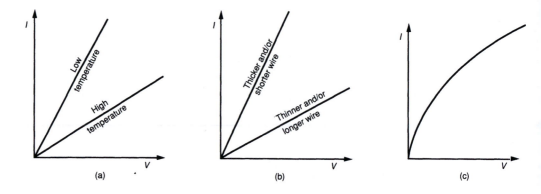

Figure 30.2

Figure 30.2(a) shows the *I–V* characteristics of an ordinary metal conductor determined at two different temperatures. As we saw in the previous topic, the electrical conductivity of a metal decreases with increasing temperature. This is reflected in Figure 30.2(a), which shows that the conductor allows less current to pass at high temperature. The figure also shows that, at constant temperature, current is proportional to potential difference. In other words, the *I–V* characteristic is linear. From this we can say that the resistance of the metal is constant, because the ratio V/I (= R) is the same at any point on the line. As Figure 30.2(a) stands, the steeper the line the lower the resistance, since the slope I/V (the reciprocal of the resistance) represents the conductance.

The proportionality between current and potential difference is described by Ohm's law (named after Georg Ohm, who discovered it early in the nineteenth century). This may be expressed in the form

$$\frac{V}{I} = \text{constant}$$

It is most important to recognise that this equation is not a restatement of Equation (30.1) ($R = V/I$), which merely defines resistance under particular current and voltage conditions. The resistance of many materials changes with voltage. Ohm's law describes the relationship between *I* and *V* only for those whose resistance remains constant.

Figure 30.2(b) simply follows from Equation (30.2) and shows the effect of changing the dimensions of a conductor, say a wire, made from a given material. A thicker and/or shorter wire will allow more current to pass than a thinner and/or longer wire under the same conditions.

Under constant conditions, metals generally obey Ohm's law. They tend to deviate from linear behaviour if their temperature is not kept constant. For example, Figure 30.2(c) shows the *I–V* characteristic of a bulb filament. The slope of the curve progressively decreases as the potential difference is increased. This is because the resistance of the filament increases as it becomes hotter.

30.4 RESISTORS

Although we have discussed resistance in terms of conductors and light bulb filaments, there are electronic components called *resistors* which are specifically designed to provide resistance in electrical circuits. Resistors are made from a variety of materials – for example, from wire wound into a coil, or from metal oxides or carbon. There are two main types – namely fixed and variable. Fixed resistors, with a fixed value, are simply provided with a connection at either end. Variable resistors consist of a track, sometimes a wire-wound coil, with a sliding contact that can be moved along its length. The resistance over the total track length is, of course, fixed (and there are normally connections at either end), but intermediate values can be tapped off by using the sliding contact to vary the length of wire or track through which current has to pass.

Resistors commonly have values of thousands or even millions of ohms, in which case the convenient units to use are kilohms (kΩ) or megohms (MΩ), respectively.

Where a number of resistors are interconnected, we can use Ohm's law to evaluate their combined effect in terms of a single resistance value.

30.5 RESISTORS IN SERIES

Figure 30.3 shows three resistances (R_1, R_2 and R_3) connected in series, so the same current I will pass through them all (as can readily be demonstrated by connecting an ammeter into any part of the circuit). Between entering R_1 and leaving R_3, each coulomb will have lost energy, as heat, equivalent to the total drop in potential V across the resistors.

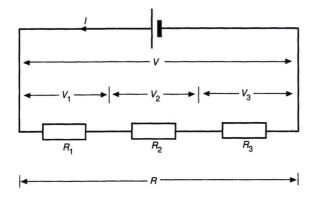

Figure 30.3

The total potential drop will be the sum of the separate drops across the individual resistors (as can readily be demonstrated with a voltmeter). In general,

$$V = V_1 + V_2 + V_3 + \ldots V_n$$

and, since $V = IR$,

$$IR = IR_1 + IR_2 + IR_3 + \ldots IR_n$$

Dividing both sides by I, which is constant throughout,

$$R = R_1 + R_2 + R_3 + \ldots R_n \tag{30.3}$$

Thus, the total combined resistance is the sum of the individual values.

Worked Example 30.1

Find the current that flows when a potential difference of 12 V is maintained across a 2 Ω resistor and a 4 Ω resistor connected in series. Hence find the potential difference across each resistor.

The circuit is shown in Figure 30.4.

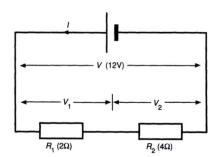

Figure 30.4

From Equation (30.3), the resistors have a combined value of 2 Ω + 4 Ω = 6 Ω. A potential difference of 12 V across a resistance of 6 Ω will result in a current I, given by

$$I = V/R = 12/6 = 2 \text{ A}$$

The potential difference V_1 and V_2 across each resistor is therefore given by

$$V_1 = IR_1 = 2 \times 2 = 4 \text{ V}$$

and

$$V_2 = IR_2 = 2 \times 4 = 8 \text{ V}$$

The example above shows that two resistors in series divide the total voltage across them in the ratio of their respective resistance values. The general case is easy to prove. The current is the same throughout the circuit; therefore,

$$I = \frac{V_1}{R_1} = \frac{V_2}{R_2}$$

so

$$\frac{V_1}{V_2} = \frac{R_1}{R_2} \tag{30.4}$$

Furthermore, if V is the total voltage across the series combination, then

$$I = \frac{V}{R_1 + R_2}$$

and, since $V_1 = IR_1$,

$$V_1 = \frac{V}{R_1 + R_2} \times R_1 \tag{30.5}$$

and similarly for V_2.

This is the basis of the *potential divider*, which is a device used to divide a voltage into given fractions.

30.6 RESISTORS IN PARALLEL

Figure 30.5 shows three resistances (R_1, R_2 and R_3) connected in parallel. In this case it is the potential difference V across them that is fixed, while the current is divided between them. The total current I is given by

$$I = I_1 + I_2 + I_3 + \ldots I_n$$

and, since $I = V/R$

$$\frac{V}{R} = \frac{V}{R_1} + \frac{V}{R_2} + \frac{V}{R_3} + \ldots \frac{V}{R_n}$$

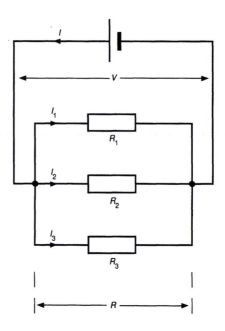

Figure 30.5

Dividing both sides by V, which is constant,

$$\frac{1}{R} = \frac{1}{R_1} + \frac{1}{R_2} + \frac{1}{R_3} + \ldots \frac{1}{R_n} \tag{30.6}$$

Thus, resistances in parallel have a combined value less than any of the individual values.

(Be very careful not to confuse these equations with the corresponding equations for capacitors. Remember that, for series combinations,

$$R = R_1 + R_2 + R_3 + \ldots R_n$$

and

$$\frac{1}{C} = \frac{1}{C_1} + \frac{1}{C_2} + \frac{1}{C_3} + \ldots \frac{1}{C_n}$$

and for parallel combinations,

$$\frac{1}{R} = \frac{1}{R_1} + \frac{1}{R_2} + \frac{1}{R_3} + \ldots \frac{1}{R_n}$$

and

$$C = C_1 + C_2 + C_3 + \ldots C_n)$$

Worked Example 30.2

Find the current that flows when a 2 Ω resistor and a 4 Ω resistor are connected in parallel across a 12 V battery.

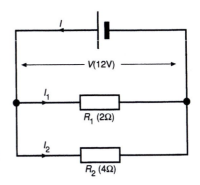

The circuit is shown in Figure 30.6
From Equation (30.6), the resistors have a combined value given by

$$\frac{1}{R} = \frac{1}{2} + \frac{1}{4} = 0.75$$

Therefore, $R = 1/0.75 = 1.33$ Ω.
A potential difference of 12 V across a resistance of 1.33 Ω will result in a current I, given by

$$I = V/R = 12/1.33 = 9 \text{ A}$$

Figure 30.6

(Alternatively, the currents I_1 and I_2 flowing through each resistor are given by

$$I_1 = V/R_1 = 12/2 = 6 \text{ A}$$

and

$$I_2 = V/R_2 = 12/4 = 3 \text{ A}$$

Therefore,

$$I = I_1 + I_2 = 9 \text{ A})$$

Worked Example 30.3

For the circuit in Figure 30.7, (a) find the potential difference between A and B and between B and C, and (b) find the current passing through the 20 Ω resistor and through the 30 Ω resistor.

The resistance R_{BC} of the parallel combination between B and C is given by

$$\frac{1}{R_{BC}} = \frac{1}{20} + \frac{1}{30}$$

which gives $R_{BC} = 12$ Ω.

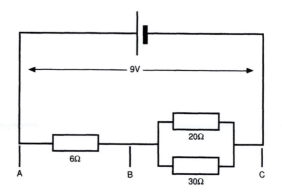

Figure 30.7

The total resistance R_{AC} is 6 Ω + 12 Ω = 18 Ω. Since the potential difference across AC is 9 V, the current drawn by the total combination is given by

$$I = V/R = 9/18 = 0.5 \text{ A}$$

(a) The potential difference V_{AB} across AB is given by

$$V_{AB} = I \times R_{AB} = 0.5 \times 6 = 3 \text{ V}$$

and the potential difference V_{BC} across BC is given by

$$V_{BC} = I \times R_{BC} = 0.5 \times 12 = 6 \text{ V}$$

(b) Since the potential difference across BC is 6 V, the current passing through the 20 Ω resistor is given by

$$I = V/R = 6/20 = 0.3 \text{ A}$$

and through the 30 Ω resistor by

$$I = V/R = 6/30 = 0.2 \text{ A}$$

30.7 E.M.F. AND INTERNAL RESISTANCE

The basic function of an electric cell or battery is to use chemical energy to raise charge through a potential difference. Equation (27.3) ($W = QV$) tells us, for example, that a 3.0 V cell gives 3.0 J of energy to each coulomb passing through it.

When a cell is not being used, the voltage across its terminals is called the *electromotive force* or *e.m.f.* (symbol E), as indicated in

Figure 30.8(a). The e.m.f. is the electrical energy supplied by the cell per coulomb. If the cell is being used to maintain a current, say through the resistance R in Figure 30.8(b), then the potential difference V across the resistor, and across the cell, is found to be less than the e.m.f. The reason for this is that the cell has its own *internal resistance* as charge moves through its interior. This means that, when a current flows around the circuit, electrical energy is consumed by the cell as well as by the resistor. It can be helpful to view internal resistance as a component of the circuit into which the cell is connected. Figure 30.8(b) is therefore redrawn in Figure 30.8(c) to show the circuit as a cell of e.m.f. E and internal resistance r connected across the resistance R.

R is in series with r, so the total resistance in the circuit is $(R + r)$. From Ohm's law, the current I flowing through the circuit will be

$$I = \frac{E}{R + r} \tag{30.7}$$

which, on rearranging, gives

$$E = IR + Ir = V + Ir$$

where $V \, (= IR)$ represents the potential difference across the cell terminals. Ir is the inaccessible voltage lost across the resistance of the cell which produces heat and which only becomes evident when a current I flows. On rearranging,

$$V = E - Ir \tag{30.8}$$

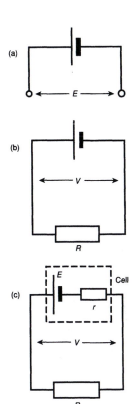

Figure 30.8

That is to say, when current passes through the circuit, each coulomb of charge passing through the cell gains E joules, but also loses Ir joules to the internal resistance. This leaves V joules for external use. Because the magnitude of Ir increases with I, the usable voltage V of the cell decreases in proportion to the current drawn from it. Equation (30.8) enables us to calculate the internal resistance of a cell or other source of e.m.f. simply by subtracting the potential difference across its terminals from the e.m.f. and dividing the result by the current.

30.8 BATTERIES

As we noted in Topic 28, a battery is simply a number of cells connected together to form a single unit. Batteries may have cells connected in series or in parallel.

If the cells are in series with positive terminals connected to negative, as in Figure 30.9(a), then their combined e.m.f. and their combined internal resistance are both obtained by adding the individual values together as follows:

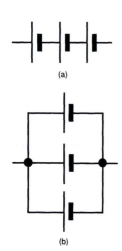

(a)

(b)

Figure 30.9

$$E = E_1 + E_2 + E_3 + \ldots E_n \tag{30.9}$$

and

$$r = r_1 + r_2 + r_3 + \ldots r_n \tag{30.10}$$

For example, six 2.0 V lead–acid cells are used in series in 12 V car batteries. (Lead–acid cells have low internal resistance, which allows high currents to be drawn briefly for starting engines.)

If identical cells are connected in parallel, as in Figure 30.9(b), then their combined e.m.f. is the same as a single cell but their *capacity* is correspondingly increased. Capacity is measured in ampere-hours, 1 ampere-hour being the charge passing a point in 1 h in a conductor carrying a steady current of 1 A. (In practice, the actual capacity generally depends on the rate at which a cell is discharged.) As we might expect from Equation (30.6), the combined resistance is given by r/n, where r is the internal resistance of each of the n identical parallel cells.

The parallel combination of different types of cell is more complex and we shall consider it in the next topic.

30.9 POWER

Equation (29.3) ($P = IV$) applies to any device converting electrical energy to another form. If the device is a resistor that obeys Ohm's law and converts the electrical energy entirely to heat, then we can modify the equation by introducing R and eliminating I or V as required. Thus,

$$P = IV$$

and, since $I = V/R$,

$$P = \frac{V^2}{R} \tag{30.11}$$

Also, since $V = IR$,

$$P = I^2R \tag{30.12}$$

Worked Example 30.4

A potential difference of 12 V is maintained across an 8 Ω and a 16 Ω resistor connected (a) in series, and (b) in parallel. Find the electrical power consumed in each case by each resistor.

The circuits are shown in Figure 30.10.

(a) The total resistance of the series combination in Figure 30.10(a) is (8 Ω + 16 Ω =) 24 Ω; therefore, the current I flowing round the circuit is given by

$$I = \frac{V}{R} = \frac{12}{24} = 0.5 \text{ A}$$

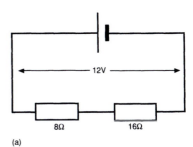

(a)

The power consumed by the 8 Ω resistor is given by

$$P = I^2 \times R = 0.5^2 \times 8 = 2 \text{ W}$$

and by the 16 Ω resistor by

$$P = I^2 \times R = 0.5^2 \times 16 = 4 \text{ W}$$

(b) The power consumed by the 8 Ω resistor is given by

$$P = \frac{V^2}{R} = \frac{12^2}{8} = 18 \text{ W}$$

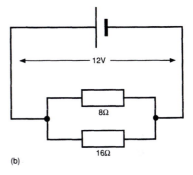

(b)

and by the 16 Ω resistor by

Figure 30.10

$$P = \frac{V^2}{R} = \frac{12^2}{16} = 9 \text{ W}$$

Questions

(Use any previously tabulated data as required.)

1. Find the resistance of a 100 m length of copper wire 1.2 mm in diameter.

2. A copper wire, 10 m long and 1.7 mm^2 in cross-sectional area, carries a current of 15 A. Find the potential drop along its length.

3. The individual resistors shown in Figure 30.11 all have the value of 3 Ω. Find the combined resistance across AB in each case.

4. What value resistor must be connected with a 20 Ω resistor to give a combined value of 12 Ω? Should it be connected in series or in parallel?

5. (a) Show that the combined resistance R of two resistors R_1 and R_2 connected in parallel is given by

$$R = \frac{R_1 R_2}{R_1 + R_2}$$

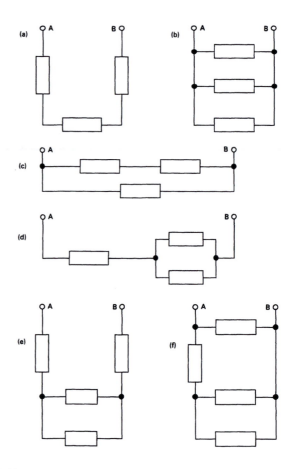

Figure 30.11

(b) Two resistances connected in parallel have a combined value of 1.2 Ω. When connected in series, they have a combined value of 5 Ω. Find their individual values.

6. For the circuit shown in Figure 30.12, find (a) the potential difference across the 20 Ω resistor, (b) the

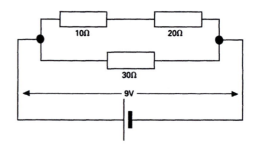

Figure 30.12

combined resistance of the three resistors, and (c) the current through the 10 Ω resistor.

7. Two resistors are connected in series across a 3 V supply. One of the resistors, with a value of 200 Ω, has a potential difference of 0.1 V across it. Find the value of the other resistance.

8. (a) A battery of 4.5 V e.m.f. with an internal resistance of 1.25 Ω is used to maintain a current of 0.2 A through a resistor. Find the potential difference across the resistor.
   (b) Find the potential difference if the resistor is changed for one that draws a current of 0.8 A.
   (c) Estimate the current which flows if the terminals of the battery are momentarily short-circuited with a very low resistance conductor.

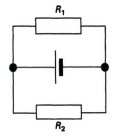

Figure 30.13

9. In the circuit shown in Figure 30.13 the cell has an e.m.f. of 6 V and an internal resistance of 0.5 Ω. R_1 and R_2 are resistors of 3 Ω and 6 Ω, respectively. Find the current through (a) the cell, (b) R_1 and (c) R_2.

10. Two identical cells, each with an e.m.f. of 4.5 V and an internal resistance of 1.5 Ω, are connected (a) positive terminal to negative in series and (b) positive terminal to positive in parallel across a 3 Ω resistance. Find the current through the resistance in each case.

11. A current of 1.2 mA flows through a resistor across which there is a potential difference of 6 V. Find the resistance and the power consumption of the resistor.

12. The maximum power rating of a particular 100 Ω resistor was given as 2.25 W. Find the maximum current that it would safely carry.

13. By considering the units on the right-hand sides of Equation (29.3) (page 283) and Equations (30.11) and (30.12), confirm that the quantity obtained in each case is power.

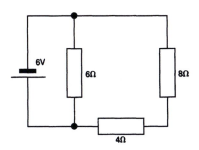

Figure 30.14

14. For the circuit in Figure 30.14 find (a) the power supplied by the battery and (b) the power dissipated in the 8 Ω resistor. (Assume that the internal resistance of the battery is negligible.)

TOPIC 31 SOME SIMPLE CIRCUITS

COVERING:

- shunts and multipliers (ammeters and voltmeters);
- potential dividers;
- the potentiometer;
- the Wheatstone bridge;
- Kirchhoff's laws.

The purpose of this topic is to develop some of the ideas that we have already met and to broaden our discussion of electrical circuits.

31.1 SHUNTS AND MULTIPLIERS

In this section we shall consider ammeters and voltmeters based on the moving-coil galvanometer. Nowadays these are tending to be superseded by digital electronic instruments; nevertheless they provide a good basis for our discussion.

The moving-coil galvanometer is an instrument that measures small electric currents. Essentially it consists of a coil mounted between the poles of a permanent magnet in such a way that the coil rotates when a current is passed through it. (We shall consider this in more detail in the next topic.) Rotation is resisted by a spring, so that the angle through which the coil moves gives a measure of the current. This is usually indicated by the position of a needle against a calibrated scale. As far as its behaviour as a circuit element is concerned, the instrument can generally be treated as a resistor because of the resistance of its internal wiring, particularly that of the coil.

Moving-coil galvanometers usually need to be modified for use as ammeters, because they are generally too sensitive for normal currents and the needle tends to go off-scale. The range of the instrument needs to be adjustable, so that the current corresponding to full-scale deflection of the needle can be varied as required. This is very easily done using a *shunt*, which is a resistance connected in parallel with the galvanometer that allows a fixed proportion of the total current to bypass the instrument itself, as in Figure 31.1

R_g is the resistance of the galvanometer G. By adjusting the shunt resistance R_s, we can vary the proportion in which the total current I is divided into I_g and I_s, the parallel currents through the galvanometer

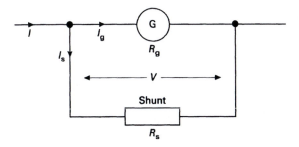

Figure 31.1

and through the shunt, respectively. Since the potential difference V across both resistances is the same, then, from Ohm's law,

$$V = I_s R_s = I_g R_g$$

Therefore,

$$R_s = \frac{I_g R_g}{I_s}$$

and, since $I_s = I - I_g$,

$$R_s = \frac{I_g R_g}{I - I_g} \tag{31.1}$$

By adjusting R_s we can adjust the range of the instrument to suit the magnitude of the current being measured. Obviously, I_g must not exceed the full-scale deflection current; otherwise the needle will go off scale.

Worked Example 31.1

A moving-coil galvanometer shows full-scale deflection with a potential difference of 75 mV across the terminals and a current of 15 mA flowing through the coil. How can the instrument be adapted to measure currents up to 2.5 A?

The resistance of the galvanometer is obtained from Equation (30.1) (on page 287) as follows:

$$R = \frac{V}{I} = \frac{75 \times 10^{-3}}{15 \times 10^{-3}} = 5 \ \Omega$$

If, in Figure 31.1, I_g is not to exceed 0.015 A, then a shunt is required which allows (2.5 − 0.015) A to bypass the galvanometer when the total current I is 2.5 A. The shunt resistance may therefore be obtained from Equation (31.1) as follows:

$$R_s = \frac{I_g R_g}{I - I_g} = \frac{15 \times 10^{-3} \times 5}{(2.5 - 0.015)} = 0.03 \ \Omega$$

Since a potential difference applied across the terminals of a galvanometer will cause a current to flow, we can use it as a voltmeter. If the wiring inside the galvanometer obeys Ohm's law, then there will be a linear relationship between V and I and it can be calibrated to measure either. But, as with the ammeter, we need to be able to vary the sensitivity of the instrument. In this case we use a *multiplier*, as in Figure 31.2.

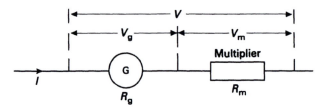

Figure 31.2

A multiplier is simply a series resistance, R_m in the figure. Together with the galvanometer resistance R_g, it divides the potential difference V in accordance with Equation (30.4) (page 293). Since the current I is the same through both resistances, because they are in series, then

$$I = \frac{V_g}{R_g} = \frac{V_m}{R_m}$$

Therefore,

$$R_m = \frac{R_g V_m}{V_g}$$

and, since $V_m = V - V_g$,

$$R_m = \frac{R_g (V - V_g)}{V_g} \tag{31.2}$$

Thus, we can adjust the range of the voltmeter by varying the multiplier resistance, remembering that V_g must not exceed the full-scale deflection voltage.

Worked Example 31.2

How can the moving-coil galvanometer in Worked Example 31.1 be adapted to measure voltages up to 12 V?

If, in Figure 31.2, V_g is not to exceed 0.075 V, then a multiplier is required across which there is a potential difference of $(12 - 0.075)$ V when the total potential difference is 12 V. The multiplier resistance may therefore be obtained from Equation (31.2) as follows:

$$R_m = \frac{R_g (V - V_g)}{V_g} = \frac{5(12 - 0.075)}{0.075} = 795 \ \Omega$$

The ideal voltmeter has infinite resistance and the ideal ammeter has zero resistance, so that they have no effect on the current flowing through the circuit where they are being used. In practice, real moving-coil instruments have finite resistance, which can affect the measurements significantly. For example, let us consider the measurement of resistance by the so-called *ammeter–voltmeter method* which we met in the previous topic (Figure 30.1 on page 288). The two arrangements in Figure 31.3 incorporate a *rheostat* to control the current through the circuit. (A rheostat is a variable resistor connected so that current flows between one of the end connections and the sliding contact.)

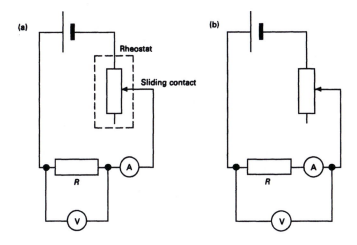

Figure 31.3

If the resistance R being measured in Figure 31.3(a) has a sufficiently high value, then the current flowing through it may be low enough for the current through the voltmeter V to be significant. This problem can be overcome by connecting the voltmeter across both the resistance and the ammeter A, as in Figure 31.3(b), so that the ammeter measures the true current through the resistance. Note that this arrangement will only give a satisfactory measurement where the resistance of the ammeter is negligible compared with the component under test; otherwise there will be a significant voltage drop across the ammeter.

31.2 POTENTIAL DIVIDERS

In the previous topic we met a potential divider in the form of two resistors in series that enable a fraction of the total voltage across them to be tapped off according to Equation (30.5) (page 293). Potential dividers take a variety of different forms, including a chain of any number of series resistors dividing the voltage across it into as many portions as there are individual resistances.

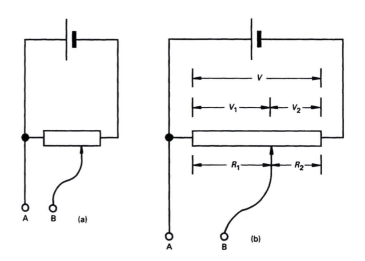

Figure 31.4

Figure 31.4(a) shows a potential divider based on a variable resistor with both ends connected across a cell. The sliding contact divides the track into two parts, corresponding to R_1 and R_2 in Equation (30.5), as shown in Figure 31.4(b). By adjusting the position of the sliding contact, the potential difference across the output terminals AB may be varied from zero up to the maximum potential difference V across the ends of the variable resistor. As the following worked example demonstrates, this voltage will fall when an external current is drawn from the output terminals.

Worked Example 31.3

For the potential divider shown in Figure 31.4, $R_1 = 200\ \Omega$, $R_2 = 100\ \Omega$ and $V = 12$ V. Assuming the battery has negligible internal resistance, find the potential difference across the output terminals AB (a) when no external current is being drawn from them, and (b) when an external resistance of 200 Ω is connected across them.

(a) From Equation (30.5),

$$V_1 = \frac{R_1}{R_1 + R_2} \times V = \frac{200}{200 + 100} \times 12 = 8\ \text{V}$$

(and $V_2 = 4$ V).

(b) With the external resistance R_e connected, the combined resistance R_{AB} across AB is given by

$$\frac{1}{R_{AB}} = \frac{1}{R_1} + \frac{1}{R_e} = \frac{1}{200} + \frac{1}{200}$$

Hence, $R_{AB} = 100\ \Omega$.

Since $R_{AB} = R_2 = 100\ \Omega$, the voltage V is divided into equal halves (i.e. $V_1 = V_2 = 6$ V). The potential difference across AB is therefore reduced to 6 V by the effect of the external resistance. (Note that the greater the external resistance the less will be the current drawn by it and the less the reduction in the potential difference across AB.)

The *potentiometer* (see Figure 31.5a) is a very simple form of potential divider. It consists of a resistance wire of uniform cross-section which has a constant potential difference maintained across its ends by means of a so-called driver cell. A sliding contact is used to tap off any fraction of the total potential difference. A scale is fixed parallel to the wire so that the lengths l_1 and l_2 (corresponding to R_1 and R_2 in Figure 31.4) can be measured. The galvanometer G has a central zero position so that it can detect current flowing in either direction when other circuit elements are connected across AB.

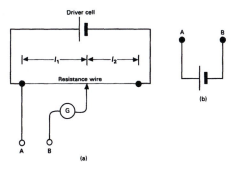

Figure 31.5

The potentiometer measures potential difference and can be used to find the true e.m.f. of a cell. The cell is connected across AB so that it is in opposition to the driver (as indicated in Figure 31.5b). Provided that the cell e.m.f. is less than the potential difference between the ends of the potentiometer, then a length of wire l_1 can be found where the potential difference across it exactly balances the e.m.f. of the cell. Under these conditions, no current flows through the galvanometer and the potentiometer is said to be *balanced*. Since no current is flowing through the cell under test, the potential difference across its terminals is equal to its e.m.f. (because $Ir = 0$ in Equation 30.8 on page 297). Thus, the balanced length l_1 is proportional to the e.m.f. of the cell. If this procedure is repeated with another cell, then

$$\frac{E_A}{E_B} = \frac{l_A}{l_B} \tag{31.3}$$

where E_A and E_B are the e.m.f. values and l_A and l_B are the respective values of the balance length l_1 for the two cells. The e.m.f. of an unknown cell can therefore be obtained by comparing it with a cell of accurately known e.m.f.

The potentiometer needs modification for measuring very small e.m.f. values, because the balance length will be very small and subject to large errors in its measurement. The problem could be solved by using a very long potentiometer wire but this would generally be impracticable. However, the same effect can be achieved by connecting a large resistor in series with the wire, as in Figure 31.6(a). The total voltage across the series pair is then divided according to Equation (30.5) (page 293), where R_1 is the resistance of the whole length of the wire and R_2 is the added series resistance.

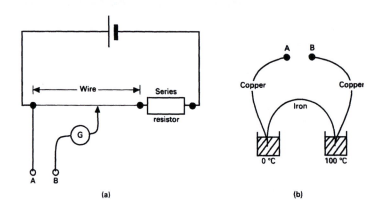

Figure 31.6

This arrangement can be used to measure the small e.m.f. produced by a *thermocouple*. Thermocouples are devices that are used for measuring temperature. For example, if a piece of iron wire is connected to the

potentiometer via two copper wires and the two iron–copper junctions
are at different temperatures, say 0°C and 100°C as in Figure 31.6(b),
then a small e.m.f. can be measured. This e.m.f. varies with the
temperature difference and the device can therefore be used as a
thermometer by keeping one of the junctions at a known reference
temperature.

Since the potentiometer draws no external current, it behaves, in
effect, like a voltmeter of infinitely high resistance and therefore does
not disturb any circuit into which it is connected. It can be used to
measure potential difference accurately, because the balance length can
be measured accurately. Being a *null method* (i.e. a balance method),
it does not rely on the accuracy of the galvanometer. It therefore has a
number of advantages over ordinary voltmeters and can even be used
to calibrate them. (*Calibration* is the determination of the true values
corresponding to the actual readings given by any type of instrument.)

A variable potential difference can be applied across the terminals
of an ordinary voltmeter by using a variable resistor as a potential
divider (as in Figure 31.4). For any given reading on the voltmeter,
the true potential difference across its terminals can be measured by
using the potentiometer. The voltmeter can therefore be calibrated over
its whole range by varying the potential difference.

The potentiometer can be used to measure current by finding the
potential difference across a known resistance through which the current
is passing and then applying Ohm's law. This may involve inserting into
the circuit an accurately known resistance of sufficiently low value not to
disturb the current. This principle can be used to calibrate an ammeter by
measuring the true current passing through it for any given reading.

The value of an unknown resistance can be obtained by connecting
a known resistance in series with it and using the potentiometer to
measure the potential difference across each resistance in turn when
they are both carrying the same current. If V_A and V_B are the potential
differences corresponding to the balance lengths l_A and l_B for the
resistances R_A and R_B, respectively, then

$$\frac{l_A}{l_B}\left(=\frac{V_A}{V_B}=\frac{IR_A}{IR_B}\right)=\frac{R_A}{R_B} \qquad (31.4)$$

(Before we leave this section, note that variable resistors used in
electronic circuits are sometimes referred to as potentiometers.)

31.3 THE WHEATSTONE BRIDGE

The Wheatstone bridge is a circuit that can be used for the accurate
measurement of resistance. It consists of a network of four resistances
R_1, R_2, R_3 and R_4, connected as shown in Figure 31.7. For the purposes
of our discussion let us assume that the voltages across them are
V_1, V_2, V_3 and V_4, respectively.

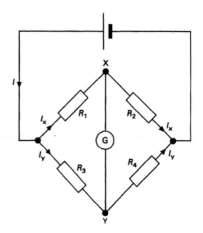

Figure 31.7

If the values of the resistances are such that no current flows through the galvanometer, then the bridge is said to be *balanced*. The potential difference across XY must then be zero and therefore $V_1 = V_3$ and $V_2 = V_4$. Furthermore, the current I entering the network divides into I_x through R_1 and R_2, and I_y through R_3 and R_4 (remembering that none flows through the galvanometer).

If $V_1 = V_3$, then $I_x R_1 = I_y R_3$ (since $V = IR$), and if $V_2 = V_4$, then $I_x R_2 = I_y R_4$.

Dividing one equation by the other,

$$\frac{I_x R_1}{I_x R_2} = \frac{I_y R_3}{I_y R_4}$$

Therefore,

$$\frac{R_1}{R_2} = \frac{R_3}{R_4} \tag{31.5}$$

Knowing the values of three of the resistances, the fourth may be found.

The *metre bridge*, shown in Figure 31.8, is a simple practical version

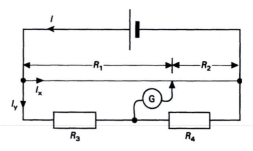

Figure 31.8

of the Wheatstone bridge. It consists of a resistance wire 1 m long which enables the ratio R_1/R_2 to be continuously varied with a sliding contact. Knowing this ratio and the value of either R_3 or R_4, the value of the fourth resistance can be found from Equation (31.5).

31.4 KIRCHHOFF'S LAWS

Kirchhoff's laws are useful when it comes to considering steady currents flowing through circuits which are too complicated to be treated as series and parallel combinations of resistances and e.m.f.s. In such cases we consider circuits in terms of *junctions* (where three or more conductors meet) and closed *loops*. We have already met the basic ideas involved. Now we shall formalise them.

The first law states that the sum of the currents entering a junction in a circuit is equal to the sum of the currents leaving it. In mathematical terms $\Sigma I = 0$ (where Σ means 'the sum of'). Currents arriving are normally treated as positive and those leaving as negative. In terms of Figure 31.9, which shows two currents arriving at a junction and three leaving it,

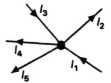

Figure 31.9

$$I_1 - I_2 + I_3 - I_4 - I_5 = 0$$

or

$$I_1 + I_3 = I_2 + I_4 + I_5$$

This is the statement of a special case of the principle of conservation of charge. It also expresses the idea that charge does not accumulate at any point under steady state conditions.

Kirchhoff's second law tells us that, in any closed loop in a circuit, the sum of the e.m.f.s is equal to the sum of the potential differences across the resistances in the loop. In mathematical terms, $\Sigma E = \Sigma IR$. This is, of course, the statement of a special case of the principle of conservation of energy. As we shall see, we have to think carefully about the signs in applying the second law. If we follow the conventional current flowing round a simple circuit, then the e.m.f. of a cell (or any source of e.m.f.) represents a rise in potential and the potential difference across a resistance represents a drop in potential. Figure 31.10 shows a more awkward case where two cells are connected in parallel to form a closed loop which is part of a larger circuit. E_1 and E_2 are the e.m.f.s of the cells and r_1 and r_2 are their respective internal resistances.

Applying the first law to the junctions at either end of the loop simply tells us that the sum of the currents I_1 and I_2 through the two branches of the loop is equal to the total current I_3.

We can apply the second law by travelling either clockwise or anticlockwise round the loop formed by the two branches. In either case we equate the sum of the e.m.f.s (i.e. the potential rises) to the

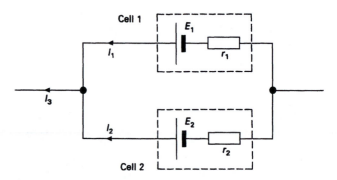

Figure 31.10

sum of the potential drops, taking into account our direction of travel. We take an e.m.f. as positive if we enter the cell via the negative terminal and exit via the positive; we take a potential drop as positive if we travel through the resistance in the same direction as the current. Thus, if we choose to travel clockwise round the loop in Figure 31.10, then E_2 and $I_2 r_2$ are positive, while E_1 and $I_1 r_1$ are negative. We have

$$\Sigma E = E_2 - E_1$$

and

$$\Sigma IR = I_2 r_2 - I_1 r_1$$

Therefore,

$$E_2 - E_1 = I_2 r_2 - I_1 r_1$$

The following worked example illustrates a simple application of Kirchhoff's laws. First, we identify the current passing through each branch. (In some cases it may be difficult to decide the current direction, but if the wrong assumption is made, then the calculated value will simply be negative.) Next we apply the first law to the junctions, then the second law to the closed loops.

Worked Example 31.4

A 10 Ω resistance is connected in parallel with two cells simultaneously, one of 3 V e.m.f. and 1 Ω internal resistance and the other of 6 V e.m.f. and 2 Ω internal resistance, with their positive terminals together. Find the current in each branch of the circuit.

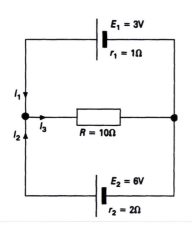

Figure 31.11

The circuit is shown in Figure 31.11.

Let us choose I_1 and I_2 as the currents in the upper and lower branches in the directions shown.

Applying the first law to the left-hand junction, I_1 and I_2 combine to give I_3, which is the current through the resistance in the middle branch.

Applying the second law to the upper loop (containing cell 1 and the resistance R), then, in the anticlockwise direction.

$$E_1 = I_1 r_1 + I_3 R$$

Therefore,

$$E_1 = I_1 r_1 + (I_1 + I_2)R$$

and, substituting the given values,

$$3 = 11 I_1 + 10 I_2 \qquad\qquad\qquad \text{(a)}$$

Applying the second law to the lower loop (containing cell 2 and the resistance R), then, in the clockwise direction,

$$E_2 = I_2 r_2 + I_3 R$$

Therefore,

$$E_2 = I_2 r_2 + (I_1 + I_2)R$$

and, substituting the given values,

$$6 = 10 I_1 + 12 I_2 \qquad\qquad\qquad \text{(b)}$$

Combining (a) and (b) to eliminate I_2 gives

$$I_1 = -0.75 \text{ A}$$

and substituting -0.75 A for I_1 in either (a) or (b) gives

$$I_2 = 1.125 \text{ A}$$

Therefore,

$$I_3 = I_1 + I_2 = 0.375 \text{ A}$$

We can check the values of I_1 and I_2 by applying the second law to the outer loop containing just the cells (but not the resistance R). Taking the clockwise direction,

$$\Sigma E = E_2 - E_1 = 6 - 3 = 3 \text{ V}$$

and

$$\Sigma IR = I_2 r_2 - I_1 r_1 = (1.125 \times 2) - (-0.75 \times 1) = 3 \text{ V}$$

(As noted above, the negative value found for I_1 simply tells us that its direction is opposite to that initially chosen and shown in Figure 31.11.)

Questions

1. A battery of 12 V e.m.f. and 15 Ω internal resistance is connected across a voltmeter of (a) 500 Ω resistance (b) 5000 Ω resistance. In each case find the potential difference across the voltmeter.

2. A 9.5 Ω resistance and an ammeter of 0.1 Ω resistance are connected in series with a cell of 1.5 V e.m.f. and 0.4 Ω internal resistance. Find (a) the current passing through the ammeter and (b) the potential difference across the cell.

3. Two 1000 Ω resistors are connected in series across a battery of 6 V e.m.f. and negligible internal resistance.

   (a) Find the potential difference across a voltmeter of 2000 Ω resistance connected in parallel across one of the resistors.
   (b) What would the potential difference be if the voltmeter had infinite resistance?
   (c) What would the resistance of the voltmeter have to be for a potential difference of 2.9 V?

4. (a) In Worked Example 31.3 (page 307), find the potential difference across AB if the value of the external resistance had been 66.7 Ω.
   (b) Find the current passing through R_2 (i) in this arrangement and (ii) in both arrangements in the worked example.

5. Twelve identical pieces of wire, each of 12 Ω resistance, are connected together to form the edges of a cube. A current of 2 A enters this network at one corner of the cube and leaves it by the corner diagonally opposite. (a) Find the current in each wire. Find (b) the potential difference between the points where the current enters and leaves the network and (c) the total resistance between these points.

6. A current of 0.2 A flows through the 50 Ω resistance in the circuit shown in Figure 31.12. Using Kirchhoff's laws, find the value of the resistance R, assuming that the battery has negligible internal resistance.

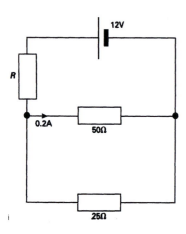

Figure 31.12

7. In the Wheatstone bridge circuit shown in Figure 31.7 (page 310) $R_1 = 10\ \Omega$, $R_2 = R_3 = 20\ \Omega$ and $R_4 = 40\ \Omega$. The battery has an e.m.f. of 6 V and an internal resistance of 2 Ω. Using Kirchhoff's laws, find the current drawn from the battery.

TOPIC 32 MAGNETIC FIELDS

COVERING:

- permanent magnets;
- fields around current-carrying conductors;
- force on a current-carrying conductor in a magnetic field;
- force on a moving charge in a magnetic field;
- torque on a current-carrying coil in a magnetic field;
- force between current-carrying conductors.

We have already seen that an electric charge gives rise to an electric field. In this topic we shall see that if an electric charge is in motion, it will produce a magnetic field as well.

32.1 PERMANENT MAGNETS

The magnetic field surrounding a permanent magnet is associated with the motion of the electrons within its constituent atoms. The earth behaves like a permanent magnet for reasons that are not fully understood but are believed to have their origins in electric currents that circulate in its molten core.

Permanent magnets have equal and opposite north-seeking and south-seeking *poles*, normally called north and south poles, corresponding to the way in which a freely suspended bar magnet aligns itself in the earth's magnetic field. The opposite poles of two magnets attract one another and their like poles repel because of the forces arising from the interaction of their magnetic fields. From this it follows that the earth's north pole is actually a magnetic south pole, because it attracts the north-seeking pole of a compass. Similarly, its south pole is a magnetic north pole. (A compass is simply a small magnet, pivoted to allow it to align itself in a magnetic field.)

Magnetic fields, like electric fields, can be represented by field lines whose concentration indicates the field strength. The field direction at any point is taken to be the direction of the force acting on a north pole placed there. The field pattern round a permanent magnet can be plotted with a small compass, but it can be revealed much more quickly by covering the magnet with a sheet of stiff paper and sprinkling iron filings on top. On gently tapping the paper, the iron filings align themselves in the field. Figure 32.1(a) represents the way in which the field varies around a bar magnet. Such fields are described as non-uniform. By contrast, Figure 32.1(b) shows straight, parallel, equally spaced field lines that are characteristic of uniform fields.

The earth behaves rather as though it contains a bar magnet along its magnetic axis. A compass needle free to rotate in any direction would set horizontally at the magnetic equator and vertically at the magnetic poles, with intermediate angles elsewhere. In the UK it would point downwards at an angle of roughly 70° below the horizontal. (This angle is called the *angle of dip* or the *inclination*.)

32.2 MAGNETIC FIELDS AROUND CONDUCTORS

Since electric current is a flow of charge, we find magnetic fields associated with current-carrying conductors such as wires. In this section we shall consider the field patterns associated with a straight wire, a flat circular coil and a solenoid (i.e. a long cylindrical coil).

Figure 32.2(a) represents the field pattern around a long straight wire running perpendicularly through the page with the current passing downwards into the paper. The cross represents the tail of a departing arrow indicating the current direction. In Figure 32.2(b) the dot represents the tip of an approaching arrow indicating that the current direction is upwards out of the paper. If the current is sufficiently large, the field will be strong enough for interference from the earth's magnetic field to be insignificant and the field lines will form concentric circles around the wire. The field direction, given by the so-called *corkscrew rule*, is the direction of rotation of a right-hand screw thread advancing in the conventional current direction. (Just think of the cross in Figure 32.2(a) as the head of an ordinary screw.)

Figure 32.3 represents the field through a flat circular coil viewed from above, with the plane of the coil set at right angles to the plane of the paper. The cross and the dot indicate the current direction through the opposite sides. Very close to the wire the field pattern takes the form of more or less concentric circles which become progressively distorted further away as the field due to current in other parts of the coil becomes more significant. The direction of the field lines in the figure is still consistent with the corkscrew rule.

Figure 32.1

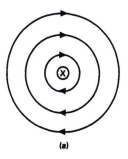

(a)

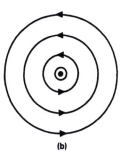

(b)

Figure 32.2

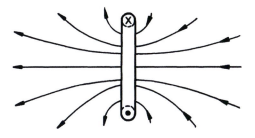

Figure 32.3

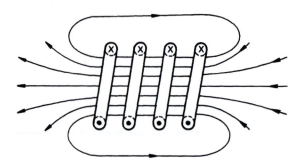

Figure 32.4

Figure 32.4 represents the field associated with a solenoid (somewhat simplified for our purposes). As we can see, there is a region of uniform magnetic field inside. The corkscrew rule still applies and the pattern is essentially an extended version of that associated with a flat coil. It is also similar to that of the permanent bar magnet in Figure 32.1(a). In fact, the solenoid will align itself in a magnetic field in just the same way as a bar magnet, and this leads us on to consider the force experienced by a current-carrying conductor in a magnetic field.

32.3 FORCE ON A CONDUCTOR IN A MAGNETIC FIELD

When a current flows through a wire suspended vertically between the poles of a U-shaped permanent magnet, the wire experiences a force and will move as shown in Figure 32.5. The direction in which it moves is given by *Fleming's left-hand rule*, which is a mnemonic involving the first and second fingers and the thumb of the left hand mutually arranged at right angles. If the *F*irst finger points in the *F*ield direction and the se*C*ond finger points in the *C*urrent direction then the thu*M*b gives the direction of *M*otion.

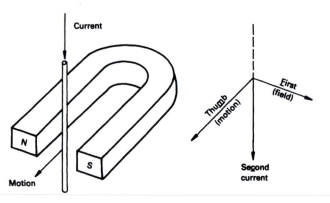

Figure 32.5

The magnitude of the force acting on the wire is proportional to the strength of the field expressed in terms of its *magnetic flux density* or *magnetic induction* (symbol B). The force F, also proportional to the current I and to the length of wire l in the field, is given by

$$F = BIl \qquad (32.1)$$

Rearrangement of this equation ($B = F/Il$) tells us that we could measure flux density by finding the force acting on a metre length of wire carrying a current of 1 A at right angles to the field. The unit of B is $N\ A^{-1}m^{-1}$ and is called the *tesla* (symbol T).

Equation (32.1) only applies when the current direction is perpendicular to the field direction. If the two directions are parallel, then the wire will experience no force at all. If the wire makes an angle θ with the field direction, as in Figure 32.6, then the force acting on the parallel component of the current ($I\cos\theta$) will be zero and the force on the perpendicular component will be proportional to $I\sin\theta$. Equation (32.1) should then be written

$$F = BIl\ \sin\theta \qquad (32.2)$$

When $\theta = 90°$, then $\sin\theta = 1$ and $F = BIl$. When $\theta = 0°$, then $\sin\theta = 0$ and $F = 0$. Remember that the direction of the force is perpendicular to the plane containing the current and field directions.

It is sometimes useful to resolve a magnetic field into components. For example, we noted that, in the UK, the earth's magnetic field dips at an angle of roughly 70° below the horizontal; the vertical and horizontal components of the earth's flux density, B_v and B_h, are therefore related by $B_v/B_h \approx \tan 70°$.

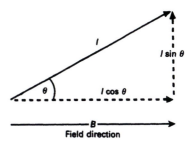

Figure 32.6

32.4 FORCE ON A MOVING CHARGE IN A MAGNETIC FIELD

Equation (32.2) can readily be adapted to obtain the force acting on a single electron in a wire carrying an electric current in a magnetic field. Equation (29.2) (page 280) gives the magnitude of the current in a metal conductor as $I = nvAe$, where n is the number of free electrons per unit volume, v and e are their drift velocity and charge, respectively, and A is the cross-sectional area of the conductor. Substituting this for I in Equation (32.2) gives

$$F = BnvAel\ \sin\theta$$

But the number of free electrons in l metres of wire is nAl; therefore, the force acting on just one electron is given by

$$\frac{F}{nAl} = \frac{BnvAel \sin \theta}{nAl} = Bve \sin \theta$$

In general, a charge q moving with a speed v at an angle θ relative to the direction of a magnetic field B will experience a force F, given by

$$F = Bvq \sin \theta \qquad (32.3)$$

If the charge is moving perpendicularly to the field, then $F = Bvq$ because $\sin 90° = 1$.

From this it follows that a magnetic field will deflect a beam of charged particles passing through it (unless the beam is parallel to the field direction, in which case $\sin 0° = 0$ and $F = 0$). The force acting on the particles will not change their speed, because it always acts at right angles to their path. (Remember that, when applying Fleming's left-hand rule to a beam of negative particles, they travel in the opposite direction to a conventional current.)

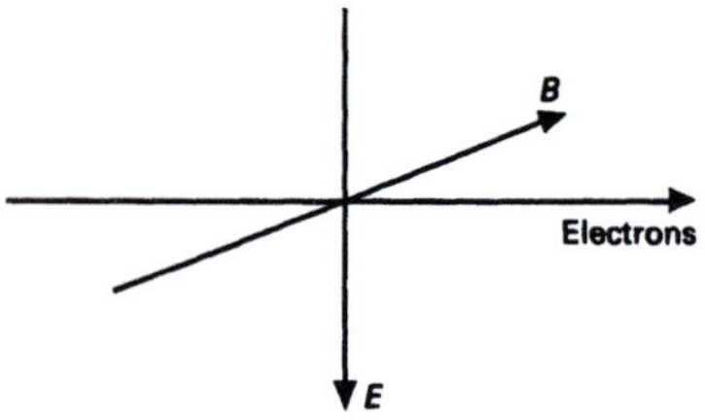

The force due to a magnetic field can be cancelled out by the force due to a superimposed electric field. For example, a beam of electrons will be undeflected when it passes through superimposed magnetic and electric fields if they provide equal and opposite forces acting at right angles to the beam. Since the force due to an electric field of strength E is given by $F = qE$ (Equation 27.1 on page 262) then, when both the fields and the beam direction are mutually perpendicular, as in Figure 32.7, electrons of velocity v and charge e will be undeflected if

$$F = Bve = eE$$

and

$$v = \frac{E}{B} \qquad (32.4)$$

Figure 32.7

Note the field directions in Figure 32.7. In the absence of an electric field, Fleming's left-hand rule tells us that the electron beam would be deflected downwards (remembering that the equivalent conventional current flows in the opposite direction). The direction of the electric field must therefore be downwards in order to deflect the beam upwards (remembering that the direction of an electric field is that of the force acting on a positive charge).

32.5 TORQUE ON A COIL IN A MAGNETIC FIELD

In the previous topic we noted that the moving-coil galvanometer relies on the principle that a current passing through a coil in a magnetic field can be used to produce rotation. The same principle applies to

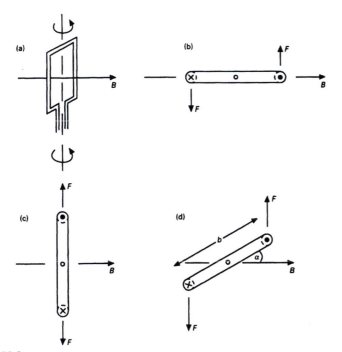

Figure 32.8

electric motors. We shall not go into the design of galvanometers or electric motors, but we need to understand the principle.

Figure 32.8(a) shows a rectangular coil that is free to rotate about its central vertical axis. The axis is at right angles to the direction of a uniform magnetic field of flux density B.

Figure 32.8(b) shows the top view of the coil when its plane is parallel to the field. The cross and dot show the current direction through the vertical sides of the coil. The direction of the force F acting on each vertical side is drawn in accordance with Fleming's left-hand rule. The current through the horizontal sides of the coil, at the top and bottom, is parallel to the field direction and therefore the magnetic force acting on them is zero. The forces on the vertical sides constitute a couple and the coil rotates until its plane is perpendicular to the field direction, as shown in Figure 32.8(c). The forces acting on the vertical sides are still the same as before but their lines of action both pass through the vertical axis, so there is no further tendency to rotate. Fleming's left-hand rule shows that there are now forces acting on the horizontal sides, but they are vertically opposed to one another and have no effect on the rotation of the coil.

Figure 32.8(d) shows the coil when it is inclined at an angle α to the field direction. From our discussion of the moments of forces in Topic 3 we can deduce that the torque T due to the couple about the axis of rotation is given by

$$T = F \times b \cos \alpha$$

where b is the width of the coil and $b \cos \alpha$ is the perpendicular distance between the lines of action of the forces F.

The vertical sides of the coil remain perpendicular to the direction of the field, whatever the value of α; therefore, F always has the value BIl (Equation 32.1), where l is the length of the vertical sides. If the coil has N turns, then, in effect, l is multiplied by N and $F = BINl$. Substituting this in the equation for T above,

$$T \, (= Fb \cos \alpha) = BINlb \cos \alpha$$

But lb is equal to the area A of the coil face; therefore,

$$T = BINA \cos \alpha \tag{32.5}$$

If $\alpha = 0°$, then $\cos \alpha = 1$ and $T = BINA$, and if $\alpha = 90°$, then $\cos \alpha = 0$ and $T = 0$, as in Figures 32.8(b) and (c), respectively.

32.6 FORCES BETWEEN PARALLEL CONDUCTORS

Two straight current-carrying conductors placed parallel to one another each experience a force because of the magnetic field due to the other. The forces are attractive if the currents are flowing in the same direction and repulsive if they are in opposite directions.

Figure 32.9 shows two parallel conductors running perpendicularly through the page, each carrying a current downwards into the paper. The current I_2 in the right-hand conductor produces a field B at the left-hand conductor (as in Figure 32.2a on page 317). The resulting force acting on the left-hand conductor pulls it towards the right (Fleming's left-hand rule). Similarly, the right-hand conductor experiences a force to the left because of the magnetic field due to the current I_1 in the left-hand conductor. Similar arguments show that the forces between the conductors are repulsive if the currents flow in opposite directions.

The ampere is defined as the steady current in each of two straight, parallel conductors of infinite length and negligible cross-sectional area, 1 metre apart in vacuum, that produces a force between them of 2×10^{-7} N per metre length.

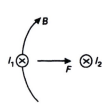

Figure 32.9

Questions

(Use any previously tabulated data as required.)

1. A straight wire 100 mm long experiences a force of 15×10^{-3} N while it carries a current of 3 A perpendicular to a uniform magnetic field.

   (a) Find the flux density of the field.
   (b) Find the magnitude of the force acting on the wire if the wire makes an angle of 60° with the field.

2. (a) Find the direction of the force acting on an electron travelling horizontally in the 9 o'clock direction when it enters a horizontal magnetic field in the 12 o'clock direction.

   (b) Find the direction of the force acting on a conventional current flowing in the same direction under the same circumstances.

3. What angle should a current-carrying conductor make with a magnetic field so that the force acting on it is half its maximum possible value?

4. An electron is travelling in a straight line at 5×10^6 m s^{-1} towards a magnetic field of 0.025 T superimposed on an electric field of 125×10^3 V m^{-1}. If the electron path and the two fields are mutually perpendicular as in Figure 32.7, find the deflection experienced by the electron 0.2 μs after entering the fields.

5. An electron is travelling in a straight line at 10×10^6 m s^{-1} midway between two parallel plates providing a uniform electric field perpendicular to the direction in which it is travelling. The plates are 10 mm apart and there is a potential difference of 1000 V between them. Find the magnitude of the magnetic field which is required to maintain the straight path of the electron.

6. A square coil with 50 mm sides is made from 125 turns of wire which has a resistance of 4 Ω per metre. Estimate the torque acting on the coil when it is connected to a 12 V supply while it is suspended from the centre of one of its sides with its plane parallel to a uniform magnetic field of 0.04 T.

7. A 12 m length of metal wire, of 8900 kg m^{-3} density and 1.7×10^{-8} Ω m resistivity, is aligned horizontally in a west–east direction. A potential difference of 890 V across the ends of the wire provides just enough support for its weight. Estimate the magnitude of the earth's magnetic field acting horizontally at that point. (Assume $g = 9.8$ m s^{-2}.)

TOPIC 33 ELECTRO-MAGNETIC INDUCTION

COVERING:

- e.m.f. and current induced in a moving conductor;
- magnetic flux;
- e.m.f. induced in a rotating coil;
- inductance;
- the transformer.

In the previous topic we saw how electric current produces motion in a magnetic field. In this topic we shall see how motion in a magnetic field *induces* electric current.

33.1 E.M.F. AND CURRENT INDUCED IN A MOVING CONDUCTOR

Figure 33.1(a) is to remind us of Fleming's left-hand rule. This is sometimes called the *motor rule*, because it concerns the motion produced by passing a current through a conductor in a magnetic field. We shall now move on to consider *Fleming's right-hand rule*, sometimes called the *generator rule*, because it concerns the current induced in a conductor that is being propelled through a magnetic field.

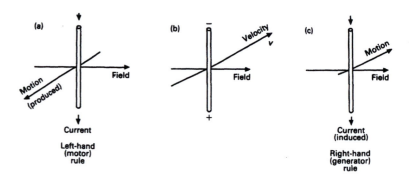

Figure 33.1

Figure 33.1(b) shows a vertical conductor, isolated from any external circuit, that is being propelled at a steady horizontal velocity v at right angles to a horizontal magnetic field. The charge carriers in the conductor are therefore being transported across the field at velocity v and, in effect, constitute a current in a similar way to charged particles in a beam. Consequently, they will experience a vertical force Bvq in accordance with Equation (32.3) (page 320) and in the direction given by Fleming's left-hand rule. If the conductor is metallic, then the charge carriers are electrons and will be displaced upwards (in the opposite direction to a conventional current). Since the positive metal ions cannot move, there will be a separation of charge, so that the top end of the conductor becomes negative while the bottom end is left positive. As charge separation continues, it creates an electric field of growing magnitude in the conductor which increasingly opposes further movement of electrons towards the top.

The electric force acting on each electron is equal to eE (Equation 27.1 on page 262), where E is the electric field strength and e the charge on the electron. Eventually this force will grow large enough to balance the magnetic force Bve (where $q = e$). An equilibrium will then be established in which

$$eE = Bve$$

and

$$E = Bv$$

(remembering that v is the velocity of the conductor).

Thus, a potential difference V is created between the ends of the conductor. If the length of the conductor is l, then, from Equation (27.4) (page 265),

$$E = V/l$$

and, substituting for E in $E = Bv$, we obtain

$$V = Bvl \qquad\qquad (33.1)$$

If the ends of the conductor are connected to an external circuit and a steady velocity v is maintained, then it will act as a generator. (In the next section we shall treat V in Equation 33.1 as an e.m.f. E. Note that e.m.f. and electric field strength have the same symbol, so be very careful not to confuse these quantities.) The direction of the induced current is given by Fleming's right-hand rule, which is illustrated in Figure 33.1(c). The field and induced current directions are represented by the first and second fingers of the right hand. These are mutually perpendicular to the thumb, which represents the direction in which the conductor is being moved.

Lenz's law tells us that the direction of an induced current is always such that it opposes the change that is causing it. This is in agreement with Figure 33.1(c), where moving the conductor as shown produces a conventional current flowing downwards. This current is just like any other, so, according to Fleming's left-hand rule, its interaction with the magnetic field produces a force that acts in the opposite direction to the movement of the conductor, as in Figure 33.1(a).

We can regard this as an example of the principle of conservation of energy. Moving the conductor induces a current which creates an opposing force that requires work to be done to overcome it. Mechanical energy is therefore absorbed by the system and electrical energy is produced.

33.2 MAGNETIC FLUX

It is helpful to think of an induced e.m.f. as the result of a moving conductor cutting through magnetic field lines or, as we shall see later, magnetic field lines sweeping across a stationary conductor.

To develop this approach, we make use of a quantity called *magnetic flux* (symbol Φ), which we shall take to represent the number of field lines. The magnetic flux through a plane is obtained by multiplying the area A of the plane by the magnetic flux density B normal to its surface. If the field direction is perpendicular to the plane, then the three quantities are related by the expression $\Phi = BA$.

The unit of magnetic flux is the *weber* (Wb) and, since $B = \Phi/A$, 1 T is equivalent to 1 Wb m^{-2}. (Now we can see why B is called the flux density.)

Let us imagine that a straight conductor of length l is moving at a steady velocity v through a magnetic field of flux density B, where the conductor, the field and the velocity are mutually perpendicular (see Figure 33.2). During a time interval Δt the conductor will have moved a distance $v\Delta t$ and swept out an area $lv\Delta t$. The flux cut by the conductor is therefore given by

$$\Delta\Phi = Blv\Delta t$$

This can be rearranged to give

$$Blv = \Delta\Phi/\Delta t$$

But, from Equation (33.1),

$$E = Bvl$$

where E is the induced e.m.f.; therefore,

$$E = \Delta\Phi/\Delta t \tag{33.2}$$

Thus, the induced e.m.f. is equal to the rate of flux change which the conductor experiences in cutting through field lines.

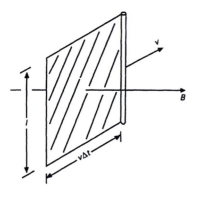

Figure 33.2

Bearing in mind that $\Phi = BA$, a flux change can also result from a change in B over a fixed area A. For example, a coil will experience change in the flux *linking* it (i.e passing through it) if it is placed in a varying magnetic field.

Figure 33.3 illustrates a case that we can interpret either way. If the coil is moved through a non-uniform field, then it will cut field lines in the process. At the same time we can see that there will be a change in the number of field lines linking the coil. Either way an e.m.f. will be induced in the coil and for our purposes it is reasonable to assume that flux cutting or a change in flux linking have equivalent effects. (Note that the weber is actually defined as the flux which, when linking a coil of one turn, and when uniformly reduced to zero in one second, induces an e.m.f. of one volt in the coil.)

These ideas are embodied in *Faraday's laws of electromagnetic induction*, which tell us that the induced e.m.f. is proportional to the rate of cutting flux or the rate of change of flux linking. Note that if a coil has N turns, then the so-called *flux-linkage* is given by $N\Phi$ and any induced e.m.f. is correspondingly increased. Putting all these ideas together gives the *Faraday–Neumann law*, which can be expressed in the form

$$E = -N\frac{\Delta\Phi}{\Delta t} \tag{33.3}$$

The minus sign, which is in accordance with Lenz's law, tells us that the direction of the e.m.f. is such that it opposes the change causing it.

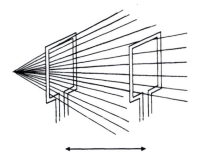

Figure 33.3

33.3 E.M.F. INDUCED IN A ROTATING COIL

We have seen that if we push a conductor across a magnetic field at right angles to the field direction, then the induced e.m.f. is given by $E = Bvl$.

Figure 33.4(a) is looking down on top of a vertical conductor that is being pushed across a magnetic field at an angle β relative to the field direction. In this case the effective velocity of the conductor relative to the field is reduced to the perpendicular component $v \sin \beta$. The induced e.m.f. is therefore reduced to $Bvl \sin \beta$ (which becomes zero if the conductor moves parallel to the field.) The negative sign at the top of the conductor in the figure indicates the charge there due to the electrons moving upwards, as in Figures 33.1(b) and (c).

Figure 33.4(b) shows the conductor as one of a pair which we shall treat at the vertical sides of a rectangular coil of width b (as in Figure 32.8 on page 321). The coil is being rotated at a constant angular velocity about its central vertical axis, so that the vertical sides cut the magnetic flux. As the positive and negative signs indicate, the e.m.f.s in the two sides act in the same direction around the coil and reinforce

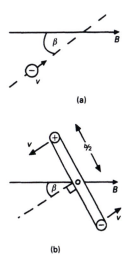

Figure 33.4

each other, so that the total induced e.m.f. (between the ends of the coil) is obtained by adding them together. Therefore, at any instant when the normal to the plane of the coil makes an angle β with the field direction,

$$E = 2 \times Bvl \sin \beta$$

From Equation (9.2) (page 69) the angular velocity of the coil is given by $\omega = \beta/t$ rad s^{-1}, where it rotates through β radians in t seconds and, if $\beta = 0$ at $t = 0$, then $\beta = \omega t$. Furthermore, if v is the linear speed of the conductors around the circumference of a circle of radius $b/2$, then, from Equation (9.3) (page 69) $v = \omega b/2$. Substituting for β and v in the equation for E above, we get

$$E = 2B \frac{\omega b}{2} l \sin \omega t$$

and since the area A of the coil is equal to bl, then

$$E = BA\omega \sin \omega t$$

Finally, if the coil has N turns, then A is, in effect, multiplied by N, so that

$$E = BAN\omega \sin \omega t \tag{33.4}$$

If required, this can be rewritten in terms of frequency of rotation f (Hz = s^{-1}), since $\omega = 2\pi f$.

Equation (33.4) tells us that if the coil rotates at a steady rate, then the induced e.m.f. across its ends varies sinusoidally with time – that is to say, it follows the pattern of a sine wave. Figure 33.5 shows how the e.m.f. between the ends of the coil alternates between positive and negative as each conductor changes direction relative to the field when the coil passes the point where its plane is perpendicular to the field direction. At this point the sides of the coil are travelling parallel to the field direction and $E = 0$.

When the plane of the coil is parallel to the field, its vertical sides are travelling at right angle to the field direction and the e.m.f. is at a maximum. At this point the normal to the plane of the coil is perpendicular to the field direction, so that ωt is either $\pi/2$ or $3\pi/2$. (Remember that ωt is in radians.) Sin ωt is therefore either 1 or -1. The e.m.f. is therefore either $+BAN\omega$ or $-BAN\omega$, where $BAN\omega$ is the amplitude of the sinusoidally alternating e.m.f.

If a resistance is connected across the coil, the alternating e.m.f. will produce an alternating current (a.c.) which periodically reverses its direction. (By contrast, direct current (d.c.) flows in one direction only.)

A major advantage of alternating current is that voltages can be stepped

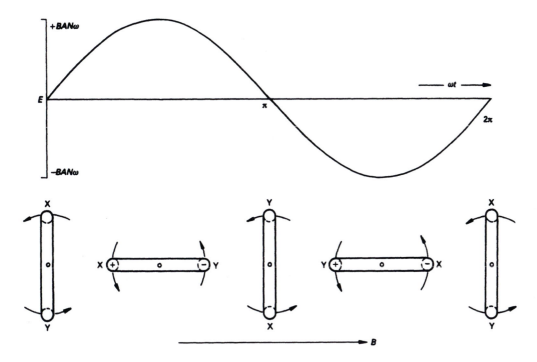

Figure 33.5

up and down very efficiently with *transformers*. However, before we can discuss transformers we must first consider *inductance*.

33.4 INDUCTANCE

Let us try to imagine what happens when a current is passed through a coil. Initially, before the current is switched on, there is no magnetic field, but as the current starts to flow, the field begins to develop. This provides a changing magnetic flux linking the coil or, if it is easier to picture them, field lines growing out from each loop of the coil which cut through their neighbouring loops. Either way, an e.m.f. is induced in the coil itself. According to Lenz's law, this e.m.f. opposes the change that is causing it, with the result that the rise in current is delayed, as indicated in Figure 33.6.

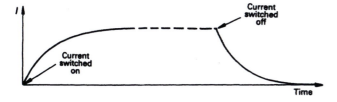

Figure 33.6

Where the curve levels off, the opposing e.m.f. is zero and the current has reached a steady value that depends on its source and the resistance of the circuit. The work done in raising the current to a steady value against the opposing e.m.f. is now stored in the magnetic field in an analogous way to the energy stored in the electric field of a capacitor.

If the current is switched off, the magnetic field collapses, thereby inducing an e.m.f. which opposes the decay of the current, as indicated in Figure 33.6. (On opening the switch contacts, this e.m.f. may be large enough to produce a visible spark as the energy stored in the magnetic field is dissipated.)

The unit of *inductance* is called the *henry* (symbol H). A coil or other *inductor* has an inductance of 1 henry if a current passing through it, while changing at the rate of 1 ampere per second, induces an e.m.f. of 1 volt. A given current change will induce a large e.m.f. in a coil with a large inductance and a small e.m.f. in one with a small inductance. Putting this into the form of an equation,

$$E = -L\frac{\Delta I}{\Delta t} \tag{33.5}$$

where L is the inductance. The equation tells us that $1\ \text{H} = 1\ \text{V s A}^{-1}$. The minus sign reminds us that the induced e.m.f. acts in opposition to the current change that is causing it (Lenz's law). Since the e.m.f. is induced in the same circuit through which the current is changing, L is often called *self-inductance*.

By contrast, the term *mutual inductance* applies when the changing current in one coil or circuit (called the primary) causes an e.m.f. to be induced in another (called the secondary) because of the changing flux linkage between them. (Or, if you prefer, because the expanding field lines from the primary cut across the secondary.)

A *mutual inductance* of 1 henry exists if a current passing through the primary, while changing at the rate of 1 ampere per second, induces an e.m.f. of 1 volt in the secondary. The e.m.f. E_s in the secondary is given by

$$E_s = -M\frac{\Delta I_p}{\Delta t} \tag{33.6}$$

where M is the mutual inductance and $\Delta I_p/\Delta t$ is the rate of current change in the primary.

If an a.c. supply is connected across the primary, then an alternating e.m.f. will be induced in the secondary. This is what happens in transformers.

33.5 THE TRANSFORMER

Figure 33.7 shows a transformer with primary and secondary *windings* side by side on a *core*. (In practice, one set of windings is often wound on top of the other.) The core, made of iron, for example, provides a very efficient magnetic linkage between the primary and secondary windings. Iron is much better than air at conveying magnetic flux (just as copper is much better at conveying electric current), so practically all the flux remains within the core.

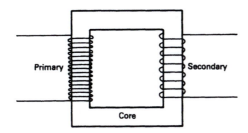

Figure 33.7

(Note that the changing flux induces *eddy currents* in the core material itself. These lead to the dissipation of electrical energy as heat. Eddy current losses can be minimised by constructing the core from laminations which are insulated from one another to interrupt the current pathways.)

The theory of transformers is beyond the scope of this book, so we shall confine ourselves to an outline discussion. An alternating voltage applied to the primary gives an alternating flux in the core which induces an alternating voltage in the secondary. Since the flux is the same through both, then the voltage V across each is proportional to the number of turns N and

$$\frac{V_s}{V_p} = \frac{N_s}{N_p} \tag{33.7}$$

where the subscripts identify the primary and secondary windings. If $N_s > N_p$, we have a step-up transformer, which increases the supply voltage; and if $N_s < N_p$, we have a step-down transformer, which reduces it.

Many transformers transfer power from the primary to the secondary with nearly 100% efficiency, in which case, using Equation (29.3) (page 283), we can write $I_s V_s = I_p V_p$, which gives

$$\frac{V_s}{V_p} = \frac{I_p}{I_s} \tag{33.8}$$

Questions

1. A straight conductor, 200 mm long and of negligible resistance, is moved at 20 m s^{-1} through a field of magnetic flux density of 5×10^{-3} T. Assuming that the conductor, the field and the direction of motion are mutually perpendicular, calculate the current if the conductor is connected across a 2.5 Ω resistor.

2. Calculate the induced e.m.f. across the ends of a straight conductor 4.0 m long under each of the following circumstances:

   (a) After having fallen freely through a distance of 10 m from a horizontal position at right angles to the magnetic north–south direction.
   (b) As in (a) but parallel to the magnetic north–south direction.
   (c) As it travels parallel to the ground at 250 km per hour in a direction at right angles to its longitudinal axis, which is orientated as in (b).

   (Assume $g = 9.8$ m s^{-2} and that the horizontal component of the earth's magnetic field is 1.9×10^{-5} T, while the angle of dip is 70°.)

3. An e.m.f. of 0.21 mV is induced across the ends of a straight, horizontal conductor 2.0 m long as it is moved vertically at 5.5 m s^{-1} at right angles to the magnetic north–south direction. The induced e.m.f. is 0.48 mV when the conductor is moved parallel to the ground at the same speed in a direction at right angles to its longitudinal axis. Calculate the flux density and the angle of dip of the magnetic field in that area.

4. A rectangular coil of 100 turns 60 mm wide and 100 mm long is rotated at 500 revolutions per minute about its central longitudinal axis, which is at right angles to a magnetic field of flux density 32 mT. Calculate the instantaneous value of the e.m.f. in the coil when its plane is at an angle of (a) 0°, (b) 60° and (c) 90° to the field direction.

5. If the coil in Question 4 is stationary and the flux density falls uniformly from 32 mT to zero in 48 s, calculate the e.m.f. induced in the coil when its plane is at an angle of (a) 90°, (b) 30° and (c) 0° to the field direction.

TOPIC 34 MAGNETIC BEHAVIOUR OF MATERIALS

COVERING:

- diamagnetism, paramagnetism and ferromagnetism;
- hysteresis;
- soft and hard magnets;
- magnetic circuits.

As Figure 32.4 (page 318) indicates, there is a region of more or less uniform magnetic field inside a solenoid when it carries an electric current. The flux density can be varied by filling the solenoid with a core of material. So-called *diamagnetic* materials slightly reduce the flux density and *paramagnetic* materials slightly increase it. On the other hand, *ferromagnetic* materials increase it greatly, some by a factor of many thousands.

Let us represent the uniform flux density in the solenoid by B_0 under vacuum conditions and by B when it is filled with different materials. From above, we can write $B < B_0$ for diamagnetic materials, $B > B_0$ for paramagnetic materials and $B >> B_0$ for ferromagnetic materials. The ratio B/B_0, which is a dimensionless quantity, gives the *relative permeability* μ_r of a material. For air, μ_r is very close to 1.

The magnetic behaviour of a material can be attributed to the orbital motion and spin of electrons in its constituent atoms. The electrons can be regarded as behaving like tiny circulating currents with associated magnetic fields.

34.1 DIAMAGNETISM AND PARAMAGNETISM

Diamagnetism is a very weak effect that is exhibited by all materials but is often swamped by the effects of paramagnetism and ferromagnetism. It results from changes in orbital motion which, in keeping with Lenz's law, tend to oppose an applied magnetic field, thereby decreasing the field in the material. Purely diamagnetic materials generally have complete electron pairing in their atomic and molecular structures.

Materials with structures containing unpaired electrons show *paramagnetic* behaviour. The unpaired electrons have associated magnetic fields which tend to become aligned in an applied field, thereby increasing the field in the material. Some metals show paramagnetic behaviour owing to the spin of conduction electrons. Paramagnetism is opposed by the randomising effect of thermal agitation.

34.2 FERROMAGNETISM

As their name suggests, *ferromagnetic* materials are epitomised by iron. They are of great importance in electrical engineering and we shall concentrate on them in this topic.

In ferromagnetic materials the effect of unpaired electron spin in incomplete inner orbitals is great enough to cause such strong interactions between neighbouring atoms that they tend to become mutually aligned. Below a certain temperature, called the *Curie point* or *Curie temperature*, the magnetic axes of neighbouring atoms are able to remain aligned and the material exhibits ferromagnetic behaviour. Above the Curie point (about 760 °C in the case of iron) there is sufficient thermal energy to destroy the alignment and the material becomes paramagnetic.

Regions of uniform alignment are called *domains*. These are typically fractions of a millimetre in size and are, in effect, small permanent magnets. In a piece of unmagnetised ferromagnetic material the domains are orientated in different directions and therefore cancel each other out. If the material is subjected to an increasing external magnetic field, then domains that are aligned in more or less the same direction as the field will tend to grow at the expense of the others. Furthermore, the magnetic axes of domains that are not aligned in the field direction may rotate if the field is strong enough.

Saturation occurs when the magnetic axes of all the domains are aligned with the external field. If the field is then reduced to zero, the material will remain magnetised. To see how this happens, let us consider the relationship between B and B_0 for a ferromagnetic core in a solenoid. To avoid complications due to end effects, it is better to consider a *toroid* (Figure 34.1), which is an endless solenoid made in the form of a ring. (B_0 is readily controlled, since it varies proportionally with the current through the toroid.)

Figure 34.2 shows the typical form of the relationship between B and B_0 for an initially unmagnetised ferromagnetic material. (Generally, $B \gg B_0$, so the vertical and horizontal scales would normally be different for a real material.) As B_0 is increased from zero at O, the material becomes magnetised, as discussed above. B increases progressively less rapidly with B_0 as it reaches its saturation value at P. If B_0 is now reduced to zero, the material retains a remanent (i.e. residual) flux density (at Q) which is called the *remanence* or *retentivity*. The value of this indicates the degree of residual distortion of the domain structure. The remanent flux density can be reduced to zero (at R) by increasing B_0 in the reverse direction. The value of B_0 required to do

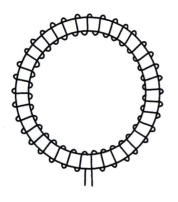

Figure 34.1

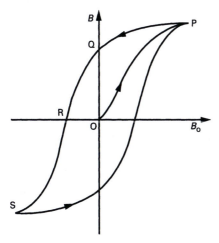

Figure 34.2

this is proportional to the *coercive force* or *coercivity*, which is a measure of the difficulty of neutralising the residual distortion of the domain structure.

On further increasing the reversed field, the material becomes saturated in the reverse direction (at S). The whole process from P to S can then be repeated in the opposite direction, from S to P, to form a closed loop called the hysteresis loop. (The word *hysteresis* describes the lagging of an effect behind its cause, in this case the lagging of B behind B_0.)

It is evident from Figure 34.2 that the relative permeability μ_r $(= B/B_0)$ is not constant for ferromagnetic materials. The maximum relative permeability, based on the largest value of B/B_0 on the initial magnetisation curve, is sometimes used to characterise a material, a high value indicating that the material is readily magnetised.

34.3 SOFT AND HARD MAGNETS

Ferromagnetic materials are described as *soft* or *hard*, depending on whether they readily lose their magnetism or tend to retain it. The use of these words is well exemplified by iron and steel, whose mechanical softness and hardness, respectively, is reflected in their magnetic properties. (The reasons for this are beyond the scope of our discussion.) Figure 34.3 shows hysteresis loops corresponding to an example of each type.

The loop for the soft magnetic material indicates low coercivity and remanence and a generally high relative permeability. This suggests a material that would be suitable for making *electromagnets*. An electromagnet is a temporary magnet, essentially a solenoid with a soft magnetic core giving a strong field, that can be controlled by varying the current. This principle is used in such diverse applications as

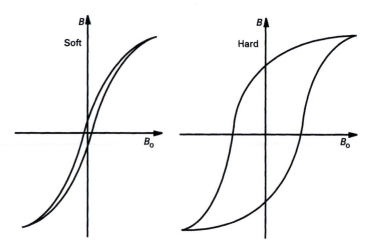

Figure 34.3

electric bells, relays, telephone receivers and for lifting heavy objects made from iron and steel.

Material used for transformer cores must be magnetically soft in order to respond to alternating current. Ferromagnetic materials consume energy as their magnetisation direction is continually changed. This energy, which is proportional to the area enclosed by the hysteresis loop, is dissipated as heat. Thus, the hysteresis loop should be narrow.

Magnetically hard materials are used to make permanent magnets. Such materials generally have high remanence – that is to say, they retain a high remanent flux density when the magnetising field has been removed. They also tend to have high coercivity, so that they are not readily demagnetised. The shape of the hysteresis loop reflects their high resistance to alteration of the domain structure.

34.4 MAGNETIC CIRCUITS

There is an interesting analogy between electric circuits and the magnetic circuit formed by the toroid in Figure 34.1. Magnetic flux Φ is driven round the toroid by a *magnetomotive force* (*m.m.f.*) in an analogous way to a current being driven round an electric circuit by an e.m.f. The magnitude of the m.m.f. is given by NI (measured in ampere-turns) where N is the number of turns and I is the current. Thus, Φ is proportional to NI.

Furthermore, the *reluctance* R_m, which is the 'magnetic resistance' of the circuit, is given by the ratio m.m.f./Φ (analogous to $R = V/I$), thus,

$$\Phi = \frac{NI}{R_m} \qquad (34.1)$$

The analogy extends further because

$$R_m = \frac{l}{\mu A} \tag{34.2}$$

where l is the length (i.e. mean circumference) and A is the cross-sectional area of the toroid. μ is the *absolute permeability* of the core material and is analogous to the electrical conductivity of a conductor (see Section 30.2).

Absolute permeability μ and relative permeability μ_r are related by $\mu = \mu_r \mu_0$ where μ_0, called the *magnetic constant*, is the absolute permeability of free space and has the value $4\pi \times 10^{-7}$ H m^{-1}. In Topic 26 we met the parallel relationship $\varepsilon = \varepsilon_r \varepsilon_0$ for the permittivity of a material in an electric field.

(The electromagnetic theory of James Clerk Maxwell, the nineteenth century scientist, showed that the speed c of electro-magnetic waves in free space depends only upon ε_0 and μ_0 as follows:

$$c = \frac{1}{\sqrt{\varepsilon_0 \, \mu_0}}$$

Although this is really beyond the scope of our present discussion, it serves to illustrate the fundamental interrelationship between electricity and magnetism that has become evident from the previous two topics.)

Question

1. An iron ring, of 420 mm mean diameter and 1.6×10^{-3} m^2 cross-sectional area, has 1000 turns of wire wound uniformly around it (as in Figure 34.1). If a current of 1.4 A in the wire produces a magnetic flux of 4×10^{-3} Wb in the iron core, estimate the relative permeability of the iron.
($\mu_0 = 4\pi \times 10^{-7}$ H m^{-1}.)

TOPIC 35 ALTERNATING CURRENT

COVERING:

- inductive reactance;
- capacitive reactance;
- impedance;
- phase angle;
- power dissipation.

The transmission of electrical power is more efficient when high voltages are used. To take an example, 100 kW of power is carried by a 1000 A current at 100 V, and by a 100 A current at 1000 V (remembering that 1 W = 1 A × 1 V). Assuming that identical cables of resistance R are used for both, the power loss P will be 100 times greater in the first case than the second because the current I is ten times greater and $P = I^2R$ (Equation 30.12 on page 298). Since transformers provide a very efficient means of stepping the voltage up or down, it makes good sense to use alternating current to transmit electrical power over long distances. In practice, enormous voltages are used for this purpose, sometimes as high as 400 kV.

Alternating current is extensively used for lighting, heating and driving machinery, and it can readily be *rectified* to direct current if a particular application demands it.

As far as we are concerned, the major difficulty with alternating current is its mathematical treatment, which is more complicated than for direct current. We have already seen that a steady direct current is opposed by the resistance of the circuit through which it flows. Alternating current is opposed not only by resistance, but also by any capacitance or inductance that the circuit may possess. But, before we move on to this, we must first understand how to quantify alternating current and voltage.

35.1 ALTERNATING CURRENT AND VOLTAGE

All we need to describe a steady direct current is its magnitude and direction. It is more difficult to describe an alternating current, because its magnitude and direction vary periodically with time. We shall confine our discussion to sinusoidal variation, although there are other types of waveform (e.g. square and sawtooth).

The *frequency* of an alternating current is measured in hertz (Hz), where 1 Hz is equal to one complete cycle per second. The ordinary mains supply in many countries is 50 Hz.

The *amplitude* or *peak value* is the maximum value, positive or negative. From our discussion in Section 33.3 we can say that the instantaneous value V of an alternating voltage at any time t is given by

$$V = V_0 \sin 2\pi ft \tag{35.1}$$

where V_0 is the peak voltage and f is the frequency. Similarly, the instantaneous value of an alternating current is given by

$$I = I_0 \sin 2\pi ft \tag{35.2}$$

For many practical purposes, we need average values that we can use in simple calculations. But alternating voltage and current have average values of zero over a complete cycle because they are positive for one half and negative for the other. We can get round the problem by considering the heating effect of a current, because this is the same in both directions. We can define the *effective* value I_{eff} of an alternating current in terms of the equivalent direct current that produces the same power dissipation in a given resistor. Although we shall not go into the reasons here, this turns out to be

$$I_{eff} = \frac{I_0}{\sqrt{2}} = 0.707\, I_0 \tag{35.3}$$

Similarly, the effective voltage is given by

$$V_{eff} = \frac{V_0}{\sqrt{2}} = 0.707\, E_0 \tag{35.4}$$

The 240 V quoted for the ordinary domestic mains supply in Britain is the effective value. Equation (35.4) gives the corresponding peak value as 339 V.

Note that, unless otherwise stated, any further reference to current I and voltage V in this topic implies their effective values.

35.2 REACTANCE

We are now in a position to consider the factors that oppose the reciprocating flow of alternating current through a circuit. As with direct current, alternating current is opposed by the electrical resistance of the materials of which the circuit is made. But, because it is continuously changing, alternating current is also opposed by the effects of inductance and capacitance. These effects are called *reactance* and are due to the occurrence of voltages that arise in inductors and capacitors which oppose the current.

We shall begin by considering *inductive reactance* X_L, which is the quantity that is used to measure the effect of an inductor such as a coil. The relationship of X_L to an inductor is parallel to that of resistance R to a resistor, and we can write

$$X_L = \frac{V}{I} \tag{35.5}$$

where V and I are the effective values of voltage and current. This expression is parallel to the definition of resistance ($R = V/I$) and, as we shall see below, reactance is measured in ohms.

A detailed mathematical analysis of reactance is beyond the scope of this book; nevertheless we need to understand it in semi-quantitative terms. As we saw in Topic 33, switching a direct current on or off through a coil (or other inductor) causes a self-induced e.m.f. which opposes the current and tends to maintain the status quo, thereby delaying current growth or decay. However, once a steady state has been reached, this e.m.f. disappears. In the case of alternating current, which is continuously changing, the opposing e.m.f. will also be continuous.

Inductive reactance increases with increasing frequency f and increasing inductance L and is given by

$$X_L = 2\pi f L \tag{35.6}$$

Since the unit of f is s^{-1} and the unit of L is $V\ s\ A^{-1}$ (from Equation 33.5), the unit of $2\pi f L$ is $V\ A^{-1}$. This is the same as the unit of resistance R (= V/I), and reactance is measured in ohms. (Note that $f = 0$ for a steady direct current, in which case $X_L = 0$ and the current is impeded solely by the resistance of the coil.)

In a purely resistive circuit, with no reactance at all, the voltage and current are in phase. That is to say, their peaks and troughs coincide as indicated in Figure 35.1. The alternating current and voltage are represented by rotating vectors, called *phasors*, which turn anticlockwise with frequency f and whose length is equal to the amplitude. Their vertical component (the projection onto the vertical axis) represents the instantaneous value.

Figure 35.2 shows that, in the case of a pure inductor, the voltage *leads* the current by a quarter of a cycle. (The opposing e.m.f., which balances the applied voltage, is at a maximum where the rate of current change is at a maximum and $I = 0$.)

Capacitive reactance X_C is the quantity that is used to measure the opposition of a capacitor to the flow of alternating current. In a similar way to inductive reactance,

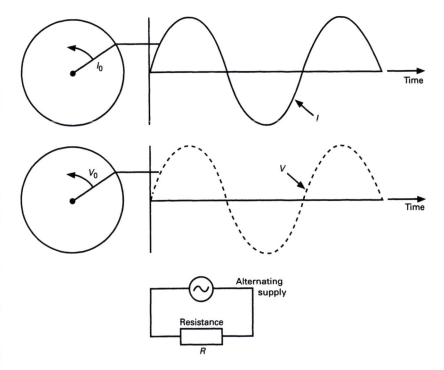

Figure 35.1

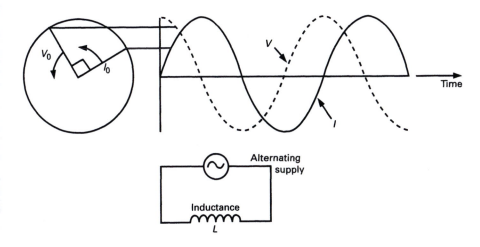

Figure 35.2

$$X_C = \frac{V}{I} \tag{35.7}$$

and, similarly, the units are ohms.

Again, we need a semi-quantitative description. In Topic 28 we saw that direct current cannot flow through a capacitor, because of the insulating gap between its plates. If it is connected across a battery, then current flows as charge builds up on the plates. But the growing potential difference across the plates increasingly opposes and eventually stops the current flow.

A capacitor does not stop alternating current, however, because charge can flow backwards and forwards from plate to plate around an external circuit without actually crossing the gap between them. However, the build-up of charge, and the resultant opposing potential difference every half-cycle, will still tend to impede the current. Capacitive reactance decreases with increasing frequency f and increasing capacitance C and is given by

$$X_C = \frac{1}{2\pi f C} \tag{35.8}$$

Remember that the unit of capacitance can be expressed as C V^{-1} (Equation 28.1 on page 270) and that 1 C = 1 A s (Equation 29.1 on page 279). The unit of X_C is therefore given by

$$\frac{1}{s^{-1} \times A \ s \ V^{-1}} = \frac{V}{A}$$

Hence, X_C is measured in ohms. (Note that for direct current $f = 0$; therefore, X_C is infinite and current cannot flow.)

Figure 35.3 shows how the voltage in a purely capacitive circuit lags behind the current by a quarter of a cycle. (The capacitor has maximum charge when the current is zero and on the point of reversing.)

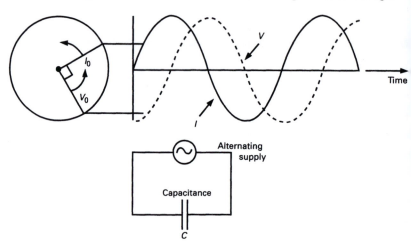

Figure 35.3

The word *CIVIL* is a useful mnemonic for phase relationships. *CIV* reminds us that, for capacitance C, I leads V, *VIL* reminds us that V leads I for inductance L.

35.3 IMPEDANCE AND PHASE ANGLE

Many circuits have inductance, capacitance and resistance. To take a very simple example, a coil will have inductance and it will have resistance due to the wire from which it is made. Figure 35.4(a) represents a simple series circuit containing all three and indicates the effective voltage across each. From the previous section we know that V_R is in phase with the current I, V_L is a quarter of a cycle ahead of it and V_C is a quarter of a cycle behind. These quantities are therefore represented as vectors, as in Figure 35.4(b), where the vector sum equals the applied voltage V as given by

$$V = \sqrt{V_R^2 + (V_L - V_C)^2} \tag{35.9}$$

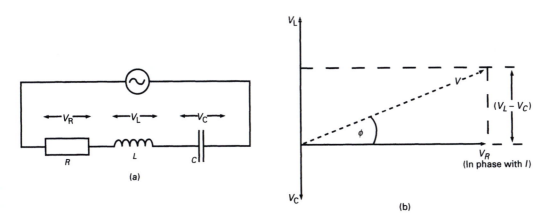

Figure 35.4

The circuit elements are in series; therefore, the current I is the same through each. From Equations (35.5) and (35.7) we know that $V_L = IX_L$ and $V_C = IX_C$; furthermore, $V_R = IR$. Therefore, from Equation (35.9), we can write

$$V = \sqrt{I^2R^2 + (IX_L - IX_C)^2}$$

and

$$V = I\sqrt{R^2 + (X_L - X_C)^2} \tag{35.10}$$

The total opposition to alternating current, called the *impedance Z*, is given by

$$Z = \frac{V}{I} = \sqrt{R^2 + (X_L - X_C)^2} \tag{35.11}$$

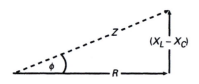

Figure 35.5

where Z is measured in ohms. (Remember that X_L and X_C depend on frequency; therefore, Z depends on frequency too.)

As the relationship between impedance and voltage suggests, we can draw a vector diagram in terms of impedance, as in Figure 35.5.

The *phase angle* Φ is the angle between the applied voltage and the current. Bearing in mind that V_R is in phase with the current, then, from Figure 35.4(b)

$$\tan \Phi = \frac{V_L - V_C}{V_R} \tag{35.12}$$

or, from Figure 35.5,

$$\tan \Phi = \frac{X_L - X_C}{R} \tag{35.13}$$

If $V_L < V_C$ (and $X_L < X_C$), then Φ will be negative, indicating that the current leads the voltage.

Although the above equations apply to series '*RLC*' circuits they can be applied to series *RL* and *RC* circuits, where the respective *C* and *L* terms are zero; thus,

$$V = \sqrt{V_R^2 + V_L^2} \tag{35.14}$$

$$V = \sqrt{V_R^2 + V_C^2} \tag{35.15}$$

and so on.

35.4 POWER

Pure inductors and capacitors do not consume power, since energy is only temporarily stored in their respective magnetic and electric fields. On the other hand, power is dissipated by resistance and its average value P is given by I^2R, where I is the effective current in a circuit containing a resistance R.

If V leads I by Φ, as in Figure 35.4(b), V_R is the component of V across the resistance in the circuit and we can write

$$V_R = V \cos \Phi$$

Hence,

$$P \ (= IV_R) = IV \cos \Phi \tag{35.16}$$

Cos Φ is called the *power factor* and, as we can see from Figure 35.5, it is equal to R/Z. If the circuit is purely resistive, then cos $\Phi = 1$

and $P = IV$. If the circuit is purely inductive, then $\cos \Phi = 0$ and $P = 0$.

Questions

(Assume effective values for current and voltage.)

1. Cables with a total resistance of 12 Ω are used to carry current from a 25 kW supply. Find the power loss and voltage drop if the supply voltage is (a) 5 kV and (b) 10 kV.

2. What is the effective value of an alternating supply of 205 V amplitude?

3. A coil has a reactance of 50 Hz that is fifty times greater than its resistance. At what frequency would you expect its reactance to be a hundred times greater than its resistance?

4. A purely inductive circuit operating at 50 Hz has an inductance of 350 mH.

   (a) Find the reactance of the circuit.
   (b) Find the current if the voltage is 110 V.

5. A purely capacitive circuit operating at 70 Hz has a capacitance of 65 μF.

   (a) Find the reactance of the circuit.
   (b) Find the voltage if the current is 2 A.

6. What is the reactance at 35 Hz of a capacitor that has a reactance of 35 Ω at 70 Hz?

7. State the phase relationship between current and voltage for a pure (a) capacitance, (b) resistance and (c) inductance.

8. Find the impedance in a series circuit where

   (a) $R = 12.5$ Ω, $X_L = 25$ Ω and $X_C = 15$ Ω
   (b) $R = 12.5$ Ω, $X_L = 15$ Ω and $X_C = 25$ Ω
   (c) $R = 3$ Ω, $X_L = 6$ Ω and $X_C = 2$ Ω.

9. Find the impedance of a series circuit in which $R = 10$ Ω, $L = 13.64$ mH and $C = 23.32$ μF at 350 Hz.

10. Find the current when a 26 V alternating supply is connected across a circuit with an inductive reactance of 6 Ω in series with a resistance of 2.5 Ω.

11. A 15 V, 2 kHz supply is connected across the combination of a 3 Ω resistor in series with a 20 μF

capacitor. Find the voltage drop across (a) the resistor and (b) the capacitor.

12. (a) Find the voltage of an alternating supply connected in series with a resistor, an inductor and a capacitor across which there are potential differences of 3 V, 8 V and 4 V, respectively.
    (b) Find the phase angle.

13. Find the impedance and phase angle where $R = 5\ \Omega$, $X_L = 14\ \Omega$ and $X_C = 2\ \Omega$. If the applied voltage is 26 V then find the voltage drops across the resistance, the inductance and the capacitance, respectively.

14. Find the impedance of a circuit in which the resistance is 25 Ω and the phase angle is 60°.

15. Find the power consumed by the circuits which, when connected to a 240 V supply, carry currents of 1.5 A which lag behind the voltage by (a) 0°, (b) 41.4° and (c) 90°.

16. Find the phase angle in a series circuit in which $R = 3\ \Omega$, $X_L = 6\ \Omega$ and $X_C = 2\ \Omega$, and find the power dissipated when a current of 1.5 A is flowing through it.

APPENDIX: CALCULATION TECHNIQUE

This appendix is addressed to those readers who are having difficulty in getting the right answers to the practice questions in the main text because they lack confidence and fluency in calculation technique.

Many of the worked examples and practice questions involve calculating the value of a quantity by substituting known values of other quantities related to it in a formula. The most common mistakes occur (1) in handling numbers expressed in standard form, e.g. 6.02×10^{23}, (2) in rearranging formulae to isolate the unknown quantity as the subject on one side, and (3) in assigning the correct units to the answer. We shall deal with these in turn.

A.1 HANDLING NUMBERS

You will need a 'scientific' calculator capable of dealing with numbers expressed in standard form. Calculator errors are easy to make, so get into the habit of rough checking your answers. For example, if 244.7 is to be multiplied by 7.8, we can say that 244.7 is roughly 250 and 7.8 is roughly 8; we therefore expect an answer of about 250 times 8, which is 2000. (In fact the precise answer is 1908.66.) The best way of avoiding errors with your calculator is to become thoroughly familiar with its operation; read the instructions carefully and practise using it at every opportunity.

A.1.1 STANDARD FORM

Engineering and technological calculations involve numbers covering an enormous range. For instance, the speed of light is very nearly 300 000 000 metres per second whereas the diameter of the titanium atom is approximately 0.000 000 000 3 metres. Writing numbers like this in full is inconvenient and makes it easy for mistakes to occur. It is much better to use powers of 10 (see Table A.1). For example, in the case of the speed of light, 100 000 000 can be expressed as 10^8 because

$$100\,000\,000 = 10 \times 10 \times 10 \times 10 \times 10 \times 10 \times 10 \times 10 = 10^8$$

This means that we can write 300 000 000 much more neatly as 3×10^8. Similarly, in the case of the titanium atom,

$$0.000\,000\,000\,1 = \frac{1}{10\,000\,000\,000} = \frac{1}{10^{10}} = 1 \times 10^{-10}$$

Therefore we can write 0.000 000 000 3 as 3×10^{-10}.

In this type of notation, called *standard form* or *scientific notation*, numerical values are expressed in the general form $a \times 10^n$ where $1 \le a < 10$ (i.e. a is equal to or greater than 1 and less than 10). a is written as a decimal number and the digits it contains are the *significant figures* (S.F.). The *exponent*

Table A.1

$$1\,000\,000 = 10 \times 10 \times 10 \times 10 \times 10 \times 10 = 10^6$$

$$1000 = 10 \times 10 \times 10 = 10^3$$

$$100 = 10 \times 10 = 10^2$$

$$10 = 10 = 10^1$$

$$1 = 1 = 10^0$$

$$0.1 = \frac{1}{10} = 10^{-1}$$

$$0.01 = \frac{1}{100} = \frac{1}{10^2} = 10^{-2}$$

$$0.001 = \frac{1}{1000} = \frac{1}{10^3} = 10^{-3}$$

$$0.000\,001 = \frac{1}{1\,000\,000} = \frac{1}{10^6} = 10^{-6}$$

or *index*, n, is an integer (whole number). Thus, for example, 2.432×10^7 represents the number 24 320 000. There are four significant figures and the exponent 7 indicates that 2.432 is multiplied by 10 000 000; in effect, the decimal point should be moved seven places to the right to write the number in full. Similarly, 1.8×10^{-4} represents 0.000 18; in this case there are two significant figures and the exponent -4 indicates that 1.8 is divided by 10 000 (or multiplied by 0.0001), thus the decimal point should be moved four places to the left to write the number in full. (Strictly speaking, the digits are moved relative to the decimal point but many people find it easier to think of it the other way round.)

It is often more convenient to work with exponents that are multiples of 3 (e.g. 24.32×10^6 rather than 2.432×10^7) because this ties in with the system of prefixes that is commonly used with units expressing particular quantities, for example *kilo-* ($\times 10^3$) and *milli-* ($\times 10^{-3}$). (See Topic 1.)

A.1.2 ACCURACY AND ROUNDING OFF

The accuracy associated with a particular numerical value should be reflected in the number of significant figures that it contains. For example, if we write 9.3×10^{-6} this implies that we know the value accurately to two significant figures (2 S.F.), whereas writing 7.6500×10^6 implies that we know the value accurately to five significant figures (5 S.F.). (Note that some formulae contain precise whole numbers, for example $A = 4\pi r^2$ where A is the surface area of a sphere of radius r.)

The large number of digits displayed by calculators tempts us to give the results of calculations to more significant figures than is justified. The accuracy of the result of a calculation is limited by the accuracy of the values used in making it, and we must *round off* the answer accordingly.

When an answer is rounded off to a given number of significant figures we use the convention that its last digit is increased by 1 if the digit following it in the original number had been 5 or more, and it is left unchanged if the digit originally following it had been less than 5. So by this convention, rounding off 1.4500 and 1.5499 to two significant figures gives 1.5 in both cases. And rounding off 326.719 508 to four significant figures gives 326.7 because the 7 is followed by a 1; rounding it off to five significant figures gives 326.72

because the 1 is followed by a 9; rounding it off to seven significant figures gives 326.7195 because the 5 is followed by a 0; and rounding it off to six significant figures gives 326.720 since the 9 is rounded up to 10 because it is followed by a 5.

How do we decide how many significant figures to use? Consider multiplying 1.5 by 1.5. These numbers are given to two significant figures so it seems reasonable to do the same with the answer, as follows:

$$1.5 \times 1.5 = 2.25 = 2.3 \text{ (to 2 S.F.)}$$

Note that the two significant figures used in the original numbers imply that their true values lie closer to 1.5 than to either 1.4 or 1.6; in other words, that they lie somewhere between 1.45 and 1.55, say 1.50 ± 0.05 (1.50 *plus or minus* 0.05). The true answer could therefore be as little as 1.45 × 1.45 or as much as 1.55 × 1.55, i.e. between 2.1025 and 2.4025 respectively. We can therefore say that the estimated *bounds* for the answer are 2.1 and 2.4 (to 2 S.F.). Similar arguments apply to division.

Although giving the answer above as 2.3 is over-optimistic as far as strict accuracy is concerned, for our purposes this is a reasonable compromise between giving answers with an unjustifiable number of digits and going to the lengths of estimating the upper and lower bounds for the result of every calculation that we perform. As a general rule of thumb for multiplication and division in this book, it is sufficient to take advantage of as many significant figures as each given value provides, then round off the answer to the number of significant figures corresponding to the least accurately known value. For example

$$\frac{9.22 \times 456.738}{73.76} = 57.092\,25 = 57.1 \text{ (to 3 S.F.)}$$

The answer is rounded off to three significant figures because the least accurately known value (9.22) is given to three significant figures.

We can use the same basic approach with numbers expressed in standard form. In effect, multiplication involves multiplying the decimal parts and adding the exponents indicating the powers of ten, as in the following example:

$$(1.7314 \times 10^6) \times (2.8 \times 10^5)$$
$$= (1.7314 \times 2.8) \times 10^{(6+5)}$$
$$= 4.84792 \times 10^{11}$$
$$= 4.8 \times 10^{11} \text{ (to 2 S.F.)}$$

The decimal part of the answer is rounded off to two significant figures because the least accurately known value (2.8 × 10^5) is given to two significant figures.

Division is similar, but the exponent in the *denominator* (on the bottom line) is subtracted from the exponent in the *numerator* (on the top); for example

$$\frac{1.7314 \times 10^6}{2.8 \times 10^5} = \frac{1.7314}{2.8} \times 10^{(6-5)} = 0.62 \times 10^1 \text{ (to 2 S.F.)} = 6.2$$

Now let us consider sums and differences. Again assuming that accuracy is reflected in the significant figures, it is meaningless to add 0.135 to 79 because the 0.135 will be swamped by the ±0.5 uncertainty implied in writing 79. On the other hand it is perfectly reasonable to add 0.135 to 79.000 to give 79.135 (although there is some uncertainty in the final digit because of the ± 0.0005 uncertainty in the original numbers). Furthermore it is reasonable to write

$$79.0 \ + 0.135 = 79.1 \qquad \text{and}$$
$$79.00 + 0.135 = 79.14$$

Differences are treated in a similar way.

For our purposes, we can use this basic approach when we add and subtract numbers expressed in standard form. In effect, we simply adjust the exponents (by moving the decimal points) so that they are the same, then we can treat the decimal parts of the numbers as above. For example

$$
\begin{aligned}
(1.55 \times 10^4) + (2.22 \times 10^3) \\
= (1.55 \times 10^4) + (0.222 \times 10^4) \\
= (1.55 + 0.222) \times 10^4 \\
= 1.772 \times 10^4 \\
= 1.77 \times 10^4 \text{ (to 3 S.F.)}
\end{aligned}
$$

Similarly with subtraction

$$
\begin{aligned}
(1.55 \times 10^4) - (2.22 \times 10^3) \\
= (1.55 \times 10^4) - (0.222 \times 10^4) \\
= (1.55 - 0.222) \times 10^4 \\
= 1.328 \times 10^4 \\
= 1.33 \times 10^4 \text{ (to 3 S.F.)}
\end{aligned}
$$

Don't forget that, in practice, this can still be over-optimistic so far as strict accuracy of the answer is concerned.

Inaccuracy in calculations is a complex subject that goes beyond the scope of this book. Nevertheless the principles outlined in this appendix should help you to give your answers to a reasonably sensible degree of accuracy. (In the main text, and in the answers to the practice questions, some values are given to a greater than strictly justifiable number of significant figures where this helps to avoid misunderstanding.)

A.1.3 ROUGH CHECKING

As mentioned earlier, calculator errors are very easy to make, so get into the habit of rough checking your answers. For instance,

$$\frac{(8.747 \times 10^5)(4.206 \times 10^3)}{(6.412 \times 10^6)} = 573.8 \text{ (to 4 S.F.)}$$

To rough check this, round off the decimal part of each number to one significant figure so that the calculation reduces to

$$\frac{(9 \times 10^5)(4 \times 10^3)}{(6 \times 10^6)} = \frac{36}{6} \times \frac{10^5 \times 10^3}{10^6} = 6 \times 10^2 = 600$$

A.1.4 EXERCISES

(Assume that the accuracy of the numbers given in these exercises is reflected in the number of significant figures used.)

1.1. Convert the following (i) into standard form, and (ii) using exponents that are multiples of 3:
    (a) 283 486
    (b) 0.000 043

(c) 183 810 397
(d) 0.000 002 31
(e) 52 921
(f) 0.000 960
(g) −0.0780

1.2. Convert the following to ordinary decimal numbers:
(a) 1.791×10^1
(b) 1.791×10^{-1}
(c) 1.791×10^5
(d) 3.49×10^{-4}
(e) 3.81×10^5
(f) 9.41×10^{-2}
(g) -9.870×10^{-3}

1.3. Evaluate the following, giving your answers to a sensible degree of accuracy:
(a) $3.46 + 7.02$
(b) $3 - 0.0007$
(c) $950 + (2 \times 10^7)$
(d) 64.34×29
(e) 64.34×29.00
(f) $144/12.3$
(g) 6.345^2
(h) $12.548/6.322$

1.4. Evaluate the following, giving your answers in standard form to a sensible degree of accuracy:
(a) $(2.47 \times 10^{-2}) + (3.6 \times 10^{-3})$
(b) $(5.8 \times 10^4) + (7.92 \times 10^5)$
(c) $(4.87 \times 10^{-3}) + (3.2 \times 10^{-8})$
(d) $(3.4 \times 10^4) - (2.47 \times 10^5)$
(e) $(4.2 \times 10^{-7}) + (3.8 \times 10^{-6})$
(f) $(1.22 \times 10^4) - (1.7 \times 10^9)$
(g) $12 \times (-2.5 \times 10^4)$
(h) $-7.0 \times (3.7 \times 10^{-9})$
(i) $64/(8 \times 10^{-3})$
(j) $(-2 \times 10^3)(2 \times 10^3)$
(k) $(53.2 \times 10^{-2})(1.2 \times 10^2)$
(l) $(2 \times 10^2)^3$
(m) $4/(8 \times 10^{-51})$
(n) $(2.4 \times 10^{-8})/(3.2 \times 10^{-9})$
(o) $(9.92 \times 10^{-5})^{-2}$
(p) $(7.81 \times 10^3) + (2.4 \times 10^8)$
(q) $(-3.47 \times 10^{-3})/(2.86 \times 10^9)$
(r) $(-2.86 \times 10^{-9}) \times (-3.11 \times 10^{12})$
(s) $(2.26 \times 10^{-7})^3$
(t) $1/(1.4 \times 10^3)$
(u) $(9.46 \times 10^{28})/(1.49 \times 10^{-7})$
(v) $(8.1 \times 10^{10})/(9.0 \times 10^{-8})$
(w) $(4.2 \times 10^{18})(3.5 \times 10^{-4})(1.47 \times 10^5)^{-1}$
(x) $(4.5 \times 10^8)/(3.7 \times 10^9)^0$
(y) $(2.79 \times 10^{-5}) \times (3.81 \times 10^4) \div (9.73 \times 10^3)$
(z) $(6.4 \times 10^3)^{1/2}$

A.2 REARRANGING FORMULAE

There is a helpful analogy between a pair of scales and a mathematical equation. The scales will balance so long as there are equal weights in each pan. Furthermore, the scales continue to balance if we change the weights, provided that we increase or reduce each side by the same amount. The same principle applies to balancing equations, and we can make good use of it in rearranging formulae. We can add or subtract and multiply or divide using any appropriate quantity to rearrange a formula, and we can raise each side to a power or take roots but, whatever we do, we must treat both sides in the same way so that the balance is preserved.

A.2.1 USING ADDITION AND SUBTRACTION

Let us begin with the following equation:

$$a + 3 = 11$$

To find the value of a we need to rearrange the equation so that we end up with a by itself on one side and the numerical value on the other. If we subtract 3 from each side then we eliminate the 3 from the left while keeping both sides equal, as follows

$$a + 3 - 3 = 11 - 3$$

hence

$$a = 8$$

To check our answer we substitute 8 for a in the original equation thus:

$$8 + 3 = 11$$

Both sides are equal, therefore our answer is correct.

Expressing the rearrangement in general terms, we can write

$$a + b = c$$

then, subtracting b from both sides,

$$a + b - b = c - b$$

hence we obtain

$$a = c - b$$

In effect, the b moves from the left-hand side of the equation to the right, changing from positive to negative as it crosses the equals sign. Similarly, if

$$a - b = c$$

then, adding b to both sides,

$$a - b + b = c + b$$

hence we obtain

$$a = c + b$$

In this case the b changes from negative to positive in crossing from left to right. The same arguments apply to crossing from right to left, so we can move any separate term in an equation from one side to the other provided that we change its sign. Also note that

if $a = b$, then

$a - b = 0$, and

$0 = b - a$

A.2.2 USING MULTIPLICATION, DIVISION, ROOTS AND POWERS

First let us consider the general case where we multiply both sides of an equation by an appropriate quantity in order to eliminate a fraction. For example, if

$$\frac{a}{b} = c$$

then, multiplying both sides by b,

$$\frac{a \times b}{b} = c \times b$$

The b's on the left-hand side cancel each other out (because $b/b = 1$) and we obtain

$$a = c \times b$$

In effect, in moving from one side of the equation to the other, the b has changed from being a *divisor* (a quantity by which something is divided) to being a *multiplier* (a quantity by which something is multiplied).

In the following worked example, both sides of an equation are divided by an appropriate number in order to isolate an unknown quantity and find its value.

Worked Example A.1

Find the value of x if

$$8x - 24 = 2x - 6$$

First we collect all the terms involving x on the left-hand side and all the numbers on the right, changing the signs of any that we have to move, as follows

$$8x - 2x = -6 + 24$$

hence

$$6x = 18$$

Now, dividing both sides by 6,

$$\frac{6x}{6} = \frac{18}{6}$$

therefore

$$x = 3$$

Thus, in general, if

$$a \times b = c$$

and we wish to isolate a on the left-hand side, then dividing both sides by b

$$\frac{a \times b}{b} = \frac{c}{b}$$

hence

$$a = \frac{c}{b}$$

The b has changed from being a multiplier on the left-hand side to being a divisor on the right-hand side.

Similarly, if

$$\frac{a}{b} = \frac{c}{d}$$

then

$$ad = cb$$

and

$$\frac{d}{c} = \frac{b}{a}$$

The general point to note is that a multiplier on one side becomes a divisor on the other, and vice versa.

In some cases it may be necessary to take roots of both sides or raise them to some power, as the following worked examples demonstrate.

Worked Example A.2

Find the value of x if

$$\frac{x}{32} = \frac{25}{2x}$$

Collecting the terms involving x on the left-hand side and the numbers on the right, we obtain

$$x \times 2x = 25 \times 32$$

hence

$$2x^2 = 800$$

and, dividing both sides by 2,

$x^2 = 400$

Taking square roots of both sides

$x = \pm 20$

Note that the answer is given as ± 20 because $20^2 = 400$ and $-20^2 = 400$.

Worked Example A.3

Find the value of x if

$$7 = \frac{21}{\sqrt{x}}$$

If

$$7 = \frac{21}{\sqrt{x}}$$

then

$$\sqrt{x} = \frac{21}{7} = 3$$

and squaring

$$x = 9$$

A.2.3 CHANGING THE SUBJECT OF A FORMULA

We shall now apply these principles to *transposing* or *transforming* a formula in order to make an unknown quantity the *subject*, isolated on one side so that it can be evaluated by substituting numerical values for the known quantities on the other side. For example, suppose that we need to find the value of T, given the values of p, V, n and R in the following formula:

$$\frac{pV}{T} = nR$$

We would transpose the formula to make T the subject as follows:

$$\frac{pV}{nR} = T, \text{ which we would normally write as } T = \frac{pV}{nR}$$

(Note that we can exchange the two sides of the formula in their entirety because if $x = y$ then $-y = -x$ and, multiplying both sides by -1, $y = x$.)

Worked Example A.4

The formula for the volume V of a sphere of radius r is

$$V = \frac{4}{3}\pi r^3$$

Find the radius of the sphere that has a volume of 11.5 cubic metres. ($\pi = 3.14$ to three significant figures.)

Let us begin by making r (the unknown quantity) the subject, on the left-hand side of the equation. Since

$$\frac{4}{3}\pi r^3 = V$$

then

$$r^3 = \frac{3V}{4\pi}$$

Taking cube roots of both sides, we obtain

$$r = \left(\frac{3V}{4\pi}\right)^{1/3}$$

and substituting the given values for V and π,

$$r = \left(\frac{3 \times 11.5}{4 \times 3.14}\right)^{1/3}$$

hence

$$r = 1.40 \text{ metres (to 3 S.F.)}$$

A.2.4 BRACKETS

Suppose that we want to make a the subject of the formula

$$s = ut + \frac{1}{2}at^2$$

First we move the term ut to the left-hand side so that

$$s - ut = \frac{1}{2}at^2$$

Then we isolate a by moving the 2 and the t^2 across the equals sign to become multiplier and divisor, respectively, of the left-hand side as follows:

$$\frac{2(s - ut)}{t^2} = a$$

The important point to note is that the multiplier 2 and the divisor t^2 operate on the whole of the left-hand side, so brackets are used to make $(s - ut)$ a single entity. Then, since the subject is normally written on the left-hand side,

$$a = \frac{2(s - ut)}{t^2}$$

Expressing this type of problem in general terms, if we want to make c the subject of the equation

$$a = b + cd$$

then

$$a - b = cd$$

therefore

$$c = \frac{(a - b)}{d}$$

Similarly, to make c the subject of

$$a + \frac{b}{c} = d$$

then

$$\frac{b}{c} = d - a$$

and, multiplying both sides by c,

$$b = c(d - a)$$

therefore

$$c = \frac{b}{(d - a)}$$

Note that brackets are often omitted when they enclose the whole of the numerator (top line) or denominator (bottom line) of a fractional term, for example

$$c = \frac{a - b}{d} \quad \text{and} \quad c = \frac{b}{d - a}$$

Worked Example A.5

Make x the subject of the equation

$$a - b = \frac{c}{x - d}$$

If we treat $a - b$ and $x - d$ as complete entities, as follows

$$(a - b) = \frac{c}{(x - d)}$$

then they can be exchanged so that

$$x - d = \frac{c}{a - b}$$

Hence we obtain

$$x = \frac{c}{a - b} + d$$

Worked Example A.6

Transpose the following formula to make ΔV the subject:

$$\frac{E}{3(1 - 2\mu)} = - \frac{h\rho g}{\Delta V/V_0}$$

(The physical significance of the quantities involved (E, μ, h, ρ, g, ΔV and V_0) need not concern us here but we should note that the symbol Δ is used to represent a change in a quantity, for example ΔV represents a change in V. Subscripts are often used to specify a particular quantity; in this case V_0 represents the value of V under specific conditions.)

In general

$$\frac{1}{a/b} = \frac{b}{a}$$

therefore

$$\frac{1}{\Delta V/V_0} \text{ can be written as } \frac{V_0}{\Delta V}$$

so that

$$\frac{E}{3(1 - 2\mu)} = - \frac{V_0 h\rho g}{\Delta V}$$

Rearranging the formula, and treating the contents of the brackets as a single entity:

$$\Delta V = - \frac{3V_0 h\rho g(1 - 2\mu)}{E}$$

Finally, a useful tip to help you avoid mistakes when rearranging formulae: *write down each operation a step at a time.*

A.2.5 EXERCISES

2.1. Find the values of the unknown quantities below and, in each case, check your answer by substituting it into the original equation:
(a) $a - 3 = 8$
(b) $b + 7 = 5$
(c) $2c - 11 = 7$
(d) $7d + 51 = 100$
(e) $7 - 3x = 5x - 57$
(f) $5y - 15 = 15 - 5y$
(g) $11(3p + 2) = 121$
(h) $5 + 3(y - 2) = 11$
(i) $2(b + 3) = -5(4 - 3b)$

(j) $d^2 - 5 = 44$
(k) $3a/5 = 9$
(l) $6/b = 1/5$
(m) $3(a - 2)/a = 2$
(n) $1/a = 1/2 - 1/3$
(o) $30(25 - T) = 270(T - 5)$

2.2. Rearrange each of the following formulae to make:
(a) m the subject of $F = ma$

(b) V the subject of $\rho = \dfrac{m}{V}$

(c) R the subject of $pV = nRT$
(d) a the subject of $v = u + at$
(e) I the subject of $E = IR + Ir$

(f) r the subject of $I = \dfrac{P}{4\pi r^2}$

(g) m the subject of $ma = mg - W$
(h) h the subject of $p = \rho g h + P$

(i) p_2 the subject of $V = \dfrac{\pi r^4}{8\eta}\left(\dfrac{p_2 - p_1}{l}\right)$

(j) v the subject of $s = \dfrac{(u + v)}{2}t$

(k) γ the subject of $V_2 = V_1(1 + \gamma\theta)$
(l) m the subject of $V = (F(m/l)^{-1})^{1/2}$

(m) m the subject of $F = \dfrac{(mv - mu)}{t}$

(n) Δl the subject of $E = \dfrac{F/A}{\Delta l/l_0}$

(o) m the subject of $Fs = \dfrac{1}{2}mv^2 - \dfrac{1}{2}mu^2$

(p) R_1 the subject of $V_1 = \dfrac{VR_1}{R_1 + R_2}$

(q) m the subject of $\dfrac{mv^2}{r} = T + mg$

(r) k the subject of $T = \dfrac{2\pi}{(k/m)^{1/2}}$

(s) R_1 the subject of $\dfrac{1}{R} = \dfrac{1}{R_1} + \dfrac{1}{R_2}$

(t) η the subject of $\dfrac{4}{3}\pi r^3 \rho_s g - \dfrac{4}{3}\pi r^3 \rho_f g - 6\pi\eta r v_t = 0$

2.3. In each of the following cases, find the value of the unknown quantity by making it the subject of the formula and substituting the known values:
(a) if $P = V^2/R$, find the value of V where $R = 64$ and $P = 4$;
(b) if $P = I^2R$, find the value of I where $R = 4$ and $P = 36$;
(c) if $V = (2gh)^{1/2}$, find the value of h where $V = 14$ and $g = 9.8$;
(d) if $V = 2(ab + bc + ca)$, find the value of b where $V = 52$, $a = 2$ and $c = 4$;
(e) if $F = Gm_1m_2/r^2$, find the value of G where $F = 49$, $m_1 = 5.0$, $m_2 = 6.0 \times 10^{24}$ and $r = 6.4 \times 10^6$;
(f) if $1/C = 1/C_1 + 1/C_2$, find the value of C where $C_1 = 3$ and $C_2 = 6$;
(g) if $a = (b^2 + c^2)^{1/2}$, find the value of c where $a = 5$ and $b = 4$;
(h) if $v^2 = u^2 + 2as$, find the value of u where $v = 60$, $a = 5$ and $s = 110$;
(i) if $T = 2\pi(L/g)^{1/2}$, find the value of L where $T = 1.0$ and $g = 9.8$. (Assume $\pi = 3.14$.)

A.3 UNITS

(Before you begin this section make sure that you have read about SI units on pages 1–3.)

It is important to recognise that we can treat units just like the symbols in a formula. For instance, if we consider a length of 2 metres, then

$$2 \text{ m} = 2 \times \text{m}$$

where m represents the standard length which we recognise as 1 metre. The volume of a cube with edges of length 2 m is therefore given by

$$2 \times \text{m} \times 2 \times \text{m} \times 2 \times \text{m} = 2^3 \times \text{m}^3 = 8 \text{ m}^3,$$

that is to say, 8 cubic metres. This example illustrates two important points: first, that units can be expressed in index form (e.g. m^3) and, secondly, that the units on both sides of a formula must balance.

A.3.1 BALANCING THE UNITS IN A FORMULA

Confirmation that the units balance can be a useful aid in checking that a formula has been rearranged correctly. For instance, consider *speed*, which is the rate of change of distance with time (see page 37). To keep things simple, we shall consider an object travelling at constant speed, then we can write

$$\frac{\text{distance travelled (m)}}{\text{time taken (s)}} = \text{speed (m s}^{-1}\text{)}$$

If we rearrange this formula to make 'distance travelled' the subject, as follows

$$\text{distance travelled (m)} = \text{speed (m s}^{-1}\text{)} \times \text{time taken (s)}$$

then the units on both sides still balance since

$$\text{m} = \text{m s}^{-1} \times \text{s}$$

The formula has therefore been rearranged correctly.

The following worked example appears to be rather complicated so go through it slowly and carefully. (It's much easier than it looks!)

Worked Example A.7

If the unit of F is kg m s^{-2}, the unit of d is m, the unit of m is kg, and the unit of u and v is m s^{-1}, then confirm that the units balance on both sides of the following formula

$$Fd = \tfrac{1}{2}mv^2 - \tfrac{1}{2}mu^2$$

Writing the units in brackets after the corresponding symbols:

$$F \text{ (kg m s}^{-2}\text{)} \times d \text{ (m)} = \tfrac{1}{2}m \text{ (kg)} \times v^2 \text{ ((m s}^{-1}\text{)}^2) - \tfrac{1}{2}m \text{ (kg)} \times u^2 \text{ ((m s}^{-1}\text{)}^2)$$

Then since

$$\text{(m s}^{-1}\text{)}^2 = \text{m}^2 \text{ s}^{-2}$$

we can write

$$F \text{ (kg m s}^{-2}) \times d \text{ (m)} = \tfrac{1}{2}m \text{ (kg)} \times v^2 \text{ (m}^2 \text{ s}^{-2}) - \tfrac{1}{2}m \text{ (kg)} \times u^2 \text{ (m}^2 \text{ s}^{-2})$$

And since

$$(\text{kg m s}^{-2}) \times (\text{m}) = (\text{kg m}^2 \text{ s}^{-2})$$

and

$$(\text{kg}) \times (\text{m}^2 \text{ s}^{-2}) = (\text{kg m}^2 \text{ s}^{-2})$$

we can write

$$Fd \text{ (kg m}^2 \text{ s}^{-2}) = \tfrac{1}{2}mv^2 \text{ (kg m}^2 \text{ s}^{-2}) - \tfrac{1}{2}mu^2 \text{ (kg m}^2 \text{ s}^{-2})$$

hence

$$Fd \text{ (kg m}^2 \text{ s}^{-2}) = (\tfrac{1}{2}mv^2 - \tfrac{1}{2}mu^2) \text{ (kg m}^2 \text{ s}^{-2})$$

Thus the units balance on both sides of the formula.

A.3.2 CONVERTING UNITS

Quantities are not always given in the units that we want, so we must be prepared to convert from one system to another. Let us take the conversion of inches (in) to millimetres (mm) as an example. Since there are 25.4 millimetres per inch, we can write

$$1 \times \text{in} = 25.4 \times \text{mm}$$

This can be rearranged to give

$$1 = 25.4 \times \frac{\text{mm}}{\text{in}}, \text{ or its reciprocal } \frac{1}{25.4} \times \frac{\text{in}}{\text{mm}} = 1$$

Multiplying a quantity by 1 has no effect on its value so if, for example, we need to convert 2.50 inches to millimetres we can use 25.4 mm/in (= 1) as a conversion factor as follows:

$$2.50 \text{ in} \times 1 = 2.50 \text{ in} \times \left(25.4 \times \frac{\text{mm}}{\text{in}}\right) = 2.50 \times 25.4 \times \frac{\text{in} \times \text{mm}}{\text{in}} = 63.5 \text{ mm}$$

Inches cancel out in the top and bottom lines, leaving the value in millimetres. If we wish to convert millimetres to inches we divide by 25.4 mm/in (or multiply by its reciprocal). For example, converting 88.9 millimetres to inches:

$$88.9 \text{ mm} \times 1 = 88.9 \text{ mm} \times \left(\frac{1}{25.4} \times \frac{\text{in}}{\text{mm}}\right) = \frac{88.9}{25.4} \times \frac{\text{mm} \times \text{in}}{\text{mm}} = 3.50 \text{ in}$$

Worked Example A.8

Given that 1 imperial gallon = 4.546 litres (to 4 S.F.), 8 pints = 1 gallon and 1000 millitres = 1 litre, find how many millilitres there are in 1 pint.

Since

$$\frac{1\ \text{gallon}}{8\ \text{pints}} = 1, \frac{4.546\ \text{litres}}{1\ \text{gallon}} = 1 \text{ and } \frac{1000\ \text{millilitres}}{1\ \text{litre}} = 1$$

we can write

$$1 = \frac{1\ \text{gallon}}{8\ \text{pints}} \times \frac{4.546\ \text{litres}}{1\ \text{gallon}} \times \frac{1000\ \text{millilitres}}{1\ \text{litre}} = \frac{4546\ \text{millilitres}}{8\ \text{pints}}$$

$$= 568\ \frac{\text{millilitres}}{\text{pint}}$$

that is to say there are 568 millilitres per pint (to 3 S.F.).

The following worked example illustrates the conversion of density (mass per unit volume) from g cm^{-3} (grams per cubic centimetre) to kg m^{-3} (kilograms per cubic metre).

Worked Example A.9

The density of a sample of concrete was found to be 2.39 g cm^{-3}. Convert this to kg m^{-3} given that 1 kg = 1 × 10^3 g and that 1 m = 1 × 10^2 cm.

If 1 m = 1 × 10^2 cm, then

$$(1\ \text{m})^3 = (1 \times 10^2\ \text{cm})^3 = 1 \times 10^6\ \text{cm}^3$$

therefore

$$1 = 1 \times 10^6\ \frac{\text{cm}^3}{\text{m}^3}$$

Since 1 kg = 1 × 10^3 g, then

$$1 = 1 \times 10^3\ \frac{\text{g}}{\text{kg}}$$

so we can write

$$2.39 \times \frac{\text{g}}{\text{cm}^3} \times \left(1 \times 10^6 \times \frac{\text{cm}^3}{\text{m}^3}\right) \times \left(\frac{1}{1 \times 10^3} \times \frac{\text{kg}}{\text{g}}\right) = 2390\ \text{kg m}^{-3}$$

As you work through the following exercises, write down each operation step by step. Remember that it is better to sacrifice a little time in getting the right answers rather than to cut corners and make mistakes. Don't hurry.

A.3.3 EXERCISES

3.1.   The units of s, t, u, v and a are as follows:

	s	t	u, v	a
Units	m	s	m s^{-1}	m s^{-2}

For each of the following equations confirm that the units balance on either side:

(a) $v = u + at$

(b) $s = \dfrac{(u + v)}{2} t$

(c) $s = ut + \dfrac{1}{2} at^2$

(d) $v^2 = u^2 + 2as$

3.2. If the unit of F is kg m s^{-2}, the unit of m is kg, and the unit of l is m, then find the unit of v in the following formula.

$$v = (F(m/l)^{-1})^{1/2}$$

3.3. If the unit of v and u is m s^{-1}, the unit of m is kg, and the unit of t is s, then find the unit of F in the following formula:

$$F = (mv - mu)/t$$

3.4. If the unit of r and l is m, the unit of V is m^3 s^{-1}, and the unit of p_1 and p_2 is N m^{-2}, then find the unit of η in the following formula:

$$V = \frac{\pi r^4}{8\eta} \left(\frac{p_2 - p_1}{l} \right)$$

(π has no units because it is the ratio of the circumference of a circle to its diameter, i.e. length/length.)

3.5. Express each of the following in metres.
(a) In ordinary decimal form:
    (i) 57.5 km;
    (ii) 1024 mm;
    (iii) 45 000 μm;
    (iv) 3.4 mm;
    (v) 250 cm;
    (vi) 72 dm;
    (vii) 2.54 × 10^5 mm.
(b) In standard form:
    (i) 0.05 μm;
    (ii) 7.9 × 10^{19} km;
    (iii) 0.07 mm;
    (iv) 0.023 dm;
    (v) 18.1 mm;
    (vi) 1.2 cm;
    (vii) 8.5 × 10^{-9} mm.

3.6. Find the number of:
(a) m^2 in 1 km^2;
(b) mm^3 in 1 m^3;
(c) mm^2 in 1 cm^2;
(d) cm^3 in 1 dm^3.

3.7. Given that 1 litre = 1 dm^3, find how many litres there are in 1 m^3.

3.8. The area A of a circle of radius r is given by $A = \pi r^2$. Find the area in m^2 of the circle whose:
(a) radius is 1.13 mm;

    (b) diameter is 2.26 mm;
    (c) radius is 5.21 cm;
    (d) diameter is 0.735 μm;
    (e) diameter is 0.41 mm;
    (f) diameter is 7.41 m;
    (g) diameter is 3.49×10^4 km.

3.9. Given that 1 mile = 1760 yards, 1 yard = 3 feet, 1 foot = 12 inches and 1 inch = 25.4 mm, convert:
    (a) 7.1 yards to metres;
    (b) 1 foot 3 inches to centimetres;
    (c) 11.6 miles to kilometres;
    (d) 0.00341 inches to μm.

3.10. Using the data from Exercise 3.9, and giving your answers to 3 S.F.:
    (a) find the area in m^2 of a rectangle measuring 30 feet $\times$ 45 feet;
    (b) find the volume in m^3 of a cube that has edges 1 yard long;
    (c) find the volume in mm^3 of a cube that has edges 1.5 inches long;
    (d) find the speed in $m\ s^{-1}$ of a car travelling at 60 miles per hour.

3.11. Given that 16 oz = 1 lb and that 1 lb = 454 g (to 3 S.F.), and using the data from Exercise 3.9, find the density in $kg\ m^{-3}$ of a cube that weighs 1 lb 4 ¾ oz and which has edges 2½ in long.

3.12. Express 150 r.p.m. (revolutions per minute) in
    (a) radians per second;
    (b) degrees per second.
    (1 revolution = 360° = 2π rad.)

ANSWERS TO QUESTIONS

Topic 1

1. (a) 89 m at 54°; (b) 102 m at 63°; (c) 178 m at 37°; (d) 127 m at 113°.
2. 9.9 m s^{-1} horizontally and 1.0 m s^{-1} vertically.
3. 146 m at 322°.
4. 3.9 m s^{-1} at 249°.
5. 58 m at 212° (approximately 8 o'clock).
6. (a) 60 s; (b) 90 m; (c) 17° west of north; (d) 62.6 s.

Topic 2

1. 7.3 × 10^{22} kg.
2. 25 mm.
3. (a) 195 × 10^6 N m^{-2}; (b) 0.0975%.
4. 0 N.
5. 98 N.
6. 104 N.
7. 400 N.
8. (a) 29°; (b)(i) 37 N; (ii) 451 N.

Topic 3

1. (a) Yes; (b) 56 N at 127°; (c) 317 N at 78°; (d) yes; (e) yes; (f) yes.
2. $\alpha = \beta = 60°$.
3. 123 N, 368 N.
4. 245 N, 1225 N.
5. (a) 735 N m; (b) 980 N m; (c) 3.3 m.
6. The forces are in equilibrium.
7. (a) 33 N, 16 N at 90°; (b) 98 N, 85 N at 0°; (c) 19 N, 53 N at 69°;
   (d) 25 N, 43 N at 60°; (e) 19 N, 34 N at 73°; (f) 28 N, 29 N at 119°.
8. 0.18.
9. 45°.

Topic 4

1. 82 kg.
2. 1.23 × 10^{-3} m^3.
3. 5500 kg m^{-3}.
4. 2620 kg m^{-3}.
5. 250 N; 1.2 × 10^6 Pa.
6. (a) 2.2 × 10^4 Pa; (b) 1.9 × 10^3 Pa; (c) 9.98 × 10^4 Pa; (d) 1.02 × 10^5 Pa;
   (e) 9.0 × 10^5 Pa.
7. 31.1 cm^3.
8. (a) 7.8 × 10^3 kg m^{-3}; (b) 0.80 × 10^3 kg m^{-3}.
9. 0.12 N.

Topic 5

1. 9 s.
2. 500 m.
3. 44 m, 103 m, 123 m, 103 m, 44 m; 10 s.
4. 0 m s^{-1}; -2 m s^{-2}.
5. -3 m s^{-2}.
6. 3.1×10^4 m s^{-2}.
7. 10 m s^{-2}; 4 s.
8. 32.0 m; 2.55 s.
9. 714 m; 122 m s^{-1} at 35° below the horizontal.
10. (a) $s = \dfrac{u^2 \sin^2 \theta}{2g}$; (b) $t = \dfrac{u \sin \theta}{g}$.

Topic 6

1. 1.7×10^{-25} m s^{-2}.
2. 1.5 m s^{-2}; 8.2 s.
3. 7.5 s.
4. 200 m.
5. 2.0 m s^{-2} in the 56° direction.
6. 360 m in the 46° direction.
7. 0.59.
8. 8.9 N.
9. 10.2 m s^{-2}.
10. 100 m s^{-1}.
11. (a) 618 N; (b) 9270 N.

Topic 7

1. 556 N.
2. 100 m s^{-1}.
3. 1.35 m s^{-1} in the opposite direction to the bullet.
4. 20 m s^{-1}; 1.8×10^5 kg m s^{-1} in both cases.
5. (a) 14 kN; (b) 2.3 kN.
6. 1.8 N.
7. 35 g.
8. 14.3 m s^{-1} in the 3 o'clock direction.
9. 6.3 m s^{-1} at 39° anticlockwise from 3 o'clock.
10. 1800 m s^{-1}.

Topic 8

1. 5 kg.
2. (a) 10 kJ; (b) 0 J.
3. 30 m s^{-1}.
4. 240 N.
5. 70 m s^{-1}.
6. 67%.
7. 1.1 m s^{-1} horizontally.
8. 180 W.
9. 8 m s^{-1}.
10. 60 W.

Topic 9

1. 2×10^{-4} m s^{-1}.
2. 0.5.

3. 2.0×10^{20} N.
4. 7.3°.
5. 3.8 rad s^{-1}.
6. 1.96 N.
7. 24.5°.
8. 2.0 N.
9. 10 rad s^{-1}; 90 rad s^{-1}.
10. (a) 250 N; (b) 1620 N.

Topic 10

1. (a) 0.025 kg m^2; (b) 12.3 J; (c) 37.0 J.
2. 3.3×10^{-4} J.
3. (a) 225 rad; (b) 30 rad s^{-1}; (c) 2700 J; (d) 360 W; (e) 2700 J.
4. 1.25 kg m^2.
5. 1.5 rad s^{-1}; 8.2 s.
6. 2.0 kg.
7. (a) 50 rad s^{-1}; (b) 37.5 kJ.
8. 33 rad s^{-1}.
9. 450 J.
10. (a) 124 J; (b) 0.28 kg m^2; (c) hoop.

Topic 11

1. (a) 0.25 m; (b) 0.99 m; (c) 3.97 m.
2. 0.45 s.
3. 1.6 m.
4. 3.5 Hz.
5. (a) 0.30 m s^{-1}; (b) 0.94 m s^{-2}.
6. 56 N m^{-1}.
7. At the top; 5 mm.
8. 0.16 Hz.
9. (a) 28 J; (b) 4.7 m s^{-1}; (c) 1.1 kN.
10. 476 g.

Topic 12

1. 825 r.p.m.
2. (a) 1.7 km; (b) 1.3 m; (c) 0.4 s; (d) 20 000–20 Hz.
3. 510 m.
4. (a) 340 m s^{-1}; (b) 255 m.
5. 1.1 W.
6. (a) 5.70 m; (b) 1.33 m.
7. 256 Hz; 512 Hz; 768 Hz.
8. 440 Hz.
9. (a) 2.0×10^{11} N m^{-2}.
10. The aluminium rod must be 4.2 times longer than the lead rod.

Topic 13

1. Light (because $\lambda = 588$ nm).
2. (a) 3.07 m; (b) 247 m.
3. (a) 656 nm; (b) 494 nm.
4. (a) 17°; (b) 17° (the light path is reversible).
5. 1.33.
6. (a) 1.23×10^8 m s^{-1}; (b) 242 nm; (c) 24.2°; (d) 33.0°.
7. (a) $2 \times i_c = 98°$; (b) total internal reflection.
8. 14.5°.

9. 588 nm.
10. (a) 5.6°–10.7° (b) 17.0°–33.7°.

Topic 14

1. Lithium (metal); nitrogen; neon; sulphur; potassium (metal); scandium (metal); manganese (metal); bromine.
2. 2; 7; 8; 8; 22; 21; 20.
3. 1.99×10^{-26} kg; 2.66×10^{-26} kg; 9.26×10^{-26} kg.
4. (a) 107.9 g; (b) 27.0 g; (c) 2698 kg.
5. 6.0×10^{26}; 9.5×10^{24}; 5.6×10^{24}.
6. 2.0×10^{22}.
7. 8.4×10^{19}.
8. 20 g.
9. 590 mm^3.
10. 92.5%.
11. (a) Iron; (b) 2.66×10^{-26} kg; (c) 1.2×10^{-29} m^3.

Topic 15

2. (a) (i) Z decreases by 2; (ii) Z increases by 1 and decreases by 1 respectively; (iii) Z decreases by 1.
   (b) (i) unchanged; (ii) decreased by 4; (iii) decreased by 4.
3. (a) $^4_2\alpha$; (b) $^0_{-1}e$; (c) ^1_1p; (d) $^{220}_{86}Rn$.
4. 8.1 days; 3.08 mg; 16.2 days.
5. (a) n; (b) α; (c) $^{14}_7N$; (d) $^{235}_{92}U$.

Topic 16

1. Ionic crystal; covalent molecule; covalent molecules joined by hydrogen bonds; atoms joined by van der Waals' forces; covalent molecule; covalent molecules joined by hydrogen bonds.
2. 30.0; 100.1; 46.0; 169.9; 119.5.
3. Methane; ethane.
4. 52.9%.
5. 13.5 g.
6. 2.275 kg.
7. $C_2H_5OH + 3O_2 = 2CO_2 + 3H_2O$; 7.5 g.

Topic 17

1. (a) 1 °C; (b) 2 °C; (c) 4.7 °C; (d) 9.1 °C; (e) 32.3 °C.
2. (a) 1650 kJ; (b) 1755 kJ; (c) 2700 kJ.
3. 812 m s^{-1}.
4. 1.75 kW.
5. (a) 293 s; (b) 575 s.
6. (a) 2.5 kg; (b) 0.5 kg.
7. 17×10^{-6} K^{-1}.
8. 30.007 m.
9. 160 °C.
10. 16 °C; 62.5 MN m^{-2}.
11. 5.0 l.
12. Steel rod 475 mm; brass rod 300 mm.

Topic 18

1. (a) 16 mm; (b) 52 mm; (c) 400 mm; (d) 20 m.
2. (a) 11.1 °C; (b) copper/steel = 8/1.
3. 69 °C.
4. 1.84 MW m^{-2}.
5. 15 W.
6. (a) 2.5 W; (b) 2.5 W; (c) 2.5 W.
7. 0.6 kg.
8. (a) Larger/smaller = 4/1; (b) larger/smaller = 1/2.
9. (a) 503 °C; (b) 315 W m^{-2}.
10. 0.5.

Topic 19

1. An increase of 240 mm^3.
2. 72.5 1.
3. 38 °C.
4. 35 1.
5. 22.4 × 10^{-3} m^3.
6. (a) 2.80 m^3; (b) 125 mol.
7. 1.0 × 10^{22}.
8. 1.3 kg m^{-3}.
9. 752 mmHg.
10. 149 mmHg.

Topic 20

2. 1.00 1 s^{-1}.
3. 2.0 × 10^{27} molecules min^{-1}.
5. 1.8 × 10^{-5} Pa s.
6. (a) Laminar (*Re* = 1475); (b) 0.30 m.
7. 18 kPa.
8. 11 × 10^{-3} N.
9. 2 × 10^{-4} J.

Topic 21

1. 50 MPa in both cases.
2. (a) 100 MPa; (b) 2.5 mm.
3. 100 MPa.
4. (a) 0.43 mm; (b) 0.23 mm; (c) 15.6 mm.
5. 4.5 mm.
6. 8 kg.
7. 99.94 mm.
8. +0.08%.
9. −0.03%.
10. (a) 77 GPa; (b) 167 GPa.

Topic 22

2. 0.139 nm.
3. 7890 kg m^{-3}.
4. 0.68.
5. 0.41.
6. 2260 kg m^{-3}.

Topic 23

1. (a) 33 GPa; (b) 0.47 GPa.
2. 1.2 mm.

Topic 26

1. (a) × 4; (b) × 1/9.
2. (a) 50 nC; (b) 5.32 mN; (c) 50 mN.
3. (a) 8.2×10^{-8} N; (b) 3.6×10^{-47} N.
4. 0.11 μC.

Topic 27

1. (a) Right to left; (b) right to left; (c)(i) 5000 N C^{-1}, (ii) 5000 V m^{-1}; (d)(i) 4.8×10^{-17} J, (ii) 4.8×10^{-17} J, (iii) 0 J; (e) 4.8×10^{-17} J; (f) 2.4×10^{5} m s^{-1}; (g) 1.0×10^{7} m s^{-1}.
2. (a) 6.4×10^{-16} N; (b) 7.0×10^{14} m s^{-2}.
3. 2.0×10^{-6} V m^{-1}.
4. 4.
5. 1 μJ.
6. (a) 8.4×10^{6} m s^{-1}; (b) 9.5×10^{-8} s.
7. (a) 2.0×10^{5} m s^{-1}; (b) 4.1×10^{-6} s.
8. 4.8×10^{-19} C.
9. (a) 1.0×10^{7} m s^{-1}; (b) 6.2 mm.

Topic 28

1. 400 μC.
2. 2.1×10^{11}.
3. 26 V.
4. 7.2×10^{-4} C.
5. 1 μF; 1.5 μF; 2 μF; 3 μF; 4.5 μF; 6 μF; 9 μF.
6. 8 V (wider-spaced plates); 4 V.
7. (a) 2 μF; (b) 2.5 mJ.
8. 15.1 °C.
9. (a) 600 μC, 1200 μC; (b) 400 μC, 400 μC.
10. (a) 45 V; (b) 13.5 mJ.

Topic 29

1. 7.5×10^{19}.
2. 5 W.
3. 4×10^{-19} J.
4. 6 W.
5. (a) 6 W; (b) 15 V; (c) 0.4 A.
6. 10 C; 2 GJ.
7. 60 V.
8. 2.5 A.
9. 192 W.

Topic 30

1. 1.5 Ω.
2. 1.5 V.
3. (a) 9 Ω; (b) 1 Ω; (c) 2 Ω; (d) 4.5 Ω; (e) 7.5 Ω; (f) 1.8 Ω.
4. 30 Ω in parallel.
5. 2 Ω; 3 Ω.

6. (a) 6 V; (b) 15 Ω; (c) 0.3 A.
7. 5800 Ω.
8. (a) 4.25 V; (b) 3.5 V; (c) 3.6 A.
9. (a) 2.4 A: (b) 1.6 A; (c) 0.8 A.
10. (a) 1.5 A; (b) 1.2 A.
11. 5 kΩ; 7.2 mW.
12. 0.15 A.
14. (a) 9 W; (b) 2 W.

Topic 31

1. (a) 11.65 V; (b) 11.96 V.
2. (a) 0.15 A; (b) 1.44 V.
3. (a) 2.4 V; (b) 3.0 V; (c) 14.5 kΩ.
4. (a) 4 V; (b)(i) 0.08 A, (ii) 0.04 A, 0.06 A.
5. (a) 0.67 A in the wires connected to the corners where the current enters
   or leaves the network and 0.33 A in the other wires; (b) 20 V; (c) 10 Ω.
6. 3.3 Ω.
7. 0.273 A.

Topic 32

1. (a) 0.05 T; (b) 13×10^{-3} N.
2. (a) Vertically upwards; (b) vertically downwards.
3. 30°.
4. Nil.
5. 0.01 T.
6. 1.5×10^{-3} N m.
7. 2×10^{-5} T.

Topic 33

1. 8 mA.
2. (a) 1.06 mV; (b) 0 V; (c) 14.5 mV.
3. 4.8×10^{-5} T; 66°.
4. (a) 1.0 V; (b) 0.5 V; (c) 0 V.
5. (a) 0.4 mV; (b) 0.2 mV; (c) 0 V.

Topic 34

1. 1875.

Topic 35

1. (a) 300 W, 60 V; (b) 75 W, 30 V.
2. 145 V.
3. 100 Hz.
4. (a) 110 Ω; (b) 1 A.
5. (a) 35 Ω; (b) 70 V.
6. 70 Ω.
7. (a) *I* leads *V* by 90°; (b) *I* and *V* are in phase; (c) *V* leads *I* by 90°.
8. (a) 16 Ω; (b) 16 Ω; (c) 5 Ω.
9. 14.50 Ω.
10. 4 A.
11. (a) 9 V; (b) 12 V.
12. (a) 5 V; (b) 53.1°.
13. 13 Ω; 67.4°; 10 V, 28 V, 4 V.
14. 50 Ω.

15. (a) 360 W; (b) 270 W; (c) 0 W.
16. 53.1°; 6.75 W.

Appendix

1.1. (a)(i) $2.834\,86 \times 10^5$, (ii) 283.486×10^3; (b)(i) 4.3×10^{-5}, (ii) 43×10^{-6}; (c)(i) $1.838\,103\,97 \times 10^8$, (ii) $183.810\,397 \times 10^6$; (d)(i) 2.31×10^{-6}, (ii) 2.31×10^{-6}; (e)(i) 5.2921×10^4, (ii) 52.921×10^3; (f)(i) 9.60×10^{-4}, (ii) 960×10^{-6}; (g)(i) -7.80×10^{-2}, (ii) -78.0×10^{-3}.
1.2. (a) 17.91; (b) 0.1791; (c) 179 100; (d) 0.000 349; (e) 381 000; (f) 0.0941; (g) −0.009 870.
1.3. (a) 10.48; (b) 3; (c) 2×10^7; (d) 1900; (e) 1866; (f) 11.7; (g) 40.26; (h) 1.985.
1.4. (a) 2.83×10^{-2}; (b) 8.50×10^5; (c) 4.87×10^{-3}; (d) -2.13×10^5; (e) 4.2×10^{-6}; (f) -1.7×10^9; (g) -3.0×10^5; (h) -2.6×10^{-8}; (i) 8×10^3; (j) -4×10^6; (k) 6.4×10^1; (l) 8×10^6; (m) 5×10^{50}; (n) 7.5; (o) 1.02×10^8; (p) 2.4×10^8; (q) -1.21×10^{-12}; (r) 8.89×10^3; (s) 1.15×10^{-20}; (t) 7.1×10^{-4}; (u) 6.35×10^{35}; (v) 9.0×10^{17}; (w) 1.0×10^{10}; (x) 4.5×10^8; (y) 1.09×10^{-4}; (z) 8.0×10^1.
2.1. (a) 11; (b) −2; (c) 9; (d) 7; (e) 8; (f) 3; (g) 3; (h) 4; (i) 2; (j) ±7; (k) 15; (l) 30; (m) 6; (n) 6; (o) 7.
2.2. (a) $m = F/a$; (b) $V = m/\rho$; (c) $R = pV/(nT)$; (d) $a = (v - u)/t$; (e) $I = E/(R + r)$; (f) $r = (P/(4\pi I))^{1/2}$; (g) $m = W/(g - a)$ or $m = -W/(a - g)$; (h) $h = (p - P)/(\rho g)$; (i) $p_2 = 8V\eta l/(\pi r^4) + p_1$; (j) $v = 2s/t - u$; (k) $\gamma = (V_2 - V_1)/(V_1\theta)$; (l) $m = Fl/V^2$; (m) $m = Ft/(v - u)$; (n) $\Delta l = Fl_0/(AE)$; (o) $m = 2Fs/(v^2 - u^2)$; (p) $R_1 = V_1R_2/(V - V_1)$ or $R_1 = -V_1R_2/(V_1 - V)$; (q) $m = T/(v^2/r - g)$; (r) $k = 4\pi^2 m/T^2$; (s) $R_1 = RR_2/(R_2 - R)$; (t) $\eta = 2r^2(\rho_s - \rho_f)g/(9v_t)$.
2.3. (a) $V = (PR)^{1/2} = 16$; (b) $I = (P/R)^{1/2} = 3$; (c) $h = V^2/(2g) = 10$; (d) $b = (V/2 - ca)/(a + c) = 3$; (e) $G = Fr^2/(m_1m_2) = 6.7 \times 10^{-11}$; (f) $C = C_1C_2/(C_1 + C_2) = 2$; (g) $c = (a^2 - b^2)^{1/2} = 3$; (h) $u = (v^2 - 2as)^{1/2} = 50$; (i) $L = T^2g/(4\pi^2) = 0.25$.
3.1. (a) $\text{m s}^{-1} = \text{m s}^{-1}$; (b) m = m; (c) m = m; (d) $\text{m}^2\,\text{s}^{-2} = \text{m}^2\,\text{s}^{-2}$.
3.2. m s^{-1}.
3.3. $\text{kg m s}^{-2} = \text{N}$.
3.4. $\text{N s m}^{-2} = \text{Pa s}$.
3.5. (a)(i) 57 500 m, (ii) 1.024 m, (iii) 0.045 m, (iv) 0.0034 m, (v) 2.50 m, (vi) 7.2 m, (vii) 254 m; (b)(i) 5×10^{-8} m, (ii) 7.9×10^{22} m, (iii) 7×10^{-5} m, (iv) 2.3×10^{-3} m, (v) 1.81×10^{-2} m, (vi) 1.2×10^{-2} m, (vii) 8.5×10^{-12} m.
3.6. (a) 1×10^6; (b) 1×10^9; (c) 1×10^2; (d) 1×10^3.
3.7. 1×10^3.
3.8. (a) 4.01×10^{-6} m²; (b) 4.01×10^{-6} m²; (c) 8.53×10^{-3} m²; (d) 4.24×10^{-13} m²; (e) 1.3×10^{-7} m²; (f) 43.1 m²; (g) 9.57×10^{14} m².
3.9. (a) 6.5 m; (b) 38.1 cm; (c) 18.7 km; (d) 86.6 μm.
3.10. (a) 125 m²; (b) 0.765 m³; (c) 5.53×10^4 mm³; (d) 26.8 m s^{-1}.
3.11. 2300 kg m^{-3}.
3.12. (a) 15.7 rad s^{-1}; (b) $900° \text{ s}^{-1}$.

INDEX

absolute permeability 337
    of free space 337
absolute permittivity 257
    of free space 257, 337
absolute pressure 30
absolute refractive index 111
absolute zero 153, 155, 179
acceleration 1, 37, 47
    angular 69
    centripetal 72
    due to gravity 42, 47
    in simple harmonic motion 86, 87
acceptors (in semiconductors) 285
accuracy (numerical values) 348
action and reaction 50
activity (radioactivity) 137
addition of vectors 3
aircraft wing, lift force on 74, 187
alkali metals 123
alloys 147, 240
alpha decay 132, 133, 135
alpha particles 131, 132, 135
alpha radiation 132
alternating current 328, 338
alternating e.m.f. 328
alternating voltage 338
ammeter 288, 302, 305
amorphous solids 204, 219
ampere 258, 279
    definition of 322
amplitude
    alternating e.m.f. 328
    simple harmonic motion 86
    waves 96
Andrews' investigation of real
        gases 180
angle
    critical 112
    of contact 195
    of dip 317
    of friction 15
    of incidence 99, 109, 110
    of reflection 99, 109
    of refraction 110
    phase 344
angular acceleration 69

angular displacement 68
angular equations of motion 70
angular momentum 80, 81
    conservation of, see conservation
        laws/principles
angular velocity 68
antinodes 104
apparent weight 51
Archimedes' principle 32
area expansivity 159
asbestos 218
atmosphere (unit of pressure) 31
atmospheric pressure 30
atomic mass 126
atomic mass unit 118, 119, 126
atomic number 119, 129, 130
atomic packing factor 209, 230
atomic structure 118
atomic weight 126
atoms, electronic structure of 120
Avogadro constant 126
Avogadro's hypothesis 180

banking (aircraft and roads) 74
bar 31
base units 1, 2
batteries 265, 270, 297
becquerel 137
bending 13, 225
Bernoulli's equation 185
beta decay 133, 135
binding energy 130
black body 170
body-centred cubic structure 211,
        212, 230, 235, 236
boiling point 154, 155, 182
bond rotation 145, 202, 204, 242,
        244
bonds, chemical 11, 141
bouncing putty 245
Boyle's law 175, 176, 179
brass 240
brittle fracture 203, 221, 231,
        232
bubble chamber 135
bulk modulus 102, 200
butane molecule 145

caesium chloride structure 213,
    214, 215
calculation technique 347
calibration 190, 309
capacitance 270
capacitive reactance 340
capacitors 270
    energy stored in 273
    in parallel 274, 294
    in series 274, 294
    parallel-plate 270
    practical 271
capacity
    cells and batteries 298
    heat 154, 155, 156
capillarity 195
carbon dating 137
cells, electric 270, 296, 297, 311,
    312
Celsius scale 155
cement 228, 229
centigrade scale 155
centre of gravity 10, 11
centripetal acceleration 72
centripetal force 72
ceramics 204, 221
    strength of 221
chain reaction 138
chain size, effect on polymer
        properties 246
chain-like molecules 145, 202,
    204, 241
charge, *see* electric charge
Charles' law 176, 179
chemical bonds 11, 141
chemical equations 150
circuits, magnetic 336
circular motion 68
    vertical 75
clay minerals 219
cloud chamber 135
coefficient of dynamic
        viscosity 192
coefficients of friction 14
coercive force/coercivity 335
coherent waves 101
coils
    magnetic field around 317
    magnetic torque on 320
    rotating, e.m.f. induced in 327
collisions 55, 65
colour 109
components of forces 19
composites, polymer based 250
compounds 144
compression 11, 13, 143, 197, 225,
    226, 228

concrete 228
conductance, electrical 289
conduction
    electrical, in metals 280
    electrical, in semiconductors 283
    thermal 163
    thermal, through successive
        layers 166
conductivity
    electrical 289
    thermal 164, 165
conservation laws/principles
    angular momentum 80
    charge 256, 311
    energy 65, 152, 186, 311, 326
    momentum 55
constructive interference 101
contact angle 195
continuity equation 185
contraction, thermal 154, 158
convection 159, 163, 169
conventional current direction 282
converging lenses 113
conversion of energy 61, 65, 88,
    223, 282
cooling 169, 170
    Newton's law of, *see* Newton
co-ordination number 209, 230
corkscrew rule 317
coulomb 258, 279
Coulomb's law 119, 257, 262
counter, Geiger 136
counter, scintillation 136
couple 22
covalent bonding 143
    contrasted with ionic
        bonding 145
    polarisation in 146
covalent crystal structures 145,
    216
crack propagation 222, 223, 224
creep 246
cristobalite crystal structure 216,
    219
critical angle 112
critical isotherm and
        temperature 180, 181
cross-linking in polymers 249
crystal structures
    body-centred cubic 211, 212,
        230, 235, 236
    caesium chloride 213, 214, 215
    covalent 145, 216
    cristobalite 216, 219
    diamond 145, 216
    face-centred cubic 208, 209,
        212, 230, 235, 236

germanium   216
graphite   217
hexagonal close-packed   208, 212, 230, 235
ice   159, 217
ionic   143, 212
metallic   146, 206
rocksalt   143, 214, 215
silicon   216
sphalerite   215, 216
zinc blende   215, 216
crystallinity, effect on polymer properties   246
cubic expansivity   159, 179
cupronickel alloys   240
curie   137
Curie point/temperature   334
current, *see*   electric current

Dalton's law of partial pressures   180
damping of oscillations   94
deceleration   37
deflection, magnetic, charged particle beams   320
deformation
    elastic   11, 197, 231, 232
    plastic   13, 203, 224, 231, 232
density   28, 102
    comparison between ice and water   159
    relative   28, 33
    water   28
derived units   1, 2
destructive interference   102
deviation of light
    by diffraction gratings   114
    by prisms   113
dew point   182
diamagnetism   333
diamond   112, 145, 216, 217, 283
dielectric constant   257, 258, 271
dielectric strength   271
dielectrics   257, 271
diffraction   100
    light   100, 114
    sound   100
diffraction grating   114
diode, semiconductor   285
dip, angle of   317
direction, field, *see*   field direction
dislocations   237, 238
dispersion of light   114
displacement   3, 37
    angular   68
    in simple harmonic motion   86
diverging lenses   113

domains, magnetic   334
donors (in semiconductors)   284
doping (semiconductors)   284
Doppler effect   102
double bond   145
drift velocity of electrons in metallic conductors   280
ductile behaviour   231, 232
dynamic friction, coefficient of   14

earth, magnetic field of the   316, 317
earthing   265
echoes   99
eddy currents   331
effective value, alternating current and voltage   339
efficiency   66, 283
elastic deformation   11, 197, 231, 232
elastic limit   13, 203
elastic modulus, *see*   modulus of elasticity
elasticity   11, 143, 197
elasticity of rubber   200, 202, 242, 245
electric cell   270, 296, 297, 311, 312
electric charge   119, 256
    conservation of, *see*   conservation laws/principles
    moving, magnetic force on   319
electric constant   257, 271, 337
electric current   257, 279
    alternating   328, 338
    conventional direction   282
    eddy   331
    in metallic conductors   280
    in semiconductors   283
    induced in a moving conductor   324
    measurement of   309 (*also see* ammeter)
electric field   261
electric field line   262
electric force   119, 257, 258, 261
electrical conductance   289
electrical conduction
    in metals   280
    in semiconductors   283
electrical conductivity   289
electrical energy, conversion of   282
electrical insulators   257, 271
electrical power   283, 298
electrical resistance   281, 287
    internal   296, 297
    measurement of   288, 305, 309

electrical resistance – *continued*
  metals, effect of temperature
    on 281, 290
electrolytes 257
electromagnetic induction 324
electromagnetic interaction 9
electromagnetic radiation 163, 169
electromagnetic spectrum 109
electromagnetic waves 108, 163, 169
electromagnets 335
electromotive force (e.m.f.) 296
  alternating 328
  induced in a moving
    conductor 324
  induced in a rotating coil 327
  measurement of 308
electron 118, 256
  behaviour in electric fields 265,
    272, 320
  behaviour in magnetic
    fields 319, 325, 327
  capture 133, 134, 135
  drift velocity in metallic
    conductors 280
  emission 133, 135
  free, in graphite 217
  free, in metals 230
  in atomic structures 120
  in beta decay 133, 135
  in charged capacitors 270, 273
  mean free path in metallic
    conductors 281
  role in behaviour of magnetic
    materials 316, 333
  role in chemical bonding 141
  role in electrical conduction
    through metals 280
  role in electrical conduction
    through semiconductors 283
  role in electrostatic induction in
    metals 259
  role in thermal conduction 164
electron gas 146
electron spin 121
electronegativity 146
electronic structure of atoms 120
electronvolt 131, 266
electropositive elements 146
electrostatic induction 259
electrostatics 257
elements, chemical 118
elongation after fracture 231, 233
emerald 218
e.m.f., *see* electromotive force
emissivity 170
energy 2, 61
  binding 130

conservation of, *see* conservation
    laws/principles
conversion/transformation 61,
    65, 88, 223, 282
  in simple harmonic motion 88
  internal 109, 151
  ionisation 121, 123
  kinetic 63, 88, 153
  levels, electronic, in the
    atom 120
  loss in hysteresis 336
  nuclear 131, 138
  potential 61, 88, 153, 264
  rotational 80, 82
  sound 98
  stored in a capacitor 273
  stored in a magnetic field 330
  strain 62, 223, 232, 273
  surface 194, 223
  thermal 153
equations
  of angular motion 70
  chemical 150
  of motion 38, 39, 41, 42
equilibrium, mechanical 19
equipotential surface 62, 264
ethane molecule 144, 145, 241,
    242, 243
ethene (ethylene) molecule 144,
    145, 241
expansion, thermal 154, 158
expansivity 158, 179
extension 11, 12
extrinsic semiconductors 284

face-centred cubic structure 208,
    209, 212, 230, 235, 236
farad 270
Faraday's laws of electromagnetic
    induction 327
Faraday–Neumann law 327
ferromagnetic materials 333, 334
ferromagnetism 334
field direction
  electric 262
  magnetic 316
field lines
  electric 262
  magnetic 316
field strength
  electric 261, 262, 267
  gravitational 10, 47, 261
  magnetic 316, 319
fields
  electric, *see* electric field
  gravitational, *see* gravitational
    field

magnetic, *see* magnetic field
uniform and non-uniform 263, 317
fired clay products 227
fission, nuclear 131, 138
flat coil, magnetic field around 317
Fleming's left-hand (motor) rule 318, 324
Fleming's right-hand (generator) rule 324, 325
floating 34
flux, *see* magnetic flux
focus, principal 113
force 1, 9, 11, 47, 54
action and reaction 50
between parallel conductors 322
between polar molecules 148
centripetal 72
components of 19
electric 119, 257, 258, 261
frictional 13
gravitational 9, 258
intermolecular 147
internal 11
lift on aircraft wing 74, 187
lines of 262, 316
magnetic, on a conductor 318
magnetic, on a moving charge 319
moments of 22
normal 13
nuclear 119, 129
parallelogram of 15
resolution of 16, 19, 24
restoring, in simple harmonic motion 87
resultant of 15, 19
van der Waals' 149
forsterite 218
forward biased p–n junction 285
free fall 42
acceleration of 42
freezing 154, 155
frequency
alternating current and voltage 339
natural 105
simple harmonic motion 89
waves 96, 97
friction 13, 72, 74, 256
angle of 15
force 13
rolling 16
fundamental 104
fundamental frequency of a taut string 105
fundamental particles 118

fusion
latent heat of 154, 157
nuclear 131, 138
galvanometer, moving coil 302
gamma radiation 134, 135
gas constant 176
gas equation 176
gases 174
gases, inert (noble) 122
gauge pressure 30
Geiger counter 136
generator rule 324, 325
germanium 283, 284, 285
crystal structure 216
glass 112, 204, 219, 224, 225
fibre reinforced polymers 250
fibres 227, 250
transition temperature 204, 244
grain boundary 238
graphite 217
gravitation, Newton's law of, *see* Newton
gravitational constant 10
gravitational field 10, 261, 263
gravitational force 9, 258
gravitational interaction 9
gravity
acceleration due to 42, 47
centre of 10, 11
Griffith's equation 224

half-life 136
harmonics 104
heat 151
latent 154, 155
heat capacity 154, 155, 156
heat transfer 163
henry 330
hexagonal close-packed structure 208, 212, 230, 235
high density polyethene, tensile properties 247
holes, positive, in semiconductors 284
Hooke's law 11, 197, 202
fundamental basis of 143
humidity, relative 182
hydraulic press 32
hydrocarbons 144
hydrogen bonding 148
hydrometer 34
hysteresis 335

ice and water, density of 159
ice crystal structure 159, 217
ideal gas 174

ideal gas equation 176
ideal (non-viscous) liquids 185
impedance 343
impulse 55
inclination 317
induced current in a moving
    conductor 324
induced e.m.f.
  in a moving conductor 324
  in a rotating coil 327
inductance 329
induction
  electromagnetic 324
  electrostatic 259
inductive reactance 340
inert gases 122
  structure 122, 141
inertia 9
inertia, moment of 78, 81
insulation, thermal 165
insulators, electrical 257, 271
intensity of sound 98
interference 101, 102
  light 114
intermediate (ionic/covalent)
    bonding 145
intermolecular forces 147
internal energy 109, 151
internal forces 11
internal resistance 296, 297
intrinsic semiconductors 284
inverse square relationships 10, 98,
    119, 257
ionic bond 141
  contrasted with covalent
    bond 145
  polarisation in 146
ionic crystal structures 143,
    212
ionisation chamber 135
ionisation energy 121, 123
ionising radiation 135
ions 141
  size of 145
isotherm 179
  critical 180, 181
isotopes 119, 136
I–V characteristics, metallic
    conductors 289

joule 2, 60

kelvin 155
kinematic equations, *see* equations
    of motion
kinetic energy 63, 88, 153
  rotational 80, 82

kinetic friction, coefficient of 14
Kirchhoff's laws 311

laminar flow 184
latent heat 155
laws/principles of conservation, *see*
    conservation laws/principles
lenses 113
Lenz's law 325
lift force on aircraft wing 74,
    187
lifts, apparent weight in 50
light 109
  deviation 113, 115
  diffraction 100, 114
  dispersion 114
  interference 114
  polarisation 116
  reflection 109
  refraction 110
  speed 108
  wavelengths 109
light emitting diodes 285
linear expansivity 158
linear momentum 54
lines, field, *see* field lines
lines of force 262, 316
liquids 184
  apparent volume/cubic
    expansivity 159
  pressure in 29, 31
  supercooled 204, 219
litre 160
lone pair (electrons) 148
longitudinal waves 97
low density polyethene, tensile
    properties 247
lubrication 16

magic numbers 130
magnetic behaviour of
    materials 333
magnetic circuits 336
magnetic constant 337
magnetic domains 334
magnetic field 263, 316
  around a flat coil 317
  around a solenoid 318
  around a straight conductor 317
  energy stored in 330
  of the earth 316, 317
  resolution into components 319
magnetic field lines 316
magnetic flux 326
  cutting 326, 327
  density 319, 326
  linkage 327

linking 327
magnetic force
  between parallel conductors 322
  on a conductor in a magnetic
    field 318
  on a moving charge in a magnetic
    field 319
magnetic induction 319
magnetic materials 333
magnetic poles 316
magnetic torque on a coil 320
magnetomotive force (m.m.f.) 336
magnets
  permanent 316
  soft and hard 335
majority carriers 284
manometer 30
mass 9
  atomic 126
  defect 130
  molecular 149
  number 119, 129
materials, magnetic 333
matter, speed of sound
    through 102
Maxwell, James Clerk 337
mean free path
  electrons in metal
    conductors 281
  gas molecules 174
mechanical waves 96
melting point 154, 155
metal conductors, current in 280
metallic bonding 146, 207, 230
metallic conductors, *I–V*
    characteristics 289
metallic crystal structures 147, 206
metals 125, 146
  alkali 123
  electrical conduction in 147, 280
  slip in 203, 221, 233
  strength and toughness 236, 238
  tensile strength 233
  thermal conductivity 147, 164
methane molecule 144
metre bridge 310
mica 219
microstrain 200
millibar 31
minority carriers 284
mixtures 144
m.m.f. (magnetomotive force) 336
modulus of elasticity
  bulk, *see* bulk modulus
  shear, *see* shear modulus
  Young's, *see* Young's modulus
molar heat capacity 156

molar latent heat 156
mole 126
molecular mass, relative 149
molecular weight 149
molecules 144
  butane 145
  chain-like 145, 202, 204, 241
  ethane 144, 145, 241, 242, 243
  ethene (ethylene) 144, 145
  methane 144
  polar 148, 259, 272
  propane 145
  water 148, 217
moment of inertia 78, 81
moments of forces 22
momentum 54
  angular 80, 81
  angular, conservation of, *see*
    conservation laws/principles
  conservation of, *see* conservation
    laws/principles
monomer 241
motion 9, 19, 37, 47, 54
  circular 68
  equations of, *see* equations of
    motion
  Newton's laws of, *see* Newton
  rotational 78
  simple harmonic 86
  thermal, *see* thermal motion
  wave 96, 108
motor rule 318, 324
moving coil galvanometer 302
multiplier 304
mutual inductance 330

neck formation
  in metals 231, 232, 233
  in polyethene 247
net force/separation curve 142,
    144, 147, 197, 198, 221
Neumann, Faraday–, law 327
neutron 118, 119, 129, 130
  number 129, 130
newton 1, 47
Newton
  first law of motion 9, 47
  law of cooling 169
  law of gravitation 9, 50, 261
  second law of motion 47, 54
  third law of motion 50
noble gases 122
nodes 104
nominal (conventional, engineering)
    stress 233
non-bonding orbitals 148
non-uniform fields 263, 317

non-viscous (ideal) liquids   185
normal force   13
n-type semiconductors   284
nuclear
    energy   131, 138
    fission   131, 138
    forces   9, 119, 129
    fusion   131, 138
    reactions   129, 137
    stability   129
nucleon   129
    number   129
nucleus   119, 129
nuclide   129
    parents and daughters   130

offset yield stress   232
ohm   287
Ohm's law   290
orbitals   120
    molecular   144
    non-bonding   148
oscillation   86, 94
    simple pendulum   91
    vertical, mass on a spring   92
overtones   104

parallel
    capacitors in   274, 294
    resistors in   293, 294
parallel conductors, force
    between   322
parallel-plate capacitor   270
parallelogram of forces   15
paramagnetism   333
partial pressures, Dalton's law
    of   180
pascal   11, 29
peak value, alternating current and
    voltage   339
pendulum
    simple   91
    torsion   94
period
    of revolution   69
    of simple harmonic motion   87,
        88, 91, 93
    of wave motion   96
periodic table of the elements
    124
permanent magnets   316, 336
permeability
    absolute   337
    absolute, of free space   337
    relative   333, 337
permittivity   257, 271, 272, 337
    absolute   257

absolute, of free space   257, 271,
    337
    relative   257, 258, 271
permittivity, unit of   273
phase angle   344
phasors   340
pitch (sound)   97
Pitot tube   187
plastic deformation   13, 203, 224,
    231, 232
plasticisation of polyvinyl
    chloride   249
p–n junctions   285
poise   192
Poiseuille's formula   192
Poisson's ratio   199, 200
polar molecules   148, 259, 272
    forces between   148
polarisation
    in dielectrics   272
    in the covalent bond   146
    in the ionic bond   146
    of light   116
Polaroid   116
poles, magnetic   316
polyethene (polyethylene,
    polythene)   149, 241, 245, 248
    high and low density, tensile
        properties   247
    neck formation   247
polymers   241
    composites based on   250
    polymerisation   241
    reinforced with glass fibres   250
polymethyl methacrylate   245, 248
polypropene   245, 248
polystyrene   245, 248
    composites with rubber   250
polyvinyl chloride   245, 248
    plasticisation   249
positive holes in
    semiconductors   284
positron emission   133, 134, 135
potential   264
potential difference   264, 265
    measurement of   308, 309
potential divider   293, 306
potential energy   61, 88, 153, 264
potential energy/separation
    curve   152, 153
potentiometer   307
pounds per square inch   31
power   2, 65
    derivation of units of   1
    electrical   283, 298
    factor   344
    in alternating current circuits   344

transmission in rotation   81, 82
prefixes for units   2, 3
pressure   29
    absolute   30
    atmospheric   30
    gauge   30
    in liquids   29, 31
    and temperature, standard   180
    transmission of   31
    units of   31
    vapour   181
pressure law   175, 176, 179
principal focus   113
principle of superposition   100
principles/laws of conservation, *see*
    conservation laws/principles
prisms   113
progressive (travelling) wave   96
projectiles   43
proof stress   232
propane molecule   145
proton   118, 119, 129, 130, 256,
    259
p-type semiconductors   284, 285
putty, bouncing   245

quantities   1

radian   68
radiation
    electromagnetic   163, 169
    heat transfer by   163, 169
    radioactivity   132
    thermal   170
radioactive decay   130, 132
radioactivity   132
radius ratio   213
reactance   339
reaction
    chain   138
    mechanical   50
    nuclear   137
real gases   180
    Andrews' investigation of   180
real (viscous) liquids   191
rearranging formulae   352
recoil (gun)   55
reflection   99
    laws of   109
    of light   109
    of sound   99
    total internal   112
refraction   99
    of light   110
refractive index   111
relative atomic mass   126
relative density   28, 33

relative humidity   182
relative molecular mass   149
relative permeability   333, 337
relative permittivity   257, 258, 271
relative refractive index   111
reluctance   336
remanence   334
resistance
    electrical, *see*   electrical resistance
    thermal   164
resistivity   288, 289
resistors
    in parallel   293, 294
    in series   291, 294
    practical   291
    variable   305, 309
resolution
    of forces   16, 19, 24
    of magnetic fields   319
    of vectors   3
resonance   105
resultant
    of forces   15, 19
    of vectors   3
retardation   37
retentivity   334
reverse biased p–n junction   285
Reynolds' number   192
rheostat   305
ripple tank   98
rocksalt structure   143, 214, 215
rolling friction   16
rotation, power transmission and
    work done in   81, 82
rotation about carbon–carbon
    bonds   145, 202, 204, 242,
    244
rotation of solids   78
rotational equilibrium   19, 23
rotational kinetic energy   80, 82
rough checking calculations   350
rounding off in calculations   348
rubber-like elasticity   200, 202,
    242, 245
rubber/polystyrene composites   250

saturated vapour   181
saturated vapour pressure   181
saturation (magnetic)   334
scalar quantities   3
scintillation counter   136
self-inductance   330
semiconductor   257, 283
    diode   285
series
    capacitors in   274, 294
    resistors in   291, 294

shear 191, 197, 199, 200
shear modulus 199, 200
shells (electronic energy
        levels) 120
shunt 302
SI units 1
side groups, effect on properties of
        polymers 247
siemens 289
significant figures 347, 348
silica 216, 225, 226
silicates 217
silicon 283, 284, 285
    crystal structure 216
silicone polymers 242, 245
simple harmonic motion 86
simple pendulum 91
sinking 32
slip in metals 203, 221, 233
        directions 235, 236
        planes 234, 235, 236
        systems 235, 236
Snell's law 111
solenoid, magnetic field
        around 318
solids 197
    structure of 206
sonic boom 103
sound 97
    barrier 103
    diffraction 100
    energy 98
    intensity 98
    reflection 99
    speed of 97, 102
    wavelengths 97
specific gravity 28
specific heat capacity 156
specific latent heat 156
spectrum, electromagnetic 109
speed 37
    light 108
    sound 97, 102
    waves 96, 102
sphalerite structure 215, 216
spin, electron 121
stability, nuclear 129
standard form 347
standard temperature and pressure
        (STP) 180
standing (stationary) waves 103
static friction, coefficient of 14
stationary (standing) waves 103
steel 240
Stefan's constant 170
Stefan's law 170
Stokes's law 193

STP (standard temperature and
        pressure) 180
strain 11, 12, 198, 199, 200
    energy 62, 223, 232, 273
    hardening 232
    transverse and longitudinal 199
strength
    ceramics 221
    field, *see* field strength
    metals 236, 238
stress 11, 12, 191, 198, 199, 200
    concentration 222
    nominal (conventional/
        engineering) 233
    proof (offset yield) 232
    thermal 160
    true 233
stress/strain plot 200, 201, 202
string, taut
    fundamental frequency of 105
    wave speed along 102
structure
    amorphous 204, 219, 225
    atomic 118
    see crystal structures
    solids 206
subshells (electronic energy
        levels) 120
supercooled liquids 204, 219
superficial expansivity 159
superposition, principle of 100
surface energy 194, 223
surface tension 194

talc 219
temperature 151
    absolute zero of 153, 155, 179
    and pressure,
        standard (STP) 180
    critical 180, 181
    effect on resistance of
        metals 281, 290
    glass transition 204, 244
    rise in metal conductors 281, 291
    scales 155
tensile strength 221, 233
tension 11, 12, 13, 143, 197
    surface 194
terminal velocity 42, 193
tesla 319
thermal
    conduction 163
    conductivity 164, 165
    contraction 154, 158
    energy 153
    expansion 154, 158
    insulation 165

motion   152, 153, 163, 174, 175, 202, 204, 242, 245, 281
   radiation   170
   resistance   164
   stress   160
   tempering   226
thermocouple   308
thermonuclear fusion   139
thermosetting polymers   249
thrust (rocket motor)   55
time-dependent behaviour of polymers   245
toroid   334
torque   22, 78, 81
   magnetic, on a coil   320
torr   31
Torricelli's theorem   186
torsion pendulum   94
total internal reflection   112
toughness   233
   glass reinforced thermosetting resins   251
   metals   236
transfer, heat   163
transformation of energy   61, 65, 88, 223, 282
transformers   329, 331, 338
transforming/transposing formulae   355
translational equilibrium   19
transmission of pressure   31
transmutation by nuclear reactions   129, 137
transverse waves   97
travelling (progressive) wave   96
true stress   233
turbulent flow   184

ultrasonic waves   99
uniform fields   263, 317
units   1, 360
universal molar gas constant   176
unsaturated vapour   182
upper yield point   240
upthrust   32

valence electrons   144
valency   144
van der Waals' forces   149
vaporisation, latent heat of   154, 157
vapour   181
vapour pressure   181
vectors   3
   addition and resolution   3

velocity   1, 4, 37
   angular   68
   electron drift in metallic conductors   281
   terminal   42, 193
Venturi meter   187
vertical circular motion   75
vibrational motion   86
viscosity   185, 191
   coefficient of   192
viscous (real) liquids   191
volt   265
voltage, alternating   338
voltmeter   288, 302, 304, 305
   calibration of   309
volume expansivity   159, 179
volume strain   200
vulcanisation   249

water
   and ice, comparison of density   159
   density of   28
   molecule   148, 217
   refractive index of   112
   waves   98, 99
watt   2, 65
wavelength   96, 97, 109
waves
   electromagnetic   108, 163, 169
   mechanical   96
weber   326, 327
weight   10
   apparent   51
wetting   195
Wheatstone bridge   309
wood   251
work   2, 60, 151
   done in moving a charge   264
   done in rotation   81, 82
   done in stretching a wire   62
   hardening   232

yield
   point   231, 232
   point, upper   240
   stress, offset yield (proof)   232
   stress/strength   232
Young's fringes   114
Young's modulus   12, 102, 198, 199, 200, 201, 224, 232
   fundamental basis of   143

zero, absolute   153, 155, 179
zinc blende structure   215, 216